St. Petersburg
Moscow
Poznan
Cracow
Warsaw
Minsk
Kazan
Perm
Yekaterinburg
Chelyabinsk
Nizhniy Novgorod
Prague
Kiev
Ufa
Omsk
Dnipropetrovs'k
Novosibirsk
Vienna
Kharkiv
Samara
Krasnoyarsk
Budapest
Kryvy Rih
Donets'k
Volgograd
Odessa
Bucharest
Tbilisi
Chengdu
Istanbul
Almaty
Sofia
Belgrade
Bursa
Yerevan
Thessaloniki
Ankara
Athens
Izmir
Adana
Baku
Haifa
Tehran
Tashkent
Hanoi
Chiang Mai
Tel Aviv
Jerusalem
Esfahan
Mashhad
Bangkok
Ho Chi Minh
Shiraz
Baghdad
Delhi
Kolkota
Bangalore
Mumbai
Chennai
Melbourne
Sydney
Sapporo
Sendai
Tokyo
Changchun
Kawasaki
Yokohama
Nagoya
Kyoto
Osaka
Shenyang
Beijing
Kobe
Hiroshima
Tianjin
Kitakyushu
Fukuoka
Pyongyang
Xi'an
Seoul
Naha
Incheon
Daejeon
Wuhan
Daegu
Shanghai
Gwangju
Nanjing
Busan
Chongqing
Hangzhou
Taipei
Guangzhou
Kaohsiung
Shenzhen
Hong Kong
Kuala Lumpur
Singapore
Manila
Jakarta
Auckland

TOWER HAMLETS
91 000 001 742 31 1

ON
THE
MAP

First published in Great Britain in 2012 by
PROFILE BOOKS LTD
3A Exmouth House
Pine Street
London EC1R 0JH
www.profilebooks.com

10 9 8 7 6 5 4 3 2 1

Design by James Alexander/Jade Design
Typeset in Bembo Book and Archer
Printed and bound in Great Britain by
Clays, Bungay, Suffolk

A CIP catalogue record for this book is available
from the British Library.
ISBN 978 1 8466 8509 5
eISBN 978 1 8476 5855 5

FSC

The paper this book is printed on
is certified by the © 1996 Forest
Stewardship Council A.C. (FSC).
It is ancient-forest friendly. The
printer holds FSC chain of custody
SGS-COC-2061

Simon Garfield

THE

Why the world looks
the way it does

PROFILE BOOKS

By the same author
Expensive Habits
The End of Innocence
The Wrestling
The Nation's Favourite
Mauve
The Last Journey of William Huskisson
Our Hidden Lives
We are at War
Private Battles
The Error World
Mini
Exposure
Just My Type

www.simongarfield.com

To Justine

Contents

Foreword by Dava Sobel
For the Love of Maps

Simon Garfield has chosen an apt double entendre as the title for his delightful paean to maps: To be *On the Map* is to have arrived. To discourse *On the Map* is to ponder cartography's course through history and throughout the cultural milieu. With pleasure, I accept the invitation he offers any reader of this book – to lose oneself in map perusal.

I love maps. I do not collect them, unless you count the ones in the box under my desk, which I've saved as souvenirs from the cities they walked me through or cross-country trips they guided. The maps I covet – early renderings of the known world before the New World came to light, mariner's portolans bearing wind roses and sea monsters – are all beyond my means, anyway. They belong where they are, in museums and libraries, and not confined to the walls (or condemned to the humidity) of my house.

I think about maps a lot. When working on a book project, I must keep a map of the territory at hand, to help the characters find their roots. Even at odd moments, say while clearing spam from the junk folders of my email accounts, it occurs to me that 'spam' is 'maps' spelled backward, and how maps, the true opposite of spam, do not arrive unbidden, but only beckon.

A map will lead you to the brink of Terra Incognita, and leave you there, or communicate the comfort of knowing, 'You are here.'

Maps look down, as I do, watching my step. Their downward perspective seems so obvious, so familiar as to make one forget how much looking-up they entail. Ptolemy's rules of cartography, written out in the second century, descended from his prior study of astronomy. He called down the moon and stars to help him align the world's eight thousand known locations. Thus he drew the tropic lines and equator through the places where the planets passed directly overhead, making his best guess of east–west distances by the light of a lunar eclipse. And it was Ptolemy who set North at the top of the map, where the pole pointed to a lone star that held still through the night.

Like everyone else these days, I rely on quick computer-generated maps for driving directions, and often find my way on foot or public transportation via the maps app on my smart phone. But for serious travel preparation I need a plat. Only a map can give me a sense of where I'm going. If I fail to see, before setting off, whether the destination is shaped like a boot or a fish tail or an animal hide, I will never gain a sense of the place once I'm there. Seeing ahead of time that streets obey a grid layout – or they circle around a hub, or follow no discernible plan – already tells me something of what wandering them will be like.

If I'm not really going anywhere, then travel by map of course provides the only possible route – to everywhere, to nowhere in particular, to the folds of the human genome, the summit of Everest, the paths of future transits of Venus for the next three thousand years. Even buried treasure, lost continents and phantom islands are all accessible by map.

What difference does it make if I never reach my map-dream destinations, when even the most admired map-makers of old stayed home? I think of Fra Mauro, immured in his Venetian monastery, spinning the thin yarns of untrustworthy travellers into his own gorgeous geography.

I revel in the visual luxury of maps. The so-called four-colour map conjecture, which defines the minimum number of pigments required for constructing a world map, sets no upper bounds on artistic licence.

The language of maps sounds no less colourful to my ear. Words like 'latitude' and 'graticule' rattle out of the mouth to cast a net around the world. And 'cartouche', the map's decorative title block or legend, whooshes off the tongue with a breeze. Some names of places yodel; others click or sing. Gladly would I go from Grand-Bassam to Tabou along the coast of the Côte d'Ivoire, if only to say so out loud.

Maps are guilty of distortion, it's true, but I forgive them for it. How could one wrestle the round world down to a flattened image on the page without sacrificing some proportion? The various methods of map projection, from the eponymous Mercator to the orthographic, gnomonic or azimuthal, all cause one continent or another to morph. Just because I grew up seeing Greenland the equal of Africa in land mass doesn't mean I believed them to be that way, any more than I fretted over the misnomer of Greenland, a place white with ice, near Iceland, green with flora. Maps are only human, after all.

Every map tells a story. The picturesque antique ones speak of quest and conquest, of discovery, claim and glory, not to mention the horror tales about exploitation of native populations. Story lines may blur in modern maps, under a welter of natural and manmade features, yet up-to-date maps make great templates for new stories: swept clean of their topographical details, and with various data superimposed, they can make a statement about the voting patterns in the latest election, say, or the spread of disease at epidemic's first threat.

The only thing better than a map is an atlas. Atlas himself, the Titan who once held the heavens on his shoulders, has lent his name to a family of rockets as well as to book-length compendia of maps. I own several of these worthy Atlas namesakes, all requiring strong arms to bear them from shelf to table.

I could enthuse about globes, too, especially the bygone ones built and sold in pairs, one orb for earth and one for sky (also depicted from *above*, with the geometry of all constellations reversed). A globe, though, is merely an inflated, reincarnated map. It starts out flat, as a series of painted or printed gores, and these need to be fitted around and pasted on a ball to make the ends of the earth meet. If maps be the fuel of wanderlust, read on.

Introduction
The Map That Wrote Itself

In December 2010, Facebook released a new map of the world that was as astonishing as it was beautiful. It was both instantly recognisable – the standard projection produced by Gerard Mercator in the sixteenth century – and yet curiously unfamiliar. It was a luminescent blue, with gauzy lines spread over the map like silk webs. What was odd about it? China and Asia were hardly there, while East Africa seemed to be submerged. And some countries weren't quite in the right place. For this wasn't a map of the world with Facebook membership overlayed, but a map generated by Facebook connections. It was a map of the world made by 500 million cartographers all at once.

Using the company's central data on its members, an intern called Paul Butler had taken their latitudinal and longitudinal coordinates and linked these to the coordinates of the places where they had connections. 'Each line might represent a friendship made while travelling, a family member abroad, or an old college friend pulled away by the various forces of life,' Butler explained on his blog. Facebook had about 500 million members at that time, so he anticipated a bit of a

mess, a crowded mesh of wires (like the back of those early computers) that would culminate in a central blob. Instead, Butler recalled, 'after a few minutes of rendering, the new plot appeared, and I was a bit taken aback. The blob had turned into a detailed map of the world. Not only were continents visible, certain international borders were apparent as well. What really struck me, though, was knowing that the lines didn't represent coasts or rivers or political borders, but real human relationships.'

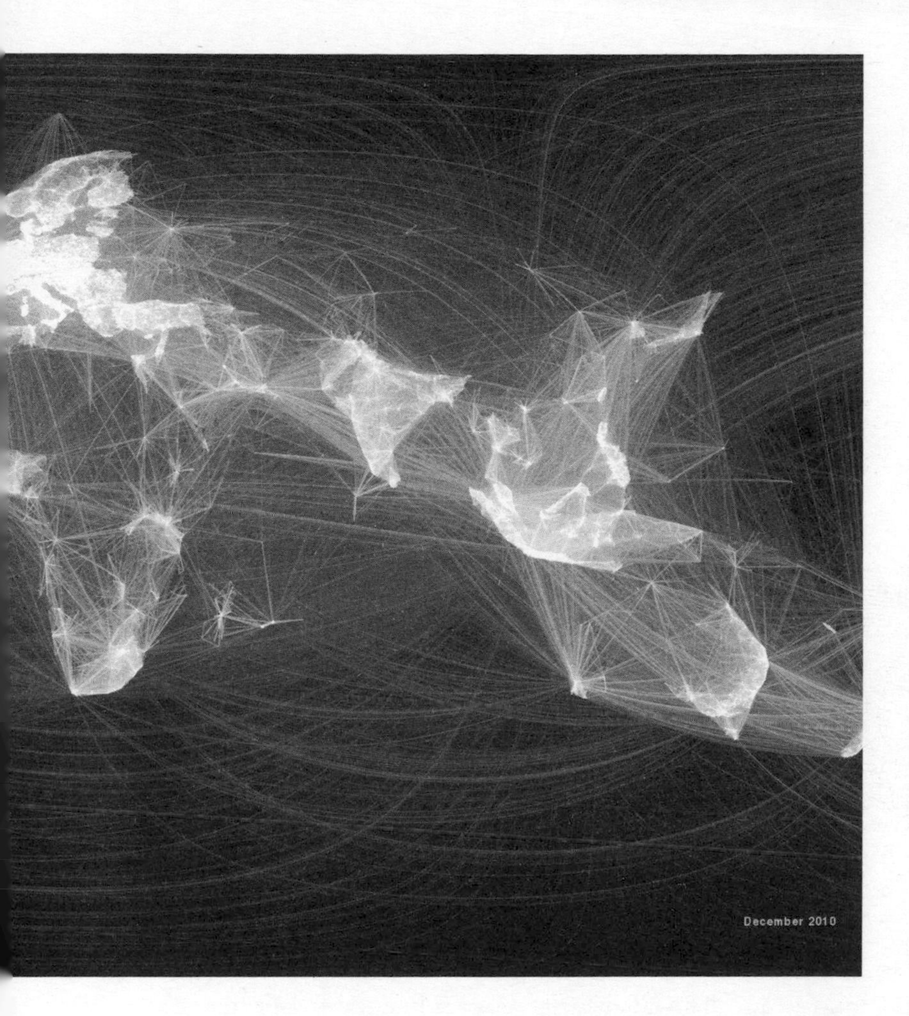

It was the perfect embodiment of something Facebook's founder Mark Zuckerberg had told me when I interviewed him the year before Butler created that map. 'The idea isn't that Facebook is one new community,' he had said, 'but it's mapping out all the different communities that exist in the world already.'

The digital revolution – so neatly encapsulated by that Facebook map – has transformed mapping more than all the other innovations of cartography's centuries. With our phone maps in our hands and Google Earth on our computers, it is increasingly difficult to recall how we managed without them. I seem to recall we used to buy maps that folded, or maps that once folded when they were new and then never again. Or that we used to pull down shoulder-dislocating atlases from shelves and thumb through their index, and perhaps wonder at how many Springfields there were in the United States.

That these simple pleasures are becoming distant memories is no small change. For physical maps have been a vital part of our world since we first began finding our way to food and shelter on the African plains as hunter-gatherers. Indeed, Richard Dawkins speculates that the very first maps came about when a tracker, accustomed to following trails, laid out a map in the dust; and a recent finding by Spanish archeologists identified a map of sorts scratched on a stone by cave dwellers around fourteen thousand years ago. Dawkins goes on to speculate as to whether the creation of maps – with their concepts of scale and space – may even have kick-started the expansion and development of the human brain.

In other words, maps hold a clue to what makes us human. Certainly, they relate and realign our history. They reflect our best and worst attributes – discovery and curiosity, conflict and destruction – and they chart our transitions of power. Even as individuals, we seem to have a need to plot a path and track our progress, to imagine possibilities of exploration and escape. The language of maps is integral to our lives, too. We have achieved something if we have put ourselves (or our town) on the map. The organised among us have things neatly mapped out. We need compass points or we lose our bearings. We orient ourselves (for on old maps east was at the top). We give someone a degree of latitude to roam.

Maps fascinate us because they tell stories. The ones in this book tell how maps came about, who drew them, what they were thinking, and how we use them. Like any map, of course, the selection is highly selective, for a book about maps is effectively a book about the progress of the world: sturdier ships in the fifteenth century, triangulation in the late-sixteenth century, the fixing of longitude in the eighteenth, flights and aerial observation in the twentieth century. And then, in this century, the Internet, GPS and sat nav – and perhaps, through them, a second reshaping of our own spatial abilities.

For the Internet has effected an extraordinary and significant change. Before astronomers faced the gallows for suggesting otherwise, our earth stood firmly at the centre of the cosmos; not so long ago, we placed Jerusalem at the centre of our maps; or if we lived in China, Youzhou. Later, it might be Britain or France, at the heart of their empires. But now we each stand, individually, at the centre of our own map worlds. On our computers, phones and cars, we plot a route not from A to B but from ourselves ('Allow current location') to anywhere of our choosing; every distance is measured from where we stand, and as we travel we are ourselves mapped, voluntarily or otherwise.

Earlier this year, a friend of mine noticed an odd thing on his Blackberry. He was walking in the Italian Alps and wanted to check out contours and elevations. When he turned on his phone his Transport for London bicycle app was open: a handy tool where you put in a London location and it tells you how many bikes are available at each docking station. It was less use in Italy, or so he thought. But, in fact, the app was still working and the map over which Transport for London had overlaid its bicycle info actually covered the entire world. The bikes were only the start of it. It could plot a route to Ravello, Cape Town or Auckland. Wherever he went, my friend *was* the map, the pivot around which the world diligently spun. And the app was no doubt tracking him, too, so that someone knew which Italian mountain he was on, as well as who was riding the bike he had docked the day before.

How on earth did we get to this point? This book is intended as an answer to that question, but it could also be viewed as a journey around an exhibition. It is by necessity an imaginary show, for it contains things that would be impossible to gather in one place: long-destroyed impressions of the world from Ancient Greece, famous treasures from the world's universities, some jaw-dropping pieces from the British Library and the Library of Congress, rare items from Germany, Venice and California. There will be manuscripts, sea charts, atlases, screen grabs and phone apps. Some exhibits are more important than others, and some are just displayed for amusement. The range will be extensive: poverty and wealth maps, film maps and treasure maps, maps with a penchant for octopuses, maps of Africa, Antarctica and places that never were. Some of the maps will explain the shape of the world, while others will focus on a street or on the path of a plane as it flies to Casablanca.

We'll need a lot of space for our guides: boastful dealers, finicky surveyors, guesswork philosophers, profligate collectors, unreliable navigators, whistling ramblers, inexperienced globe-makers, nervous curators, hot neuroscientists and lusting conquistadors. Some of them will be familiar names – Claudius Ptolemy, Marco Polo, Winston Churchill, Indiana Jones – and some will be less well known: a Venetian monk, a New York dealer, a London brain mapper, a Dutch entrepreneur, an African tribal leader.

You hold in your hand the catalogue to this show, and it begins in a library on the coast of Egypt.

Chapter 1
What Great Minds Knew

Maps began as a challenge of the imagination and they still perform that role. So imagine yourself in your bedroom. How good would you be at mapping it? Given a pencil and pad, could you draw the room well enough so that someone who's never been there would get a fair picture? Would the size of the bed be in proportion to the door and the bedside table? Would the scale be right in relation to the height of the ceiling? Would your kitchen be harder or easier to map than the bedroom?

This shouldn't be too hard really, because these are places you know well. But what about the living room of a friend? That would be partly a test of memory – would you get it right or would you be struggling? But what about your first school: would you remember where your classroom was in relation to others? Or the world? Could you draw that? Could you correlate the relative size – and geographic relationship – of Mongolia and Switzerland? Would you get the oceans even half right in the southern hemisphere? And what if you'd never seen another map before, or a globe, and you'd never been to any of these places yourself? Could you construct a map of the world based purely on what people had told you, and what people had written down? And if you did manage this, would you be happy for it still to be used as the principal map of the world some 1350 years after you had drawn it?

Only, I imagine, if your name was Claudius Ptolemy.

Considering his impact on the world, and beyond the fact that we should regard the P in his surname as silent, we know curiously little about Ptolemy. But we do know where he worked – at one of the greatest buildings in ancient Egypt, lying just a little way inland on a small cloak-shaped port on the banks of the Mediterranean.

The story of the vanished Great Library of Alexandria is one of the most romantic of the ancient world, and it appeals partly because we are unable to imagine a modern equivalent. Today's British Library is a library of record, receiving a copy of each new work in the English language, but it has no ambitions to house a complete collection of the world's manuscripts, nor to contain the sum of human knowledge. The same with the Bodleian in Oxford, and the New York Public Library. But the Great Library of Alexandria did aspire to such ambitions, and it existed at a time when such a thing was broadly achievable.

From its inception in around 330 BC, the Library was intended as a place where every scrap of useful information found a home. Other private libraries were commandeered for the common good; manuscripts arriving in the city by sea would be transcribed or translated, and only some were returned; often the ships would sail away again with the originals replaced by copies. At the same time, Alexandria became Europe's principal supplier of papyrus, from which the majority of its Library scrolls were made. And suddenly the supply of papyrus for export dried up: some claimed that all the papyrus was required to supply the Great Library, though others detected a plot designed to inhibit the growth of rival collections – an elitism, passion and quest that all obsessive book and map collectors will recognise.

The Great Library was the legacy – like the city itself – of Alexander the Great. During a journey along the western reaches of the Nile Delta, Alexander had come across a site that, according to the Roman historian Arrian, he predicted would be 'the very best in which to found a city.' Its subsequent foundation signalled the shift of governmental and cultural power from Athens.

Alexander had been tutored by Aristotle in the ways of morality, poetry, biology, drama, logic and aesthetics, and it was through Aristotle that he became devoted to Homer, taking the *Iliad* into battle and living by its teachings. His conquest of the Persian Empire was followed by the destruction of Tyre and the rapid capitulation of Egypt, and it was here that he became afflicted with immortal ambitions: he wanted his legacy to be a symbol of learning rather than destruction, a place from where the Hellenistic worldview would be spread through the empire and beyond. And so he laid plans for a city marked by a devotion to scholarship, high ideals and good governance, and its vast Library was to be its pantheon.

The Library, completed several decades after Alexander's death in 323 BC, was in effect the world's first university, a place of research and colloquy, whose scholars included the mathematician Archimedes and the poet Apollonius. They discussed scientific and medical principles as well as philosophy, literature and political administration. And they were responsible for drawing up the first accomplished maps of the world: a role for which, living in a port city at the heart of both western and eastern trade routes, and with first-hand testimonies from travellers and sailors, they were ideally placed.

If we stumbled across a map of ancient Alexandria today, we would see an orderly place, a grid system of boulevards and thoroughfares. A heavily populated Jewish Quarter lies to the east, while the Library and Museum stand in the Royal Quarter in the centre. The city is surrounded by water, with the Great Harbour (home of the royal palaces) on small islands in the north. At the city's northern harbour rises the Pharos lighthouse, one of the Seven Wonders of the World, more than a hundred metres tall, with a flame at its top reflected by a mirror and visible some thirty miles out to sea. It would be difficult to miss the metaphor: Alexandria was a beacon city, a landmark both liberated and liberating in a city pulsing with illuminated thinking.

But the world beyond Alexandria – how did that look at the beginning of the third century BC?

Despite the Great Library's accomplishments in science and mathematics, the study of geography was still in its infancy. Its first scholars constructed an important proto-map of the world, based largely on the writings of the Greek historian Herodotus. His nine-volume *Researches* had been completed a century and a half earlier but his description of the rise and fall of the Persian empire and the Greco-Persian wars remained the most detailed source on the known world. Homer, too, was regarded as an important source for geographical knowledge, not least through the travels depicted in the *Odyssey*.

It is thought that this Alexandrian map depicted the world as round, or at least roundish, which by the fourth century BC was commonly accepted. It is possible that Herodotus shared this view, though he may have seen it as a flat disc floating on water. Homer, certainly, was a flat-earther, back in the eighth century BC, believing the earth was a place where if you continued sailing you would eventually fall off the end. But by the fifth century BC, Pythagoras had argued persuasively that the earth was a sphere. (The myth that the earth was still considered flat until the time of Columbus is an oddly enduring one. Why should this be? A combination of general ignorance and our love of a good story: the image of Columbus returning home with the news that his fleet did not in fact topple into a great abyss is madly appealing.)

Herodotus upheld the common wisdom that the world was divided into three sections – Europa, Asia and Libya (Africa) – but argued against a widespread belief that they were the same size and made up the whole of the earth. Neither Britain nor Scandinavia featured in his accounts, and the Nile ran throughout Africa to Morocco's Atlas Mountains. Only a small section of Asia was examined, and it was dominated by India. Herodotus admitted to uncertainty over whether Europe was surrounded wholly by water, but he suggested Africa might be. He also saw the Caspian Sea – accurately – as a vast inlet, unlike many of his successors.

As the Great Library developed its collections, the variety and reliability of its sources yielded a vast collection of fragmentary information about the world – and the possibilities of creating maps to reflect this. Eratosthenes of Cyrene (in modern-day Libya) was one of the first scholars able to marshall the city's new geographical knowledge into the art of cartography. Born in 276 BC, he studied mathematics and astronomy in Athens, combining the disciplines to form the first primitive armillary sphere (or astrolabe), a series of metal rings arranged as a globe that showed celestial positions with the earth at its centre.

At the age of forty, Eratosthenes became the third Librarian at Alexandria and began his great treatise *Geographica* shortly afterwards. There was no study of geography comparable to that of medicine or philosophy (indeed, Eratosthenes is believed to have coined the word 'geography' from the Greek words *Geo*/Earth and *graphien*/writing) but at the Great Library he would have encountered an abstract map created in the sixth century BC by Anaximander of Miletus for his treatise *On Nature*. This map, long extinct, showed the world as a flat disc with named parts for the Mediterranean, Italy and Sicily. He may also have benefitted from an inventory of countries and tribes – a 'Circuit of the Earth', but in truth more a circuit around the Mediterranean – provided in the same period by Hecataeus of Miletus. (Miletus, in modern-day Turkey, was something of a Classical geographical hothouse. In the fifth century BC it was also home to Hippodamus, a forefather of urban planning responsible for some of the earliest civic maps).

But Eratosthenes' own geographic study was to be on an altogether grander scale, making fullest use of the Library scrolls, the accounts of those who had swept through Europe

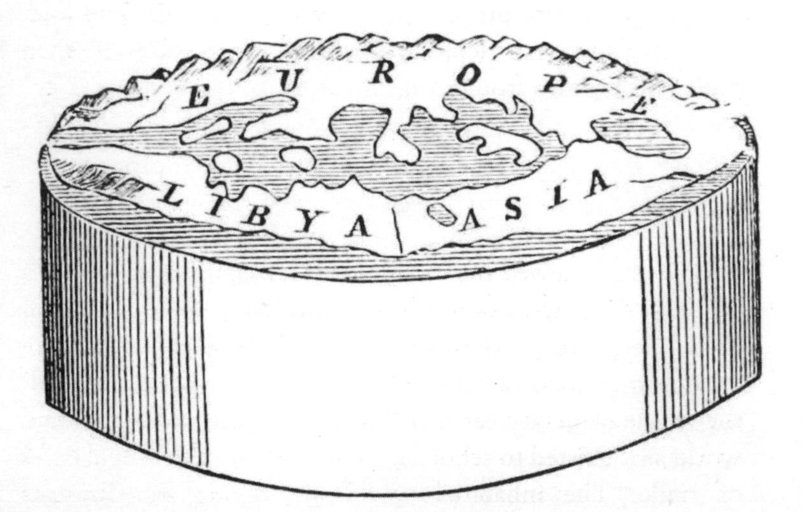

Three continents in a fountain: Anaximander imagines a disc-like earth surrounded by water in sixth century BC.

and Persia in the previous century, and the pertaining views of the leading contemporary historians and astronomers. His world map was drawn in about 194 BC. No contemporary version exists, but the cartographer's descriptions were interpreted for a Victorian audience, and this remains the generally accepted and widely used reproduction. It peculiarly resembles a dinosaur skull. There are three recognisable continents – Europe to the north-east, Africa (described as Libya and Arabia) beneath it, and Asia occupying the eastern half of the map. The huge northern section of Asia is called Scythia, an area we would now regard as encompassing eastern Europe, the Ukraine and southern Russia.

The map is sparse but sophisticated, and noteworthy for its early use of parallels and meridians in a grid system. Eratosthenes drew a main parallel running east-west through Rhodes, and a main meridian running north-south, again with Rhodes at its centre. His map was then divided into unequal rectangles and squares, which appear to the modern eye as locational grids but served the Greek geographer more as an aid to achieving accurate proportions. They affirmed the common belief that the earth's length from west to east was more than double its breadth from north to south.

Eratosthenes viewed the earth in the contemporary way: as a sphere at the centre of the universe with the heavenly bodies in full rotation every twenty-four hours. In his view, there were two distinct ways of interpolating and depicting the world: the whole planetary earth as it hung in space, and the known world as it existed to scholars, navigators and the beneficiaries of trade. The inhabited world (something the Romans would later call 'the civilised world') was believed to occupy about one-third of the northern hemisphere and was wholly contained within it. The northernmost point, represented by the island of Thule (which may have been Shetland or Iceland), was the last outpost before the world became unbearably cold; the most southerly tip, labelled enticingly as Cinnamon Country (Ethiopia/Somaliland) was the point beyond which the heat would burn your flesh.

In Eratosthenes' map the oceans are interconnected, the Northern Ocean covering the top of Europe and Scythia, the Atlantic propping up the coasts of Libya, Arabia, the Persian empire and a square-shaped India. There are giant inlets of the Caspian Sea and the Persian Gulf, both of which erroneously flow into the oceans. Brettania, vaguely accurate in shape but excessive in scale, is sited to the far north-west, sitting in good proportion to both Ireland and Europe. All three give the impression of being loosely connected, separated only by navigable inland waters or mountain ranges. And they appear purposely huddled together, as if the huge encroaching oceans and the vast areas of the unknown world are joining forces against them. There is no New World, of course, no China, and only a small section of Russia.

Nonetheless, in its reliance on scientific principles, the map made great methodological strides over its predecessors. And although Eratosthenes consciously elongated the continents to fit his workings, he set the template for a new goal – the formulation of a precise and consistent map of the world.

If it were just for his descriptive map, Eratosthenes would now be regarded as a minor character in the story of ancient cartography (indeed his colleagues referred to him as a 'Beta' talent, compared to the 'Alpha' virtues of Aristotle or Archimedes). But this judgement should be revised, for he did one great thing which goes beyond mapping: he made ground-breaking calculations as to the earth's measurements, and his working principles, based on the large Babylonian pole known as a gnomon (a forerunner to the classical vertical sundial), are rightly considered a timeless and fool-proof technique, if rather a clumsy one.

His eureka moment, reported subsequently by the Greek scientist Cleomedes, has now taken on the mythical weight of a Newtonian apple, but it may be true. Eratosthenes had observed that on midsummer's day the sun shone directly overhead at the Nile settlement at Syene, a fact demonstrated by its reflection in a deep well at noon. He knew, by the time it took to journey between the two towns by camel, that Syene (modern Aswan) lay roughly 5,000 *stades* (about 500 miles) due south of Alexandria (on the main meridian on which he had plotted Rhodes). By measuring the angle of the sun's elevation from the Great Library at the same moment (7°) he could plot a circumference of the earth. Assuming the earth to be spherical and made up of 360°, his 7° difference between 500 miles worked out at 1/50th of the whole sphere. Eratosthenes thus declared that the earth had a circumference of 250,000 *stades* (roughly 25,000 miles), a calculation he increased to 252,000 to fit his desire for a pleasing symmetrical division by 60.

Eratosthenes came remarkably close to the true figure. We now accept the earth's circumference as 24,901.55 miles (40,075.16 km). By some estimates his figure was only two per cent over, although much depends on the definition of the *stadion*, his unit of measurement, which has both an Attic definition and an Egyptian one. But given that Eratosthenes was operating with such primitive estimates (Syene was not precisely due south, the earth is not perfectly spherical but bulges slightly at the Equator), we may marvel not only at his accuracy, but also at what the great distances said about the size of the unexplored world around him. Was there ever a greater

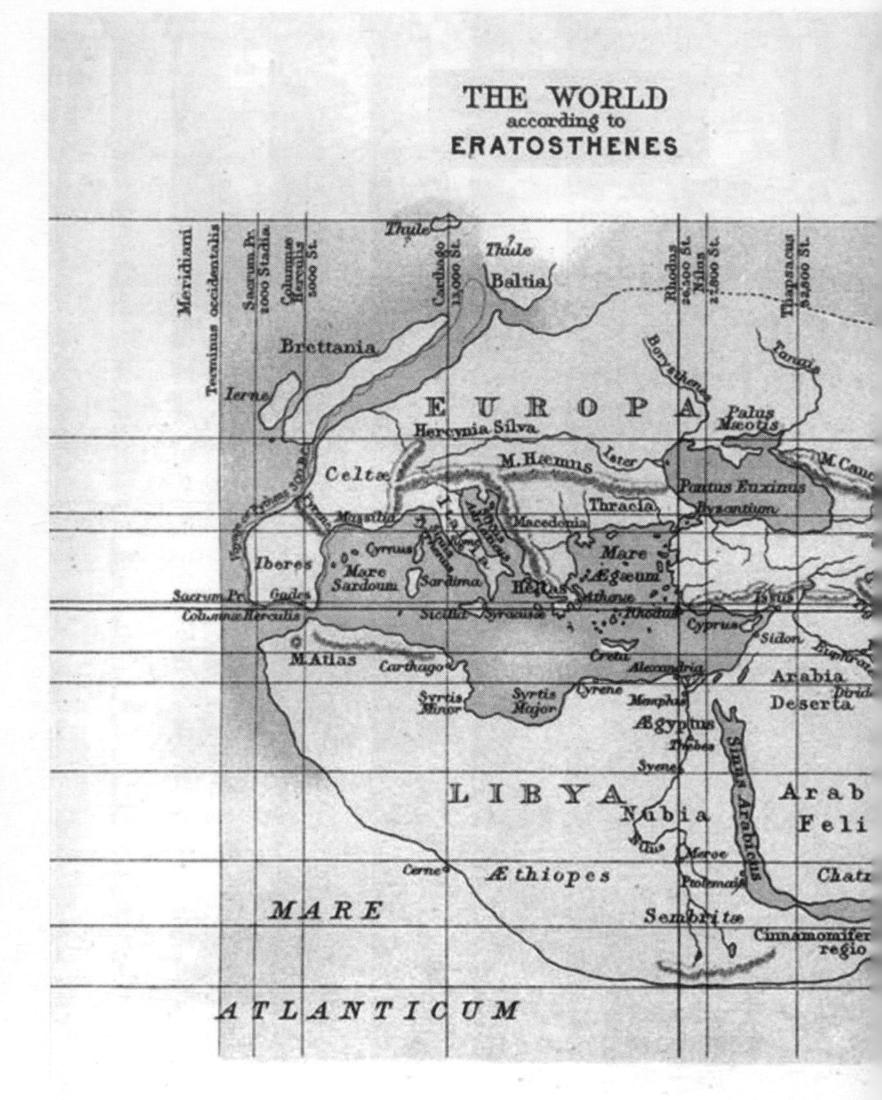

A skull-like vision of the world from Eratosthenes, with the equator through Rhodes, and Cinnamon Country spicing up the southern tip of Africa in this Victorian recreation.

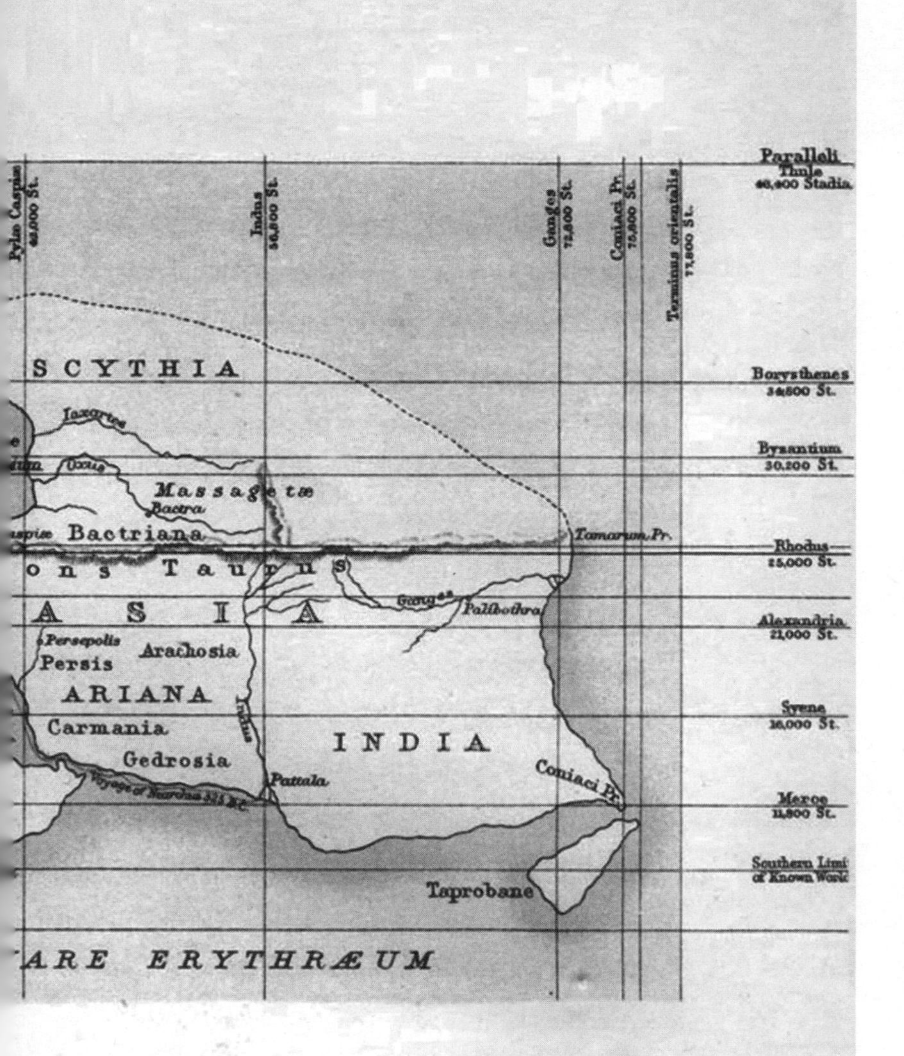

Paralleli
Thule
40,400 Stadia
Pylæ Caspiæ
42,000 St.
Indus
56,600 St.
Ganges
72,600 St.
Coniaci Pr.
75,600 St.
Terminus orientalis
77,800 St.
SCYTHIA
Borysthenes
34,800 St.
Iaxartes
Oxus
Byzantium
30,200 St.
Massagetæ
Bactra
Bactriana
Tamarum Pr.
Rhodus
25,000 St.
Taurus
Ganges
Palibothra
ASIA
Alexandria
21,000 St.
Persepolis
Persis
Arachosia
ARIANA
Indus
Syene
16,000 St.
Carmania
INDIA
Gedrosia
Coniaci Pr.
Pattala
Voyage of Nearchus 325 B.C.
Meroe
11,800 St.
Taprobane
ARE ERYTHRÆUM

invitation to explorers and geographers to map what was yet unknown?

The destruction of the Great Library by fire in 48 BC (conceivably an accident caused when Julius Caesar's troops set ablaze their own ships in an attempt to thwart the invading force of Cleopatra's brother Ptolemy XIV) was only the first to afflict it. It was destroyed or ransacked at least three times more, though each time succeeded in re-establishing itself, either on the same site or to the south-west of the city. Mark Antony replenished the library's stocks in 37 BC by raiding the library of Pergamum and donating some 200,000 volumes as a wedding present to Cleopatra.

Several years after the first firestorm, something remarkable happened to our understanding of the world: the emergence, in seventeen volumes, of the *Geographica*, the most comprehensive account of the world yet written. Its author, the historian and philosopher Strabo, was born in 63 BC in Amasia by the Black Sea, and survived long enough to straddle the Common Era.

Strabo was almost sixty before his first volume emerged about 7 BC; the last appeared a year before his death at the age of eighty-five. He was one of the world's first great travellers and much of the value of his geography lay in the descriptive passages of areas he himself had seen. He was not modest about these travels: in his second volume he boasts of a journey westwards from Armenia to Sardinia and to the south from the Euxine Sea to the borders of Ethiopia. 'Perhaps there is not one among those who have written geographies who has visited more places than I have between these limits.'

All but one of the volumes of Strabo's *Geographica* survive. Their stated purpose was to show how knowledge of the

idea

Idea Store Whitechapel
Self service

Friday, December 02, 2016 - 00:02
Borrower number: ******5175

You have borrowed 1 item

Title	Due back
On the map: why the world looks the way it does	24/12/16

You have 3 other items on loan

Title	Due back
Japan	17/12/16
Pocket Tokyo: top sights	17/12/16

inhabited world had developed in line with the expansion of the Roman and Parthian empires, and the volumes (divided into geographical regions) are invaluable in our understanding not only of cartography, but also of how the civilised world saw itself at the time of Julius Caesar and the birth of Christ. No physical map survives but it seems likely that Strabo was writing with a large manuscript map in front of him, or perhaps a selection of maps from which he drew a mental composite.

Intriguingly, Strabo's world is smaller than that described by Eratosthenes, his predecessor by two centuries. The earth's width is reduced to 30,000 *stades* (compared with Eratosthenes' 38,000), while its length is 70,000 *stades* compared to Eratosthenes' 78,000. Or that at least is his inhabited world, which he describes as 'an island' floating in a sea in the northern hemisphere. He believed that the world he knew and described took up about a quarter of the earth.

Strabo was no mathematician, and he distrusted the scientific advances in measurements and map projection made by Eratosthenes. Accordingly, he described his world in the most literal of ways, akin to the conceits of astrology. Taken as a whole, the inhabited world resembled a *chlamys*, a short tapering cloak worn by Greek soldiers and hunters. Britain and Sicily were triangular, while India was a rhomboid. He compared the northern part of Asia to a kitchen knife; Iberia to an ox-hide; the Peloponnese to a leaf on a plane tree; while Mesopotamia had the profile of a boat with the Euphrates as its keel and the Tigris the deck.

We read Strabo's *Geographica* now with a mix of awe and bemusement: awe at the scale of the enterprise, bemusement at some of its assumptions. Britain is thought not worth conquering, described as wretched and uninhabitable on account of its climate (Strabo notes that the sun hardly shines in Britain, particularly not in the region we now call Scotland). Ireland is full of cannibals. Ceylon, an island seven days' sailing from India, has an unusual crop: 'It produces elephants.'

Although Strabo is a geographer rather than mapmaker, he acknowledged the limitations of his descriptions, instructing that his prose should be visualised on a flat surface. For this he suggests a simplified grid of parallels and meridians on a parchment seven foot long and three foot wide. But he also envisaged a far better method of representing his research: a globe.

He mentions a sphere constructed by the philosopher Crates of Mallus in the previous century that was ten feet in diameter and showed the world divided into four clear regions, all islands, all of roughly equal size, two above the 'torrid zone' dividing the northern and southern hemispheres and two below.* Only one of these islands – his own – was definitely inhabited, but Crates, drawing on a combination of Eratosthenes and Homer for much of his information, believed that the other three might also be temperate and populated, with at least one other region below the equatorial ocean cultivated by 'Ethiopians' who had no connection with other Ethiopians in Cinnamon Country.

Strabo suggested that his own globe should also be at least ten feet in diameter in order to capture sufficient detail. But he appreciated that most of his readers would find the construction of such a thing beyond them.

The Great Library of Alexandria had one more defining contribution to make to the history of cartography, and although it built on the gains of Eratosthenes and Strabo it was such a momentous piece of individual scholarship that it set the tone

* This is the first terrestrial globe we know of, although it no longer exists. Crates of Mallus, a leading literary critic, was believed to be the librarian at Pergamum, Alexandria's biggest rival. But the fleeting accounts of his life in the history books remember him for one other thing, too – breaking his leg while examining a sewer in Rome.

and look of map-making in the European and Arab worlds for hundreds of years. This wasn't a map itself, but a descriptive atlas, and its originator could be said to be the world's first modern cartographer. It was a book of instructions, in Greek, that changed the way we looked at the world so fundamentally that – almost 1,350 years later – it was, in modified form, one of the main navigational tools Columbus carried with him when he departed for Japan in 1492.

The atlas was the work of Claudius Ptolemy, who lived between 90 and 170 AD, studied at Alexandria for the majority (if not all) of his life, and had earlier produced a highly influential treatise on Greek astronomy called the *Almagest*. This contained detailed star charts and a multilayered model of the earth's position in the cosmos, with the earth, stable in the centre, playing host to the daily revolution – in order of proximity – of Moon, Mercury, Venus, Sun, Mars, Jupiter and Saturn, and a sphere of fixed stars sparkling on the outer edge. Ptolemy also wrote a scientific investigation into optics, examining the process of seeing and the role of light and colour.

But the work we are interested in is Ptolemy's *Geographia*. This was a two-part interpretation of the world, the first consisting of his methodology, the second of a huge list of names of cities and other locations, each with a coordinate. If the maps in a modern-day atlas were described rather than drawn they would look something like Ptolemy's work, a laborious and exhausting undertaking, but one based on what we would now regard as a blindingly simple grid system. In the seventh section of *Geographia* (there were eight in all), Ptolemy provided detailed descriptions for the construction of not just a world map, but twenty-six smaller areas. No original copies have survived, and the closest we can get to it is a tenth-century Arab description of a coloured map – though whether that was an original or merely inspired by his text is unknown, and at any rate, it no longer exists.

The modern winds of change: Ptolemy's classic map of the world, beautifully rendered in 1482 by the German engraver, Johannes Schnitzer of Armsheim.

As one would expect, Ptolemy had a skewed vision of the world. But while the distortion of Africa and India are extreme, and the Mediterranean is too vast, the placement of cities and countries within the Greco-Roman empire is far more accurate. Ptolemy offered his readers two possible cylindrical projections – the attempt to project the information from a three-dimensional sphere onto a two-dimensional plane – one 'inferior and easier' and one 'superior and more troublesome'. He gives due credit to a key source, Marinus of Tyre, who had advanced the gazetteer listings system a few decades earlier, assigning his locations not merely a latitude and longitude, but also an estimated distance between them. (Marinus had another claim, too: his map data was the first to include both China and the Antarctic.)

Ptolemy boasted that he had greatly increased the list of cities available to the cartographer (there were about 8,000), and also disparaged the accuracy of Marinus's measurements. But he had his own flaws. Indeed, the map historian R. V. Tooley suggests that Ptolemy stood apart from his predecessors not just in his brilliance but in his disregard for science. Where earlier cartographers were willing to leave blanks on the map where their knowledge failed, Ptolemy could not resist filling such empty spaces with theoretical conceptions. 'This would not have mattered so much in a lesser man,' Tooley contends, but so great was his reputation 'that his theories assumed an equal validity with his undoubted facts.' As we shall see, this had the uncanny ability to send ambitious sailors, Columbus among them, to places they had no intention of seeing.

There were maps of the world before these Alexandrian advances – a clay tablet here, a papyrus shroud there – but they were unique and random objects.* By contrast, the maps by Eratosthenes, Strabo and Ptolemy spawned at the Great Library were logical and disciplined. The reputation of the library as the most important the world has seen has some grounding here – and it is a legend made more romantic by the various destructions that befell it over the centuries.

The Library's ultimate destruction occurred nearly half a millenium after Ptolemy's death, in 641, when Alexandria fell to the Arabs. At the time, the Library had again been replenished, and although it was not the powerhouse of learning it once was, it still contained many hundreds of thousands of volumes. But its new captor apparently had no use for books. When asked about the fate of the Library, the Caliph Omar is said to have replied: 'If the contents of the

* The Babylonian clay tablet that sits proudly in the British Museum (and at the beginning of many pictorial histories of maps) is believed to date from the Persian Period of 600 to 550 BC, and is a mystical and brilliantly imaginative item, the sort of thing that inspires conspiracy theories and blockbuster novels. We only have a damaged portion of the whole, believed to have been no more than 12.5 x 8cm when made. Its purpose is unclear, but it does conform to the general pattern of ancient world maps in so far as its creator placed his own world at the centre of it. So Babylon sits in a sea surrounded by seven unnamed circles, which may be either cities or countries. Around these sits an encircling ocean named Bitter River, into which flows the Euphrates, and on the edge of this lie seven triangular islands. We glean what we can from the damaged text above the map and on the back of the tablet: the islands are only seven miles apart from the Babylonian world, and are described principally in terms of light. One, due north, lies in complete darkness, and may betray knowledge of polar regions, while others lie 'where the morning dawns' or in light brighter than stars. Yet another contains a horned bull that 'attacks the newcomer'. The text also describes a Heavenly Ocean ringed by a constellation of animals, some of which we would recognise today as Leo, Andromeda and Cassiopeia.

books are in accordance with the book of Allah, we may do without them, for in that case the book of Allah more than suffices. If, on the other hand, they contain matter not in accordance with the book of Allah, there can be no need to preserve them. Proceed, then, and destroy them.'

But there is one more improbable thing. We have seen that Ptolemy's *Geographia* appeared in about AD 150, and we could logically have anticipated a steady stream of cartographic progress. The coordinates and projection that he employed were a universal system, something to be employed and expanded as our knowledge of the world itself grew over the centuries. It was like an enormous net, able to catch new information and spread out accordingly. But it didn't happen. The steady cartographic advance one might have anticipated failed to materialise. Where was the Ptolemy of the fourth or fifth century? Why do we not know what Harold thought of the shape of the world when he trotted out to Hastings in 1066? Or how Saladin saw the Middle East? Because there are no maps to show us.

Neither the Romans nor the Byzantines progressed Ptolemy's work. There were some fine localised beauties – the Peutinger Table from the fifth century (a long, schematic roadmap showing the key settlements of the Roman Empire), and the sixth-century Madaba map (a mosaic of the Holy Land, preserved in a church in Jordan, that includes street plans of Jerusalem and other cities). But they show little curiosity about the world beyond, and neither of them advances the science of mapmaking.

Instead of progress, the world appeared to fall into the cartographic dark ages for about a thousand years. Did our ambitions towards exploration, conquest and the pursuit of wealth suddenly disappear like so much candle smoke? And what about globes? They too spun backwards. The concepts of latitude and longitude, the emergence of the graticule and the prime meridian – all these were put back in the box, only

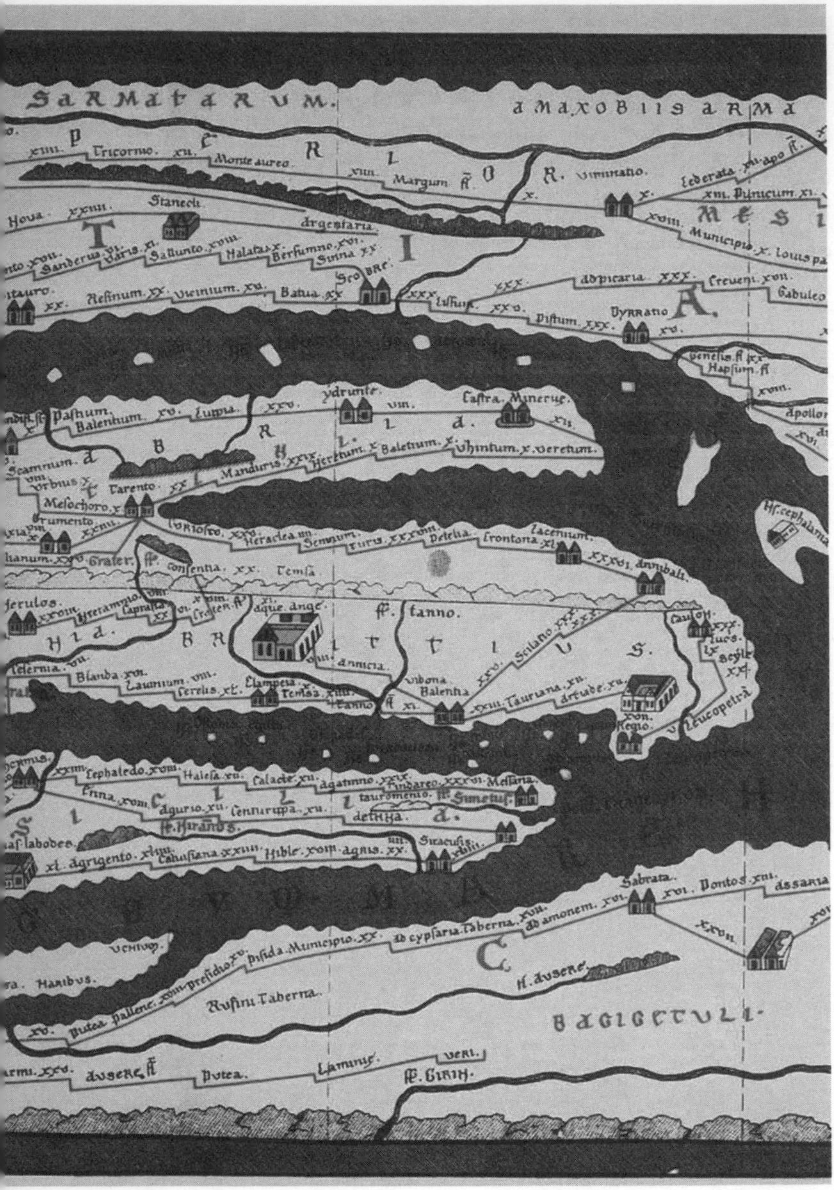

The long and winding empire: a detail from the Peutinger Table, a fifth-century Roman roadmap stretching from the Dalmatian Coast to the African Med.

really to emerge into the sun again in teeming Venice and Nuremberg in about 1450.

And what was it that did actually emerge at the height of the Renaissance? Some great new picture of the world? The discovery of new continents? Something to do with America? No, what emerged was the translation from Greek into Latin of a volume that had been thought lost since the glory days of Alexandria. It was Ptolemy's 'atlas', and its rediscovery – matched with the boom in European printing – heralded the birth of the modern world.

But let's stay awhile in the dark ages. Or, more precisely, Hereford in the winter of 1988.

Chapter 2
The Men Who Sold the World

On Wednesday 16 November 1988, the Dean of Hereford, the Very Reverend Peter Haynes, and Lord Gowrie, a former Arts Minister who was now Chairman of Sotheby's, stood outside Hereford Cathedral in suits and posed for photographs beside a framed facsimile of a large brown map. The map, almost as tall as the pair holding it, was due to be auctioned the following June, and Sotheby's had agreed a reserve price of £3.5m, which would make it the most valuable map in the world. Later that day it would be described by Dr Christopher de Hamel, Sotheby's expert on medieval manuscripts, as 'without parallel the most important and most celebrated medieval map in any form.'

Lord Gowrie regretted that such an important object might soon be leaving the country to the highest bidder, but said that all attempts to save it for the nation had failed. He had been trying for almost a year to keep the map in the UK, but now needs must. The Dean explained that his eleventh-century cathedral, one of the most impressive Norman constructions in England, was in need of £7m to prevent it from crumbling to the tiled floor, and disposing of the map was the only way forward. After their announcement, the two men handed the frame to the cathedral staff, and departed, Gowrie back to London, the Dean back to his troubled place of worship.

Mass unhappiness ensued.

The map in question was Hereford's Mappa Mundi, *c.*1290, and it wasn't a beautiful thing to look at. A large shank of tough hide – measuring 163 cm by 137 cm – it has a murky rendition of the world that, at fist sight, is hard to fathom with its faded colours and indistinct lettering. It is also a map that if you had been transported from the Great Library at the time of Ptolemy, would have come as quite a surprise. Gone is the careful science of coordinates and gridlines, longtitude and latitude. And in their place is, essentially, a morality painting, a map of the world that reveals the fears and obsessions of the age. Jerusalem stands at its centre, Paradise and Purgatory at its extremes, and legendary creatures and monsters populate the faraway climes.

And this is very much its conception. The *mappa* (the word meant cloth or napkin rather than map in medieval times) had a lofty ambition of metaphysical meaning: a map-guide, for a largely illiterate public, to a Christian life. It has no reservation in mixing the geography of the earthly world with the ideology of the next. Its apex displays a graphic representation of the end of the world, with a Last Judgement showing, on one side, Christ and his angels beckoning towards Paradise, and on the other the devil and dragon summoning to another place.

But it seems likely that those who saw it first at the end of the thirteenth century would have done what we do now and looked for the 'You Are Here' spot. If so, they would have

A scandal in the making: Hereford Cathedral's Dean, Peter Haynes (left), and Sotheby's chairman Lord Gowrie announce the sale of the Mappa Mundi.

found themselves in the south-west region of the giant circle, with Hereford one of the few places mentioned in England, and England itself a fairly insignificant part of the global story. Around them is a world crowded with cities, rivers and countries teeming with human activity and strange beasts. Ancient and brilliant cartographic theories have been replaced by something else: the map as story, the map as life.

Such a thing had never required a reserve price in an auction catalogue before. But now, according to God's current earthly representatives, a judgement had to be made. The timing of this can be fixed precisely, to February 1986, when an expert in medieval artefacts from Sotheby's arrived at the cathedral to appraise its most prized possessions. At the time the Mappa Mundi was not on Hereford's sale list. The cathedral's great treasure was thought to be the books and manuscripts of its Chained Library, a collection of theological works secured by chains to their cases to enable study but not theft. While ascending a winding stone staircase to the library, the assessor saw the dimly lit Mappa Mundi below, and asked how much it was insured for. He was astonished at the answer: £5,000. He suggested it might be worth a little more.

The furore that ensued once the sale was announced took the cathedral entirely by surprise. Britain's National Heritage Memorial Fund expressed 'outrage that one of the most important documents in the world is to be put into an auction room'; the British Library complained that it had not been consulted on a possible sale (though Lord Gowrie claimed that it was 'talking rot'). *The Times* composed a leader that concluded: 'The Mappa ought to remain in England, on public display, and preferably in Hereford. Its ancient and original link with that city is part of the Mappa's identity. As a work of art, it gains from being in Hereford. It is, so to speak, the only proper frame for it.'

The following day, amid resignations from the Hereford Cathedral's fundraising committee, several offers of a private purchase emerged, meeting the £3.5m reserve. But Canon John Tiller, the cathedral chancellor, announced that the auction would proceed regardless to gain the best possible price: 'Our first priority is the future of the Cathedral.'

Other funding schemes were raised but came to nothing. Then some months later came one that stuck: the Mappa Mundi Trust was established with a £1m donation from Paul Getty and £2m from the National Heritage Memorial Fund, amid plans for a new building to house the map, to which the public would be charged admission. And in this way the map was saved for the nation. While these plans were laid, the map was loaned to the British Library in London where it was seen by tens of thousands who had only recently been made aware of its existence.

❦ ❦ ❦

What exactly did the British Library visitors see? They saw what pilgrims arriving in Hereford around 1290 would have seen, but with less colour, better footnotes and tighter security. The Mappa Mundi provides a masterful cartographical insight into medieval understanding and expectations. What appears at first glance like wonderful naivety is on more informed inspection an extraordinary accumulation of history, myth and philosophy as it stood at the end of the Roman Empire, with a few medieval additions.

The map is frantic – alive with activity and achievement. Once you grow accustomed to it, it is hard to pull yourself away. There are approximately 1,100 place names, figurative drawings and inscriptions, sourced from Biblical, Classical and Christian texts, from the elder Pliny, Strabo and Solinus to St Jerome and Isidore of Seville. In its distillation of geographical, historical and religious knowledge the mappa serves as an itinerary, a gazetteer, a parable, a bestiary and an educational aid. Indeed, all history is here, happening at the same time: the Tower of Babel; Noah's Ark as it comes to rest on dry land; the Golden Fleece; the Labyrinth in Crete where the Minotaur lived. And surely for contemporaries – locals and pilgrims – it must have constituted the most arresting freakshow in town. With its parade of dung-firing animals, dog-headed or bat-eared humans, a winged sphinx with a young woman's face, it seems closer to Hieronymus Bosch than to the scientific Greek cartographers.

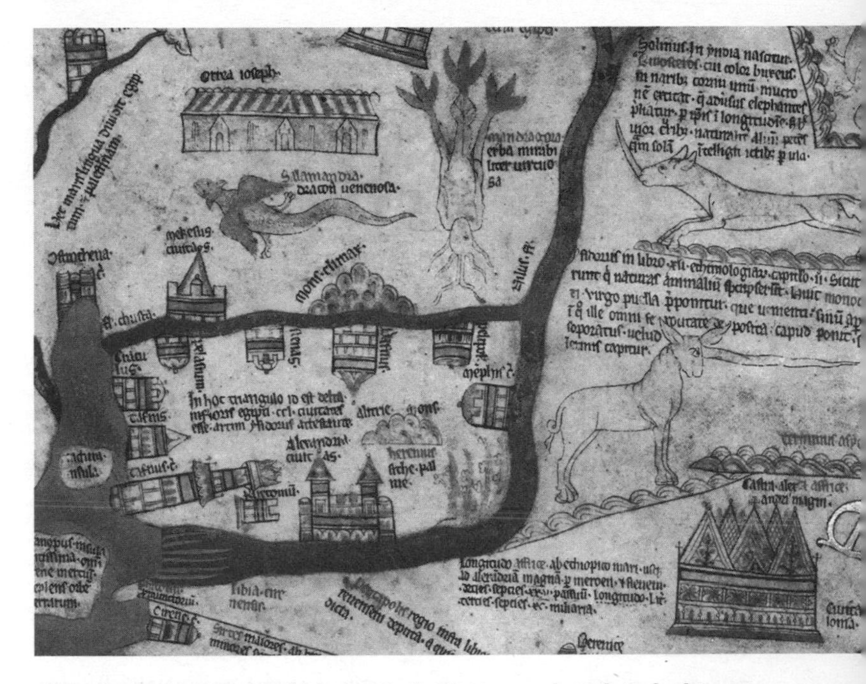

These days you'd be tested for chemical substances: the Nile Delta bisects a magical world of unicorns, castles and a peculiar mandrake man.

We are about ninety years from Chaucer, and although there is much to be read on the map in clear gothic Latin and French script, most visitors to Hereford would have obtained their knowledge from the pictures. A century and a half before the printing press, these drawings – primitive and devoid of perspective, with almost every turreted building indistinguishable from the one next to it – would have been the first big storyboard they had ever seen, and its images would surely haunt their dreams.

Observed through modern eyes, the map is also a sublime puzzle. Things are not where we might expect them to be. What we regard as north lies to the left, while east is at the top, a placing that has given us the word 'Orientation'. There are no great oceans, but instead the map is surrounded by a watery frame and floating islands of malformed creatures abound.

There are dismal transcribing errors in which Europe is labelled *Affrica* and Africa appears as *Europa*. Cities and emblematic structures seem to be selected by a curious mixture of importance, hearsay, topicality and whim: the Colossus of a misplaced Rhodes occupies more space than cities more valuable for trade or learning, such as Venice. Both Norway and Sweden appear on the map, although only Norway is named.

The British Isles appear at the northwest corner of the map, on their side, to fit the space. North-east England is populated with names, while the south-west is almost ignored. Edward I's new castle at Caernarfon makes an appearance – only a few years after its construction began – which not only helps with the dating of the map, but also confirms that new local landmarks are judged as important as those of antiquity.

There are many further anomalies. Moses appears with horns, a familiar medieval confusion between *cornu* (horn) and the intended *cornutus* (shining). The monster Scylla (labelled *Svilla*) appears twice, once in the familiar pairing with the whirlpool Charybdis, and once where the Scilly Isles lie, possibly a mishearing by a scribe. And there is another story going on here, as with many *mappae mundi*. The wilderness – the scary stuff, the unknown lands – sends a message to the viewer about the glories of civilisation and order and (self-)control. For contemporaries, it is another Christian doctrine: follow the laid-down path. To the modern viewer, though, it is the weirdness that most enthrals: the rich, demonic and comical features, like the sciapod – a man deploying his single swollen foot to shield himself from the sun.

To cartographic historians, the Hereford Mappa Mundi is

The basic form of a T-O map, crudely dividing up Asia, Europe and Africa. This one is from a twelfth-century Spanish manuscript.

categorised as a 'T-O' (or 'T within an O') map. This is a form developed at the time of the Roman Emperor Agrippa (after 12 BC) and described a simple way of separating the spherical earth into three parts. The known continents of the ancient world – Asia, Europe and Africa – are divided in the middle by the horizontal Don and Danube rivers and the Aegean Sea (on the left) and the Nile (on the right) as they flow into the wide vertical Mediterranean.

But the Hereford map is more than a circle. Determined to use every inch of available space on the hide, it encompasses monumental scenes above and below it. So the world is crowned by the Last Judgment, while at the base a scene on the left depicts the Emperor Augustus instructing surveyors to 'Go into the whole world and report to the Senate on each continent.' A scene to the right of this is less obvious – and may perhaps be the map's equivalent of the 'and this just in...' news bulletin. It shows a horseman and hunter in conversation, the message (in French rather than the map's general use of Latin) being 'Go ahead'. Who these figures are is unclear; one interpretation has them as the subject of a court case over hunting rights that took place in Hereford as the map was drawn.

In the January 1955 edition of the English periodical *Notes and Queries*, a scholar called Malcolm Letts analysed the more vivid drawings on the map, and divined their meaning. Letts was easily shocked, but his breathless prose has never been bettered in conveying sheer wonder. He describes the exploits of gold-digging ants in great detail, and devotes a whole paragraph to a salamander. He admires a picture of the Gangines 'busily collecting fruit from a tree. These creatures were said to live on the smell of apples which they always carried about with them, otherwise ... they died anon.' Close by 'comes the lynx whose urine congealed itself into the hardness of a stone (most vividly presented by the artist.)' Then, in the centre of the map, near Phrygia, he sees the Bonacon, 'whose method of defence was to discharge its ordure over three acres of ground and set fire to everything within reach. The creature is shown in its customary posture of defence ...' Reviewing the drawings on the extreme right of the map, Letts observes, 'two men embracing each other.' They are the Garamantes of Solinus, of whom little is known 'except that they abstained from war and disliked strangers.'

How could such a marvellous thing ever be sold to fix a leaky roof?

To meet the man who almost sold the Mappa Mundi one must drive up a hill from Hereford town centre, past vineyards, hop fields and many barns, until one eventually arrives at the beamed home of the Very Reverend Peter Haynes, who retired from his cathedral in 1992 and now spends time indulging his passion for model trains. I made this pilgrimage in the summer of 2011, to be greeted with tea, lemon cake and a slim cuttings file. The file included newspaper articles, a press release and a Mappa Mundi Plc share prospectus which didn't quite fly.

Haynes is eighty-seven but remains Dean Emeritus of the cathedral, and when he visits each Sunday many people still address him as Mr Dean. He served in the RAF during the war, was ordained after theological college at Oxford, became Vicar of Glastonbury at the time of the first festival in 1970, moved to Wells as Archdeacon in 1974, and, after a personal request by Margaret Thatcher, took up as Dean of Hereford in 1982.

He says that the first thing he did was send copies of the cathedral accounts in confidence to an old accountant friend, the financial director of Clarks, the shoe firm. 'He nearly had kittens. He sent them back to me and said, "You're in for terrible trouble."' There had been a continuous operating deficit for years, and an overdraft at the bank of more than £150,000. 'I realised that what the congregation was raising each year, about £17,000, and they fondly imagined was paying for the clergy, was actually going over to Lloyds Bank to service the overdraft.'

Worse was forecast: the staff pension scheme was inadequate, a building survey had revealed large and dangerous cracks, the choir required an endowment, and the cathedral's historical treasures were inadequately cared for and poorly presented. An appeal was launched in April 1985 by the Prince and Princess of Wales, but its target of £1m to restore the fabric of the building was soon deemed inadequate. It was estimated that the cathedral needed a capital injection of £7m to secure its long term security and endowments, and so the Mappa Mundi would have to go (this would also ensure that the Chained Library would not be broken up). At the time, the Dean believed that it wouldn't be much missed. 'I would often welcome visitors when they came to the cathedral, just to keep my hand in. I would tell them, "Oh, there's a very ancient map in the North Choir Aisle if you'd like to see it," but nobody thought anything of it.'

When Haynes examined the map himself he noticed damp around the perimeter. So he contacted a man called Arthur David Baynes-Cope at the British Museum. 'He was a world authority on mould,' Haynes told me, although Dr Baynes-Cope, a chemist, was also an expert on paper and book conservation. Not long before he died in 2002 he said that he was particularly proud of his forensic work that had exposed Piltdown Man as a fake. 'I got him up here,' Haynes continued, 'and he had a look at it, and he said, "Oh, I think I know what we can do with that." He came up about a fortnight later, and he put this cord around the side, and I said, "What's that then?" And he said, "Oh, this is pyjama cord from Dickins & Jones".'

Haynes says that he supervised a great amount of other work while at the cathedral, but knows that his time there will mostly be remembered for the mappa saga. He seems not entirely unhappy with this. His eyes sparkle as he tells me of an idea they had for a share issue to raise funds: 'At the beginning it was important that we kept it all secret. So there was a codename for it, an anagram: Madam Pin-Up.'

These days we can all buy a pretty good Mappa Mundi of our own. The copy that Peter Haynes and Lord Gowrie posed with outside the cathedral in 1988 was a lithograph from

1869, at that time the best facsimile available. But in 2010 the Folio Society brought out a spectacular version, 9/10ths in scale, which benefitted not only from digital reproduction (a phrase which too often spells the death of art), but also from the imagination and learning of twenty-first-century experts, including Peter Barber, the head of maps at the British Library and a trustee of Hereford's Mappa Mundi. The map was not only visually crisped up, but rendered with vivid colours thought to be close to the original – lustrous reds, blues, greens and golds. It was printed on a vellum-like material called Neo-bond, mounted on canvas, topped and tailed by wooden struts of Hereford oak, and accompanied by erudite essays; limited to an edition of 1,000, it would set you back £745.

Academic study of the Mappa Mundi has had a resurgence of late – another benign result of thc abortive sale – and now tends towards the investigative and forensic rather than

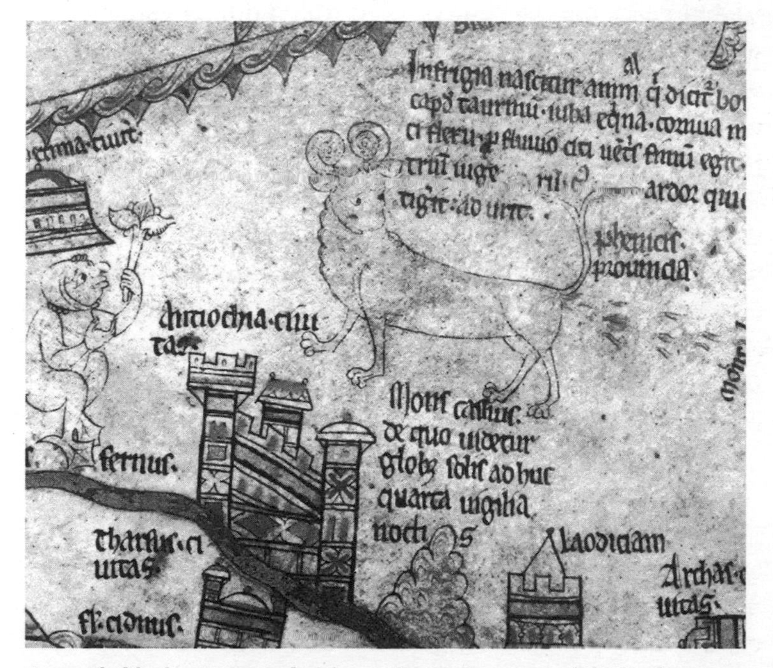

You probably don't want to live here: the Bonacon spreads his 'ordure'.

the interpretive. But many fundamental questions remain definitively unanswered, such as exactly who made it.

The left-hand corner reveals the main clue. There is a plea to all those who 'hear, read or see' the map to pray for 'Richard de Haldingham e Lafford' who 'made and laid it out'. The place names can be transposed to Holdingham and Sleaford in Lincolnshire, but who was this man, and what did he 'make' exactly? In 1999, a symposium in Hereford attracted the leading Mappa Mundi scholars, most of whom agreed upon the hand of a man called Richard of Battle, known in Latin as Richard de Bello, a canon of Lincoln and Salisbury, a prebendary of Sleaford who may have lived in Holdingham; but they were unsure if this was one man, or a Richard of Battle Snr and Jnr, related or not.

Four years later, the map historian Dan Terkla presented a paper on the Mappa Mundi at the twentieth International Conference on the History of Cartography at Harvard. He speculated that four separate men were directly responsible for the map's design, three of them named Richard – Richard of Haldingham and Lafford, Richard de Bello, Richard Swinfield – and the other, Thomas de Cantilupe. The second Richard, he believes, was a younger relation of the first, who worked on the map at Hereford after moving from Lincoln; the third was a friend working as a financial administrator and bishop at the cathedral; while Thomas de Cantilupe was Swinfield's predecessor and may also have been the mounted huntsman on the map's fringe.

Terkla asserted that the map formed a part of what he called the Cantilupe Pilgrimage Complex, a collection of possessions and relics associated with the bishop, who was canonised in 1320. De Cantilupe's shrine in the north transept had proved a huge draw on the pilgrim trail even before his sainthood, after word spread of its uncanny associations with miracles. Almost five hundred miraculous acts were recorded by the custodians of the shrine between 1287 and 1312, with seventy-one miracles listed in April 1287 alone – the year the royal family visited.

In 2000, a rather different Mappa Mundi scholar, Scott D. Westrem, travelled from the United States to examine the map without the glass, and his forensic report displays Sherlockian sleuthing. 'The vellum on which it was drawn came from the hide of a single calf, probably less than one year old when it was slaughtered,' he surmised. He observed that the map was drawn on the inner side of the skin, noting the silvery fleshy membrane. He described how the cured hide was carefully scraped to remove hair and residual fat, and suggested that the skinner's knife slipped only once during the process, severing the skin near the tail end, probably when it encountered scar tissue. 'The skin's quality is very high and apparently of even thickness; there is almost no sign of the rippling that results from the impression of the rib cage and other bones, indicating that the beast was consistently well fed.'

In May 2011, Dominic Harbour, the cathedral's commercial director, is taking another pair of visitors around the Mappa Mundi. I'm one of them. Two weeks before, the map had been installed in a new frame, something that has brought it about a foot down the wall so that Jerusalem now coincides with most people's eye level. 'Before, it was presented within a space in an architectural way,' Harbour explains. 'But people relate to it in an ergonomic way. It needs to be related to your height and what you can see and what you can touch.'

The touching is important. Although not actively encouraged, visitors are not reprimanded for doing what comes naturally – putting their fingers to the glass as a primitive way of navigating around it. At the end of each day, the glass is cleaned of fingerprints, and there are some marks over

Jerusalem, and others over Europe, reflecting the origin of that day's visitors. Americans keep their fingers to themselves. But the greasiest spot on the glass is around Hereford. 'The fingerprints tell exactly the same story they did when the map was new,' Harbour says. 'For a long time people thought that "Hereford" had been added to the map after it was originally made, suggesting that it may have begun life elsewhere. Now the thinking is that Hereford was indeed added to it later, but only after the original inscription of Hereford had been worn away from people touching it.'

Harbour is in his late-thirties, and has been working with the map since he was twenty-two. He arrived fresh from art college in 1991, helped design an explanatory booklet and a facsimile map with an English translation, and soon realised that his six-month contract might have to be extended. He began thinking about how the map could be presented in a more cohesive and effective way, and helped plan its current, impressive new exhibition space in a cathedral cloister, designed to accommodate both the Mappa Mundi and the Chained Library. The space, completed in 1996, is essentially a glorified lean-to, albeit a lean-to where a fifteenth-century stone wall adjoins an eleventh-century one.

Harbour takes his visitors to the spots where the map has been displayed or concealed through the centuries – the Lady Chapel, various transepts, the vestry where it lay under floorboards. He says he once drew a map of the map's movements, and 'ended up with scribbles all over the place'. He recalls that when he was first taken to see the map, at the age of eight, he saw 'this really strange brown thing in a case, otherworldly, magical, like some scientific sample in a jar. I don't think it was explained at all, or at least nothing that was accessible to me. Just, "This is Mappa Mundi and it is very important."'

As we look at the map together, I find myself nodding as Harbour suggests that 'it is still delivering new things'. No modern traveller can look at it and not feel desire. It is one of the most appealing features of large maps, and world maps in particular, that all journeys are feasible. On the Hereford map, everywhere except Paradise seems reachable in sturdy vessels, and even the fiercest beasts look biddable. And then it struck me: in 1290, unlike today, there seemed to be little left to explore, and no great wilderness or sea to detain you long. Unfathomable sea monsters and great white polar silences only came later. The simple message here is: we've done our work in this place, for the inhabitable world is laid down on the back of a calf. So what remains for us mere mortals? Only miracles, a higher calling, and things forever beyond our grasp. Spread the word, pilgrims.

Pocket Map
It's 1250, Do You Know Where You Are?

These days we employ the term 'road map' as a political phrase, to denote the prospect of progress. A situation may seem hopeless, but at least we have a plan: if we get to staging post A, we then have a chance of getting to staging post B. Of course, the phrase is occasionally used by people not going anywhere, notably, in 2002, by George Bush, Tony Blair and others engaged in the Middle East peace process.

Back in the thirteenth century, a few decades before the Mappa Mundi was created, a monk called Matthew Paris (*c*.1200–59) was involved in constructing an authentic road map to the Middle East – a map, in fact, that ended at Jerusalem, then under relatively benign Muslim rule and attracting large numbers of Christian pilgrims.

As part of his monastic duties, Paris worked as a manuscript illuminator and historian in the abbey at St Albans, north of London. The abbey possessed the attributes of an early university and Paris was keen to distil his learning in both visual and textual form. The result was his *Chronica Majora*, an ambitious history of the world from the Creation to the present. Writing in

Jerusalem this way (or maybe that ...). Matthew Paris's interactive road map suggests several ways to salvation.

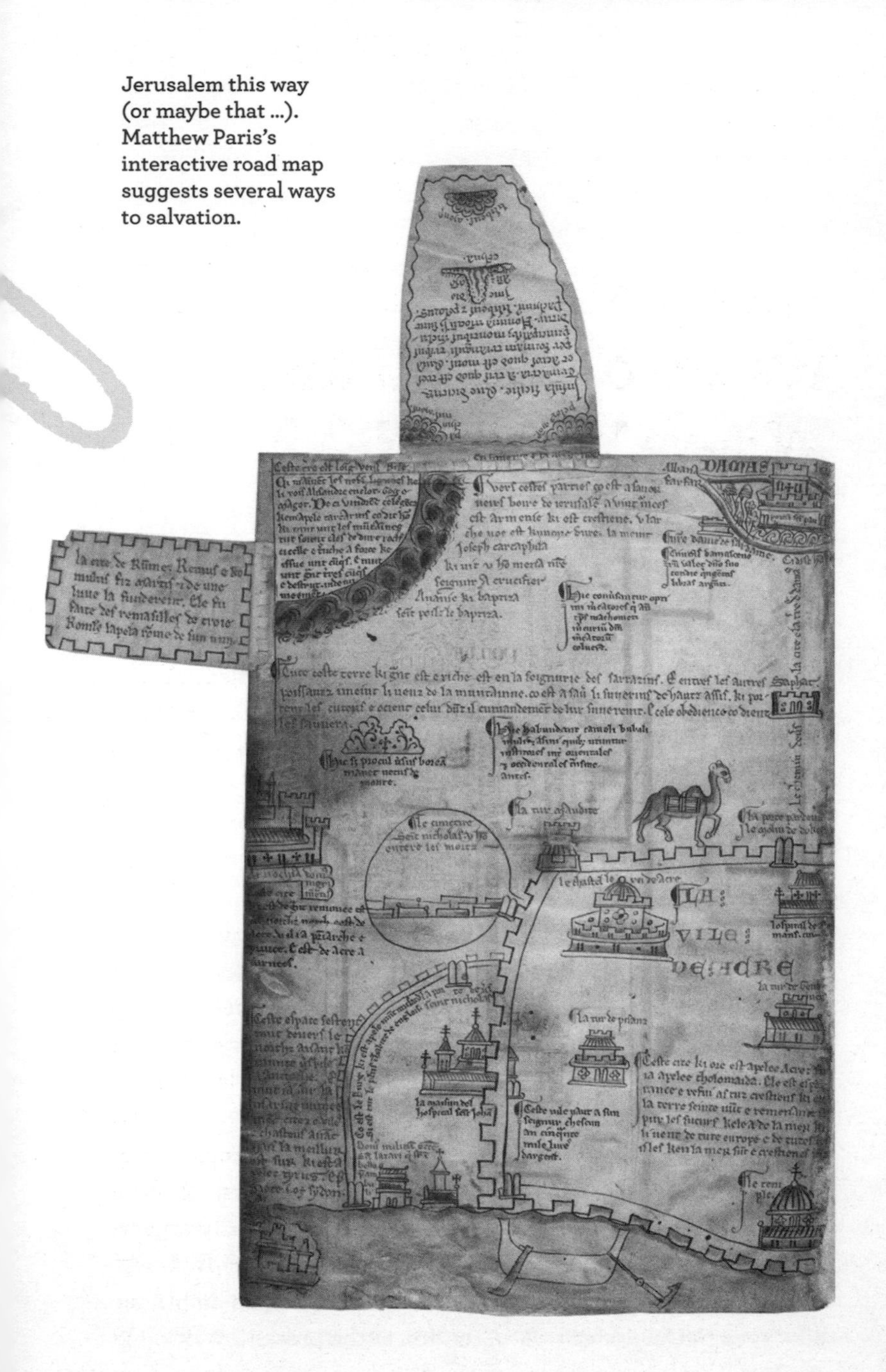

Old French and Latin, Paris combined the work of Roger Wendover, an immediate predecessor at St Albans, with his own experience. This included his own extensive travels in Europe and visits to the court of Henry III, as well as the stories gleaned from visitors to his own abbey.

His strip map from London to Jerusalem occupies seven pages of vellum at the beginning of his *Chronica Majora* and is an enchanting manuscript, full of little diversions and side-panels to keep its readers entertained, with glued-on pop-up leaves extending a journey or explanation over or round a page. One flap attached to the top of a page features Sicily and Mount Etna, described as the Mouth of Hell. Interactivity went further still, as the viewer was often presented with a choice of alternative routes, heading through France and Italy at different angles. Was this the first map with movable parts? It was certainly the first route map that we know of with such a laissez-faire attitude to the actual route.

Paris's readers may not have been particularly attracted to the prospect of actually embarking upon the itinerary laid before them (quite probably the reverse), but they would have revelled in the imagined, spiritual journey – the virtual crusade. And they would have been engaged, as we are still, with its pictorial evocation of the route.

But the map is significant for another reason. Paris refers to it as an 'itinerary', and it is where we get the word journey, from the Old French *jornee* or *jurnee*, the estimate of a reasonable 'day's travel' on a mule. The word appears between many destinations on Paris's map, and on one occasion, with nothing else of importance en route, it is stretched out for dynamic effect into 'ju-r-r-r-n-ee'.

Each page, which has two columns and flows from bottom to top and left to right, contains about a week's expedition. We begin in London, displayed as a walled city with the 'River de Tamise', 'Audgate' and 'Billingesgate' marked against a selection of crenellated buildings and steeples, dominated by St Paul's.

From there it's a day to Rochester, another day to Canterbury, one more to Dover and northern France before heading through Reims, Chambery, and Rome. The map's accuracy declines as Paris leaves Paris, but to criticise the appearance of Fleury as coming just after Chanceaux rather than just after Paris would be to misjudge the map's intentions; it was clearer, for the narrative of the story, to have the site of Saint Benedict's bones there rather than earlier.

Another entry, on another flap, suggests a certain weariness with the map-making craft: 'Toward the Sea of Venice and toward Constantinople on this coast,' Paris writes, 'are these cities which are so far away.' But the strip eventually reaches Jerusalem, the final destination, which Paris depicts with the Dome of the Rock and church of the Holy Sepulchre, and a

Are we there yet? Jerusalem finally in view as we near the end of the journey.

reasonably coherent coastline, from the gateway port of Acre, and beyond it, Bethlehem.

Paris does seem to have been troubled by problems of scale. On another map, one of four he drew of Britain, he writes regretfully (across an illustration of London) that 'If the page had allowed it, this whole island would have been longer,' which is not necessarily the ideal cartographic lesson to pass on to impressionable minds. Despite the squeeze, Scotland is permitted a generous and bulbous display, a rarity in this period. But Paris also produced another map in which Britain has uncannily accurate proportions, not least in Wales and the West Country, and it is difficult to argue with the British Library's claims that it is the earliest surviving map of the country with such a high level of detail.

One other significant Paris strip map has also survived, from London to Apulia, but it is less elaborate than the Jerusalem version and crammed onto one page of his *Book Of Additions*, an addenda to his larger history that contains such oddities as a map of the main Roman roads with Dunstable at the centre and a map of the world's leading winds (with the earth at the centre).

With all these maps, Matthew Paris achieved one other thing easily overlooked by detailed cartographic analysis: some fifty years before the Hereford Mappa Mundi, he made objects that provided a highly engaging and unique viewing experience, and he was uncommonly prescient in showing how maps may delight by their beauty and intrigue. His maps nourish the imagination, and they prompt interaction and engagement. They have an uncanny resemblance to the maps we draw as children.

Chapter 3
The World Takes Shape

The Mappa Mundi at Hereford is unique, but it is also part of a genre. Between the twelfth and fifteenth centuries, *mappae mundi* were drawn up frequently throughout the western and Arab world, and the ones that survive provide remarkable evidence of how the medieval world viewed itself. They took many different forms, some more recognisable than others, many distinctly bizarre. Most, however, share a common purpose: they were not intended for use, at least not for travel use. Rather, they were statements of philosophical, political, religious, encyclopaedic and conceptual concerns.

These qualities apply to the European cartographers, like the creators of the Hereford map. But they are applicable, too, to the Arab scholars, based in the cultural whirlpool of Baghdad – the heart of the Arab caliphate – which had inherited an Alexandrian appetite for amassing the world's knowledge, and drew upon Ptolemy as well as first-hand reports from Arab sailors and Chinese explorers.

Strangely, though, the finest – and most modern – of all the medieval cartographers was an Arab geographer based in Europe – one Muhammad al-Idrisi. Al-Idrisi traced his lineage back to the Prophet and came from a noble Arab family from

Muslim Andalusia. He travelled widely as a young man – in Spain, North Africa and Anatolia – before settling at the court of Roger II, the Norman King of Sicily. Here, around 1150, he completed his finest work, producing what was essentially an early atlas, combining various regional maps to form a global picture. Perhaps because of his position as a Muslim working for a Christian king, his mapping was notable for the absence of religious symbolism and fable. Instead, drawing on Islamic maps and travellers, as well as the accounts of Norman sailors, it showed what was possible if geographical accuracy was given its head. His view of the world looks more familiar to us today than most of the maps that followed it for hundreds of years.

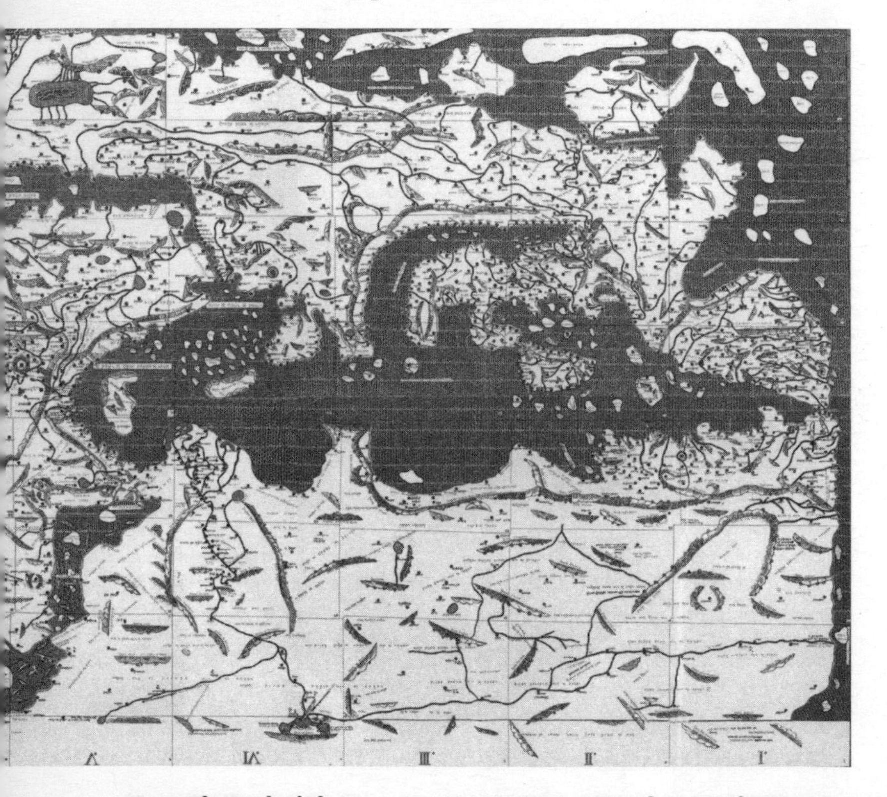

From *The Book of Pleasant Journeys to Faraway Lands:* an Arabian geographer plots North Africa and Europe.

Indeed, his mapping of the Nile and the lakes that form it was unimproved until Stanley's expeditions seven centuries later.

In the 1360s, a map of the world produced by a Buddhist monk in Japan displayed an entirely different view of life's priorities. It was another map for pilgrims, this time heading elsewhere than Jerusalem. The elderly cartographer inscribed a plain explanation: 'With prayer in my heart for the rise of Buddhism in posterity, I engaged myself in the work of making this copy, wiping my eyes which are dim with age, and feeling as if I myself were travelling through India.'

His map looks like a lantern floating over an ocean. The Sumeru mountain dominates the centre, with a series of named shrines following the Silk Route. Many of these are indebted to the Chinese Buddhist scholar, Hsuang-Tsang, whose seventh-century topography, *A Record to the Regions of the West of China*, describes a fifteen-year journey in India. Those travels, marked on the map with a red thread, had taken seven hundred years to find such lavish cartographic form, and would take about half that again to make an impact in the west.

Not that Christian ideology didn't produce anything of significance, or at least beauty, in this time. Some of the most abstract examples of divine geography were spread over facing pages in medieval books, and the most remarkable suggested exciting places on the map that wouldn't actually be viewed by explorers for many centuries. The Beatus Map, based on the writing of the Spanish Benedictine Monk Beatus of Liebana, was among the most notable. It was created towards the end of the eighth century and faithfully recreated on numerous medieval manuscripts (fourteen European examples survive from the twelfth and thirteenth centuries).

One highly stylised edition from 1109, housed in the British Library, measures 43x32cm, and looks like a giant oval fish platter (indeed it has fish swimming upstream within its framing ocean). It is a wonderfully fanciful rendering of the world, perhaps all the more attractive for its complete disregard for anything we might recognise as geographical accuracy. Again, allegory and fearful biblical learning dominate, with the Revelation of St John and the work of Saint Isidore of Seville to the fore. Adam and Eve, almost anatomically drawn next to a serpent, lie in Eden at the top, with India to their left. Alexandria rather than Jerusalem occupies the centre (with the Pharos lighthouse close by), with Africa directly below. Ravenna, a crucial cultural and political centre of the Byzantine and then Lombard empire, is given equal importance towards the north. Several locations, including Britain and the Fortunate Isles, are placed in boxes at the base of the map, like Scrabble tiles awaiting placement. Mountain ranges resemble piles of steaming dung.

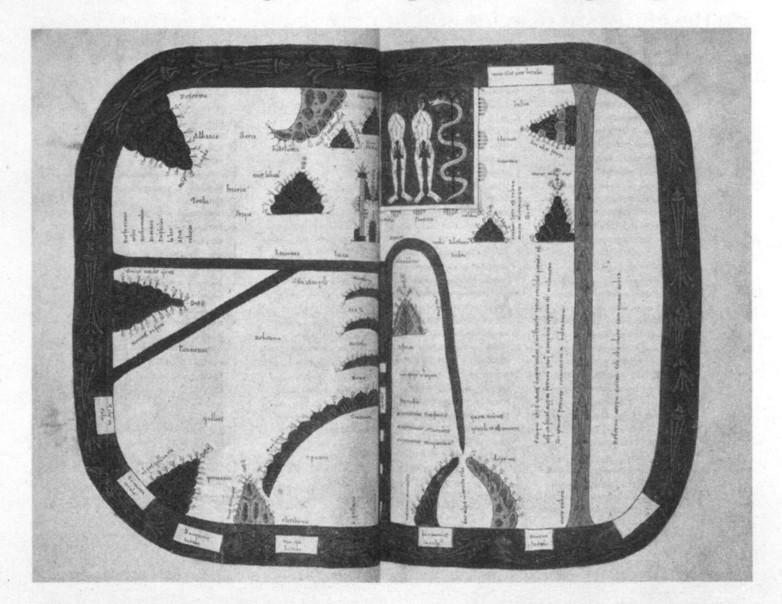

Adam and Eve at the helm of the Beatus Map in the British Library. But what is that mysterious fourth continent on the right?

But the map is remarkable for geographical reasons, too. The Red Sea, which spans the entire length of the map to the south, divides the three known continents from the possibility of a mysterious fourth. There is only a brief description of this land – we learn it is unknown torrid desert – but it is a feature of most of the Beatus interpretations.

Many western *mappae mundi* were designed as wall decorations and have since crumbled or been painted over, while many that do survive are later copies on paper, and on a much smaller scale. One of the most intricate and significant is the so-called Psalter Map, drawn sometime after 1262 and perfectly preserved within a small prayer book. It is only 15x10cm but shows many of the elements that were to appear on the Hereford map less than thirty years later. Jerusalem is at the centre, east is at the top, the trading cities are prominent, the Red Sea is massive and dripping with colour, and newly created saints (in this case Richard of Chichester) make a last-minute appearance.

It also contains much detail for its size: the Thames and the Severn are visible, Adam and Eve appear mournfully stranded in a walled Eden. The fearful Gog and Magog, supposed forces of the Antichrist, also make an appearance in the east, where they appear to be contained by a wall built by Alexander the Great, one of the many legends of Alexander's travels reflected in *mappae mundi* (the fortification has also been identified, contrary to received wisdom, as the Great Wall of China).

The cartographic historian Peter Whitfield has called the Psalter world and others like it 'maps of the religious imagination', and there is no finer or more stunning example than a map discovered in a Benedictine abbey in Ebstorf, lower Saxony, in 1832. At about 3.5m in diameter it was more than twice the size

SAMARIA
SICILIA
LIGURIA

of the Hereford map, equally encyclopaedic and just as bizarre. Its origins are uncertain but it is believed to date from as early as 1234 and to have been the work of the abbey's nuns, pooling their own knowledge with library manuscripts and reports from visitors. (A rival theory links it to Gervase of Tilbury, an English legal scholar teaching in Bologna.)

Its fame today is partly associated with its allegorical representation of Christ, whose body is splayed out (as if on the cross) so that it grasps the whole of the world – head at the top by Paradise, hands at the extremes of north and south, legs at the bottom due west, Jerusalem as his navel. The concept of Christianity embracing all humanity is echoed within the map itself, which is nothing less than a Bible class.

The map is packed with story and text, with one inscription hoping it will provide 'directions for travellers and the things on the way that most pleasantly delight the eye'. But time again has stood still, and the traveller may find themselves searching in vain for such sights depicted here as Noah's Ark and the Golden Fleece, or the unnerving possibilities of Africa, including a race yet to discover fire, a people without nose or mouth that speaks only in gesture, a tribe that is always falling on its face as it walks, and a nation with upper lips so elastic that they may pull them over their heads for disguise and shelter. But, alas, these wonders may now only be seen on photographs, as the map was destroyed in an allied bombing raid in 1943.

There may be a simple reason why these medieval maps contained many of the same features and landmarks: they may have been drawn from the same manual. In 2002 the French historian Patrick Gautier Dalche reported on his discovery of two manuscripts called *Expositio Mappe Mundi* that represented a template of how to make a map of the

The world comes alive on the Ebstorf map, not least Christ's feet at the bottom.

world. The manuscripts were copied in Germany in the middle of the fifteenth century, but their origin is believed to be English and much earlier, possibly dating back to the Third Crusade between 1188 and 1192 (their author may have been part of Richard the Lionheart's army). The guides contain not only a list of place names, but their spatial relationship to each other: places were 'above', 'below' or 'opposite' others, while regions were 'demarcated' and rivers 'spanned'. Of the 484 items detailed in the *Expositio*, some 400 made it accurately onto the Hereford Mappa Mundi.

Another clue to vague but emerging global knowledge surfaces in another mappa, from around 1436, named after the sailor-cartographer who drew it on vellum, Andrea Bianco. As befits its origins and current home in Venice, the earth resem-

The Andrea Bianco map – here come the poles

bles a giant fish in a bowl, a flounder perhaps, with Europe, Asia and Africa surrounded on all sides by ocean. A dark blue rim encircles the whole, sparkling with heavenly constellations, the earth again a symbolic marble spinning in a broadly philosophical sphere.

The British Isles, Spain and France are relatively accurate depictions, and there is the gentle introduction of the newly discovered Azores in a greatly enlarged Atlantic Ocean. But then there is real geographical news, as astonishing to us today as it must have been when the ink was fresh: the southerly base of Africa is encircled by what may be the south pole, enlivened by a merman and a man dangling perilously on a rope, while a similar circular section intrudes in the northern sphere, where the text tells us of frozen tundra. Elsewhere there are familiar figures – elephants, camels, enthroned monarchs within marquees, the Virgin Mary with child – while less familiar lands contain winged beasts and men with heads of dogs. But was the suggestion of the two polar regions sheer guesswork, tutored intuition or evidence of a greater awareness of regions beyond the habitable? There is no way of knowing, but it is clear that the map is edging towards something distinctly modern.

Pocket Map
Here Be Dragons

In the *New York Times* of 29 February 2012, the erudite columnist Thomas L. Friedman began his article about the fall-out from the Arab Spring with the following lines: 'In medieval times, areas known to be dangerous or uncharted were often labeled on maps with the warning: "Beware, here be dragons." That is surely how mapmakers would be labeling the whole Middle East today.'

Nice historical parallel, not quite true. The phrase 'Here Be Dragons' has never actually appeared on a historic map. There have been lots of ironic, nostalgic and fearful uses in literature, but try finding those three words on a map from the medieval or golden ages – the big Dutch atlases, say, or the fanciest German or British maps anytime from the fifteenth century to the twentieth – and you'll look in vain. 'Here Be Dragons' is a map myth as mythical and wonderful as dragons themselves, and we may wonder how such a thing has come about.

Blanks on maps make it look like vital information is absent. So we put in something to hide our shame: very big curving country names (M-E-X-I-C-A-N-A), chunks of text about the unusual flora of a country, a proud message from the mapmaker about his new projection. We once used another phrase to describe the unexplored: *Terra Incognita*. But, romantic though that sounds, you can't beat animals, or better still, imaginary beasts. On the earliest medieval maps, which, as we have seen, have a tendency towards the fearful in their morality lessons, the trend was to portray the angriest, scaliest and toothiest fish that sailors had ever seen, and the biggest, ugliest winged monsters that intrepid colonisers had been warned about by canny natives. Sometimes this was like Chinese whispers: the animal would begin as an elephant, become a mammoth, and by the time the maps of Africa or Asia were drawn up back in London or Amsterdam, transform into a nightmare. In China, the Chinese whispers would, one assume, naturally take the form of one of the country's sacred cultural symbols, the dragon.

'Here Be Dragons' may, however, have once appeared on a globe, but this rather depends on interpretation and translation. The Lenox Globe is believed to have been made around 1505, although its origin and maker is unknown. It's a small thing, an engraved hollow copper ball of less than 12cm in diameter, and is proudly displayed at the New York Public Library as the earliest

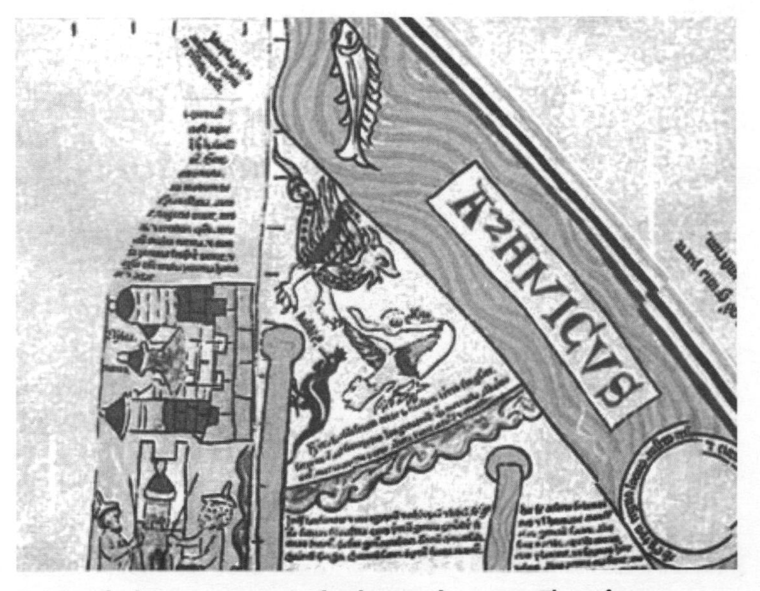

Some early dragon activity on the thirteenth-century Ebstorf map

known example of a globe showing 'Mundus Novus' – the New World. The feature we are interested in – the Latin phrase *HIC SUNT DRACONES* – is to be made out just below the Equator in 'East India' (China). It might be a reference to Chinese dragons, assumed to be real. But scholars also suggest, spoilingly, that it could be translated as 'Here are Dagronians', referencing the cannibals of the kingdom of Dagronia mentioned in the travels of Marco Polo.

Pictures of dragons on maps are another matter. They are many and magnificent. The American historian Erin C. Blake and her friends have compiled a scholarly list of early maps and globes which contain pictures of dragons (or almost-dragons, like scorpions with tongues), which include the previously-mentioned Psalter Map from c.1262, which has a dragon in the area beneath the world. She also notes that the Ebstorf Map has the word 'Draco' in south-east Africa.

Blake has also recorded the phrase's appearances in literature, where the earliest confirmed finding is surprisingly late: a Dorothy L. Sayers short story about a treasure hunt, 'The Learned Adventure of the Dragon's Head' from 1928, in which a character reports seeing 'hic dracones' on an old map.* Perhaps he did. We are the poorer for having failed to find it.

* Dorothy L. Sayers was clearly a fan of the saying, and the possibilities it contained. In 1918 her poetry collection *Catholic Tales and Christian Songs* contained the following phrase, a world-view we may all aspire to:
'Here be dragons to be slain, here be rich rewards to gain;
If we perish in the seeking, why, how small a thing is death!'

Chapter 4
Venice, China and a Trip to the Moon

Some maps get the location they deserve. The Hereford Mappa Mundi is still in its cathedral; the first globe (and maps) to mention America have made their way to the United States. But for one of the most elaborate and important maps of the world to be hung in a dimly-lit corridor above a chilly Venetian stairwell? That makes sense too.

If you walk to the western corner of St Mark's Square, climb up the stone steps of the Museo Correr, pay €16, saunter through nineteen rooms of marble, coins and globes, you end up at a glass door. This is the Biblioteca Marciana, the civic library built in the 1530s to house a vast collection of Greek and Roman manuscripts and later a copy of every book printed in Venice. And there, between the museum and the library, visible through the glass door but accessible only with special permission from an attendant, is the work of a Venetian monk named Fra Mauro, who somehow, in 1459, knew more about what was where in the world than anyone else.

Mauro lived and worked on the Venetian island of Murano, already famous for its glass by the time he established his cartographer's studio there in the 1440s. He had travelled more than most, and some of his early naval and trade charts were drawn from experience. His circular world map (coloured ink on parchment, about two metres in diameter) was constructed for King Alphonso V of Portugal, and although the original no longer survives, we are fortunate that a copy was made for a Venetian lord.

The map contains almost three thousand place names and a vast amount of explanatory text, and although it contained the usual misplaced rivers and regions, it was a geographical masterpiece. It is also – almost definitively – transitional, hovering between the old world and the new, and between the medieval depiction of the earth as one round 'planisphere' and the dual-hemisphere projection that emerged in the sixteenth century. It is the last great map of a former age, history as soon as it was framed.* Venice's role as 'the hinge of Europe' was beginning to come to an end, and Mauro's vision of a world enclosed within a fishbowl would also lose its dominance. Columbus would set sail within a couple of decades, and Mercator would chart his voyages on a map enticingly open to the navigable oceans.

The map's location in Venice is apt for another reason. Its ground-breaking depictions of China, Japan and Java derived from the journals of the most famous of all Venetian travellers, Marco Polo. Polo recounted his travels during his year-long incarceration in a Genoese jail (how he got there is uncertain: one theory runs that he had sponsored a Venetian war vessel for its attack on Genoa in 1298 and his opponents regarded him as a prize catch). His Boswell was a fellow prisoner named Rustichello da Pisa, and although the veracity of some of his journeys has been

* The Fra Mauro map of the world was not just a great cartographic milestone; it also signalled the death of Paradise. The space between its round world and the frame of the map is occupied by cosmological notes and drawings, including descriptions of the distance of the stars, the flow of the tides and theory of the elements. Paradise is here, too, but it is off the map: a Garden of Eden set apart from the inhabited world.

Somewhere between the old and the new: Fra Mauro's map of the world, shown inverted with Britain, Ireland and Europe clearly recognisable on the left. The mass of grey on the land is text.

challenged (Polo was famed for his storytelling, and in da Pisa, a romantic novelist, he had found his perfect amanuensis), there is no doubting the great impact of his account. The book was influential on cartographers after its first printing in Old French in 1300, but when the Venetian presses got hold of it 150 years later it became the most popular travel book of its day, examined by traders on the Rialto much as one might consult a train timetable today.

The travels of Marco Polo had little corroboration, and in many ways their worth should be measured not in terms of discovery – others had made spectacular journeys to the East before him – but in terms of their recording. As with maps, it is not the exploration itself we encounter, but its historical footprint.

The story begins with the Venetian brothers Niccolò and Matteo Polo leaving their trading base in the Crimean port of Soldaia around 1260 to trade in jewels with Mongols in present-day Volgograd. Their journey was extended by war, and while residing in the Central Asian city of Bukhara they met an envoy of the Great Khan Khubilai, who invited them to his court. The Great Khan asked them to return one day with oil from the Holy Sepulchre in Jerusalem and a hundred skilled educators from Rome to act as missionaries. The brothers returned to Venice after an absence of some fifteen years, with Niccolò now seeing his son Marco for the first time. Two years later the three of them ventured back to the East.

Marco Polo's (or rather da Pisa's) descriptions of their journeys begin in the Holy Land and eventually reach the summer palace of the Great Khan in Cathay (northern China). Polo boasts of becoming indispensable to the court, travelling back and forth to India, and learning of Java and Japan before returning to Europe via Sumatra and Persia. But his *Book of Travels* is very far from the sort of travel literature we read today; there is very little route detail, and great areas are covered without mention of sea or land, merely of trade; we learn as much about the abundance of sapphire, amethyst, silk, perfume and spices as we do about geography.

On his deathbed, Polo is supposed to have said he had recounted but half of what he had seen. But the jury remains out, even today, as to whether he really did take the Silk Route, or sail close to Japan. Nonetheless, if his routes are uncertain, the existence of these mystical lands in such a popular account did more than anything to expand the fifteenth-century European view of the world. For Columbus, who is known to have treasured his own copy of Polo's *Travels*, it was to serve as goal and inspiration.

But no one was as heavily influenced by Polo's *Travels* as Fra Mauro, and he refers to it several times in the text that covers his map. Almost two centuries had elapsed between Polo's expeditions and Fra Mauro's visual representation of it, yet the traveller's accounts in Cathay had not yet been superseded by further Western explorations. Fra Mauro and his collaborator Andrea Bianco used Polo's place-names and legends, perhaps also drawing on a mural of his travels at the Doge's Palace that was later destroyed by fire. In 1550, the geographer Giovanni Ramusio wrote that Mauro had also relied on a map that Polo himself drew while in Cathay, although no such map has survived.

Even today, Fra Mauro's enthusiasm for the novelty of Polo's discoveries is infectious. He describes how the great city of Cansay (Quinsay) was built, like Venice, upon water. He mentions its 12,000 bridges and vast population of 900,000. He wonders too at the splendour of the Yangtze, and its exotic trade in porcelain, ginger and rhubarb.

It was inevitable that the western cartographic view of China was determined by prospects of exploitation; the attractions of 'opening up' such an enormous and lucrative market has

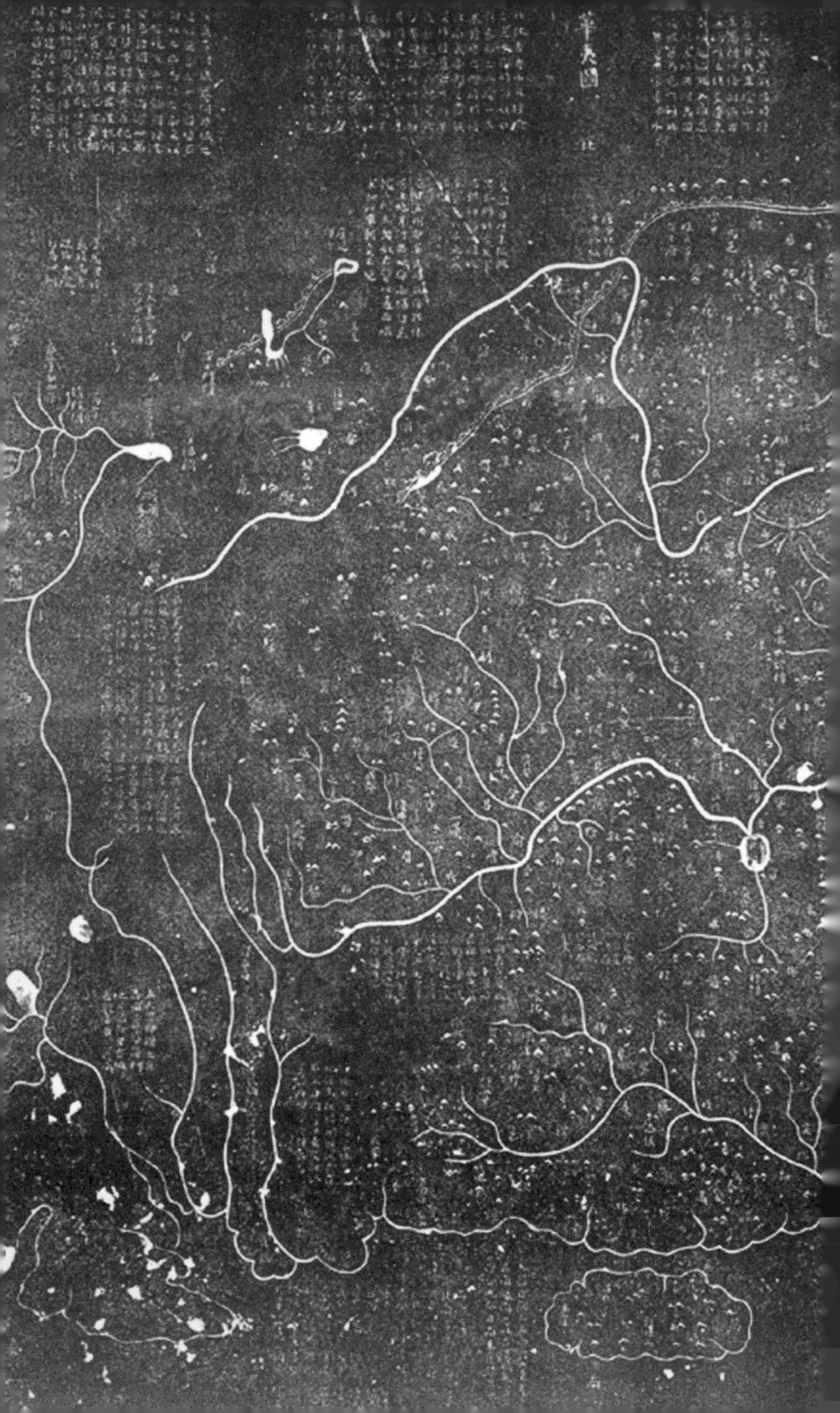

changed little from the earliest silk route to the present. It was also inevitable that Fra Mauro's interpretation of Marco Polo's travels was not quite how the Chinese saw themselves.

Our knowledge of the mapping tradition in the Far East goes back to the second century, when a man called Chang Heng drew up a mathematically based system of size and scale. His guidelines were advanced by P'ei Hsiu, an official of the royal court who constructed an eighteen-sheet atlas upon the principles outlined in the *Ch'in Shu*, the official history of the Ch'in dynasty. The world was shaped by a construct of angles, curves and straight lines upon a rectangular frame, with the intention that the result 'can conceal nothing of their form from us.' Neither Heng or Hsiu's maps, alas, have survived.

The earliest maps from China that we do have are dated 1137, and reside on two large stone slabs at the Pei Lin Museum in Xi'an. Measuring almost a metre square, they represent the two traditional styles of Chinese map-making, although one of them, with a strictly etched grid system, would not look out of place on a modern draughtsman's drawing board. The first, more classical, is translated variously as 'The Map of China and Foreign Lands' or – in a precursor to its strict isolationist policy – 'The Map of China and Barbarian Lands'. It is a heavily annotated and highly politicised representation of China's place in the world, with the administrative centres of the country receiving prominence over distant rivers and coastlines. This was not a map for plotting trade, but a teaching aid apparently designed for those preparing for civil service examinations, reinforcing the message that China was the 'Middle Kingdom' in which the Emperor ruled 'all under Heaven'.

The second and more contemporary-looking map dispensed with the rest of the world altogether. Called 'Maps

Geopolitics carved in stone: the Xian 'Map of China and Barbarian Lands' placing China firmly at the centre of the world.

of the Tracks of Yu The Great', its purpose is again believed to have been instructional and historical rather than practical, although the location and flow of rivers are uncannily accurate for a map from the twelfth century. The grid scale maintains its integrity throughout the map, with the side of each tiny square representing about thirty miles (and there are hundreds of such squares). As such, it invites generous comparisons not only with the mathematical cartography of Ptolemy, but with the far less accurate sea charts used by western explorers for centuries to come.

China did, in fact, have a Ptolemy of its own, albeit one with blinkers on. In a country where maps were tools of

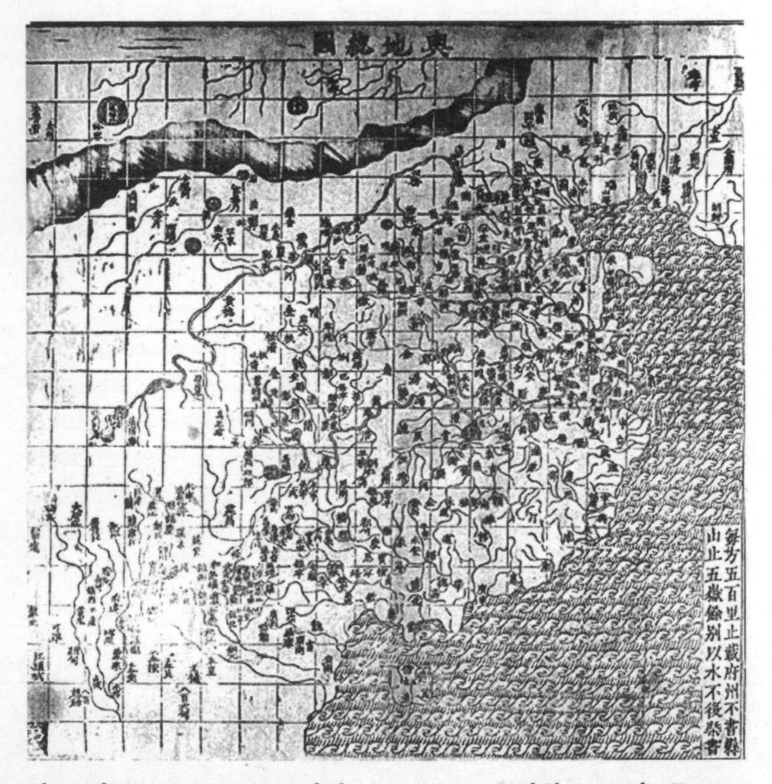

The earliest surviving copy of Chu Ssu-Pen's map of China, in the *Kuang Yü T'u* (Enlarged Terrestial Atlas) of about 1555.

power like nowhere else, the work of Chu Ssu-Pen (1273–1337) was to be the official basis not only of its maps for five hundred years (first in traced and rubbed copies of his manuscripts and then printed), but also indicative of the fears and wilful ignorance its citizens held towards the outside world. Ssu-Pen's 'Earth-Vehicle Map', inevitably focussed its attention on the cartographer's homeland, with the Great Wall symbolically dominant. Knowledge of the wider world derived predominantly from Arab traders, but Chu Ssu-Pen remained wary of their attributes. 'Regarding the foreign countries of the barbarians south-east of the South Sea and north-west of Mongolia,' he reasoned, 'there is no means of investigating them because of their great distance, although they are continually sending tribute to the court. Those who speak of them are unable to say anything definite, while those who say something definite cannot be trusted; hence I am compelled to omit them here.'

Back in northern Italy, the tradition of Venetian exploration was extending well beyond Marco Polo. Forty years after Fra Mauro made his world map, John Cabot (Giovanni Caboto), the Venetian commissioned by Henry VII of England, planted the flag of Saint Mark beside an English one when he made landfall in the New World, while his son Sebastian, who also claimed the city as his own, explored uncharted parts of South America and may have found one of the earliest trade routes via the Northwest Passage. Alvise de Mosto, another native, made great inroads into West Africa in the 1450s and is credited with the discovery of the Cape Verde islands.

All of these explorations soon found their way onto the city's maps. Andrea Bianco, who was largely responsible for the Venetian copy of Fra Mauro's world map, also made important navigation charts for wealthy patrician merchants. And Giacomo Gastaldi, who worked in Venice for most of his life, was the first to map many areas of the New World in the mid-sixteenth century. He also drew large murals of Asia and Africa in the Doge's Palace, and in 1548 produced what is recognised as the first pocket atlas when he published a version of Ptolemy that included the new and old world. He made a major technical innovation in the printing of maps, too, using copper plates in place of woodcuts, which allowed much greater detail.

What made the Venetians such purposeful cartographers and their maps the envy of Europe? Power, in the main. The Serene Republic wanted a solid and irrefutable display of governance and fiscal strength, not only over Venice but over all dominions under their control. Maps provided the documentation. But how did the famous cartographers of Venice see their own home? With the same sort of wonder that renders us breathless today. Not long before Marco Polo set forth, the writer Boncompagno da Signa had seen it as 'incomparable ... its floor is the sea, its roof is the sky, and its

Jacopo de Barbari's woodcut 'aerial photograph' of Venice, centuries before such a thing was possible.

walls are the flow of its waters; this singular city takes away the power of speech because you cannot nor will ever be able to find a realm such as this.' But perhaps maps could help define what words could not.

The most famous map of the city, and the one that cemented its teeming, intriguing, Byzantine image, was published in 1500 by the painter and engraver Jacopo de' Barbari. It was a huge six-woodblock undertaking, an aerial view that reflected the ease of trading with the city's merchants and established the defining (and true) image of the city as a pair of interlocking hands, or – more famous still – a monstrously sized flounder. His work was all the more remarkable because Venice is seen from above – a bird's eye view four centuries before aerial photography made such images possible and commonplace.

Above all, though, de' Barbari's map suggested that Venice was a place of the imagination, a civic notion both mythic and unfathomable. It was a place where the tourist – from 1500 or 2012 – would get lost no matter how good the map they brought, the tiny *calli* and confusing *sestieri* only an invitation to further disorientation. The sat nav is hopeless here, as is the digital map on your phone. You just walk and hope and ask and point, and you may still never get your bearings beyond the four corners of St Mark's Square (the only grid in the city). Long after you have ceased to notice the lack of cars, long after you have been fleeced in a gondola ride, and long after you have stumbled across a church with the glorious Giorgiones, you will continue to get lost and profit from it. It is this, its charming and ancient unmappability, quite as much as its Bellinis and Carpaccios, that will draw you back into history.

But there is one more place that became important to this story, and it lies far beyond Cathay. It is the Moon. On 5th

February 1971, Apollo 14 astronauts Alan Shepard and Edgar Mitchell landed on the near side of the moon by a 1000-foot diameter crater formed by a meteorite. In those rare intervals between hitting golf balls and jumping around, they began gathering rock samples, and the rocks they brought back were a little younger than researchers at Caltech in Pasadena had expected – only about 3.9 billion years old rather than an anticipated 4.5 billion.

The area where the spacecraft landed was called the Fra Mauro Formation, named after the Fra Mauro crater, one of the largest on the moon at 80km across. This didn't quite have the mellifluous ring of Sea of Tranquillity, the site of the first moon landing in 1969, and there is no official record of why members of the International Astronomical Union nominated this particular fifteenth-century Venetian monk as a small but significant addition to its Planetary System Nomenclature. But it may be safe to assume that they were mapheads, and they dug his work.

Chapter 5
The Mystery of Vinland

If they know what they're doing, rare book and map dealers can make a lot of money, but they hardly ever make a lot of history. Laurence Claiborne Witten II was an exception to the rule – a dealer who found a map with the power to change our fundamental understanding of the world.

Witten was based in New Haven, Connecticut, but he often travelled to Europe to increase his stock. In the autumn of 1957 he was browsing in a fellow dealer's shop in Geneva when he came across something that set his heart racing – a crude map on vellum that suggested that North America had been discovered and settled by Norse travellers some five hundred years before Columbus.

Viking voyages to that part of the globe had long been an accepted part of geographic folklore, but there had never before been cartographic evidence. And now, perhaps, there was. The map showed what may have been part of Newfoundland or Labrador, which would have been the earliest European document to show any part of the New World. But there was also colossal controversy. Could everything we had been taught in school be wrong? Or was the map an elaborate forgery that would go on to fool some of the most brilliant

The Vinland Map (with Vinland itself on the far left). A true voice from the past, or one of the most elaborate hoaxes ever created?

Thule ultima
Insule Sub aquilone
Magnum mare Tartarorum
Postreme Insule
mare Indicum
Ethiopi

map historians in the world? And if it was a fake, who was the faker?

The story of the Vinland Map – so called because Vinland (or Wineland) was, by AD 1000, the name given by the Vikings to North America – is one of the most important and compelling in the history of cartography. Its story also demonstrates to perfection the romantic and mysterious allure of maps, amplifying the impression that they are seldom what they appear to be on the surface.

Larry Witten was born into a family of wealthy Virginian tobacco farmers who later moved into furniture manufacture. He opened his rare book store in New Haven in 1951, and swiftly established a reputation for selling fine medieval and Renaissance manuscripts. He had entered the market at the perfect time: many European libraries and collectors were selling prized possessions to make ends meet, and collections that had been ransacked during the war were finding their way to dealers at knockdown prices. One of Witten's most trusted suppliers was a Swiss bon viveur called Nicolas Rauch, who not only arranged currency deals at a time of heavy European restrictions, but also a salon in Geneva for the rare book trade to meet and exchange information; his gala dinners before a big sale were unmissable fixtures.

One of Rauch's regular suppliers, occasionally from unspecified and uncertain sources, was one Enzo Ferrajoli de Ry, a former Italian army officer who had taken to 'running' rare books and manuscripts around Italy, Spain, Portugal and Switzerland in search of a swift profit. In September 1957, Witten happened to be with Rauch when Ferrajoli drove up to his door in his Fiat Topolino and began unloading his new wares.

One of these rarities was what we now know as the Vinland Map. The map measured 27.8cm by 41 cm (10.9in x 16.1in) and had been folded vertically in the middle. It appeared in a slim volume alongside a version of a manuscript called *The Tartar Relation*, a handwritten description on vellum and paper of the travels of the Franciscan friar John de Plano Carpini to Mongolia in 1247–48. Both works were thought to date from around 1430 to 1450, although when Witten first saw them they were in a modern binding.

If genuine, the map would show two stunning things. The first was that there was European knowledge of Norse sailings to North America some fifty years before Columbus. And the second was that, according to an explanation on the map, tightly written in a gothic script above the depiction of Vinland, the discovery itself had occurred between 985 and 1001, uncovering a great island of impossible promise.*

What did Larry Witten make of it all in his friend's Swiss shop in the middle of the twentieth century? He was excited but sceptical. The map felt right to him (map dealers pride themselves on instinct) but he was aware that there were many people far more expert in medieval maps than he was. What he didn't know at that point was that several of them had already viewed the map and felt nervous about it.

Witten thought about buying the map and the accompanying manuscript for several hours. 'My reasons for ruling out

* The full inscription reads: 'By God's will, after a long voyage from the island of Greenland to the south toward the most distant remaining parts of the western ocean sea, sailing southward amidst the ice, the companions Bjarni and Leif Eiriksson discovered a new land, extremely fertile and even having vines, the which island they named Vinland. Eric, legate of the Apostolic See and bishop of Greenland and the neighbouring regions, arrived in this truly vast and very rich land, in the name of Almighty God, in the last year of our most blessed father Pascal, remained a long time in both summer and winter, and later returned northeastward toward Greenland and then proceeded in most humble obedience to the will of his superiors.'

The dating of the discovery between 985 and 1001 comes from a contested account of these journeys in a fourteenth-century collections of traveller's tales, although it is thought that Bjarni and Eiriksson may have made separate rather than joint voyages.

forgery are unchanged today,' he wrote thirty years later. He explained that such forgeries are usually instantly obvious, and the hurdles were large: you needed the right vellum, the correct writing instruments, ink made of the right ingredients, a perfect command of the current writing style and language, and a firm grounding in cartographic knowledge and practice. It would be rare for one person to possess all these skills, and even a team of forgers would struggle to meet the material requirements.

Then there was motive: why would anyone go to the trouble of making such a thing so well? What would be the financial rewards? And what of its provenance? It hadn't featured in any public auction, and no dealer had any record of the map changing hands before. It was the equivalent of finding a Rembrandt in the attic. It was improbable because it was improbable. But it wasn't impossible.

The map looked deceptively simple. But it reflected almost all the cartographic knowledge of the day, much of it undoubtedly copied from at least one other medieval map. It showed the three continents of the medieval world (Europe, Asia and Africa), north at the top, loosely oval in shape and surrounded by oceans. But it was drawn crudely and with relatively sparse labelling; there are only five marked cities – Alexandria, Rome, Jerusalem, Mecca and Cairo. Asia is most populated continent, with names of rivers, mountains and other features, many of them also mentioned in the *Tartar Relation*. The sketch of England, with its westward outcrop of Somerset, Devon and Cornwall, is distinctly recognisable, as is the shape of Ireland (Ibernia) and the Isle of Wight. Scotland is not yet imagined, although there are small floating islands above it, which may represent the Shetlands and Faeroes. There was no border, no illumination or illustration, no display of allegory or fable. But the map had some startlingly accurate new details, almost certainly the result of original exploration.

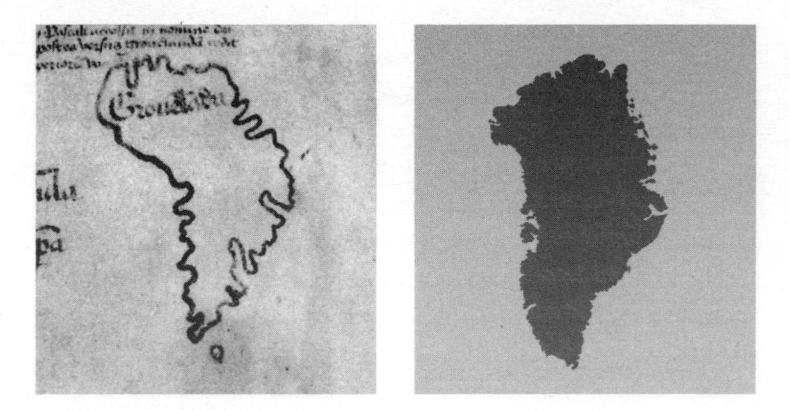

The Vinland map's outline of Greenland – a suspiciously good shot for a fifteenth-century cartographer

The outline and scale of Greenland, for example, although depicted as an island, could be traced over a map today and present an uncanny fit; the match would later be one of the strongest cases against its authenticity. The depiction of Vinland was much smaller than the North America we know now (only twice the height of England), but the shape of the east coast could find modern parallels. Its position west and slightly south of Greenland is representational, the most westerly point available on the vellum; the cartographer was limited by the size of his surface, and like most of his contemporary medieval map-makers probably abhorred blank space. It is shown with two deep inlets, possibly suggesting that Vinland is really three islands combined – Vinland, Helluland and Markland.

But there was one thing that nagged at Witten more than anything else: worm holes. These existed on both map and document, but the placing didn't match up. The suggestion was that if they weren't created at the same time (and the cursive script looked as if it was written by the same hand on both), could it be that they were forged on two separate bits of parchment that had already been riddled with worm holes before they were written on with modern ink?

Despite these doubts, Witten bought the items from Ferrajoli for a considerable price: $3,500. He told no one of his purchase until he was on the plane home two weeks later, when he couldn't help telling the passenger next to him, an American engineer. He noted that the circuitous journey back to the US – a fuel stop in Iceland, a flight-path over the southern tip of Greenland and Labrador, a second stop in Newfoundland – may have been dramatically close to the passage of his Norse explorers.

On his return, Witten showed the map and the manuscript to friends at Yale, but they too were worried about the discrepancy with the worm holes and lack of provenance. And there was one more disconcerting puzzle: an inscription on the back of the map that read 'Delineation of the first part, the second part [and] the third part of the Speculum'. What could this possibly mean? It would turn out to be the clue to everything.

Witten always knew that his map would be controversial, and now felt he could do little to prove its authenticity or otherwise. Enzo Ferrajoli refused to provide any more details as to its history, and Witten feared he might be ridiculed as obsessional if he pursued the matter. So after all the excitement, he decided to let the issue drop for a while, and get on with other work. He didn't offer the map in his sale catalogues, but instead gifted it to his wife Cora. And so, for the next couple of years, the map stayed in their New Haven home, occasionally being shown to dinner party guests, more of an enticing curio than an explosive document of history.

As the years passed, Witten found that his map, far from being 'discovered' by him on a chance find in his friend's shop, had in fact previously undergone an intense period of examination by at least two of the world's leading cartographic scholars; had he known of this earlier, he might have believed he was being set up.

A few months before Witten's visit to Geneva in September 1957, the Vinland Map and *Tartar Relation* were at the British Museum being analysed by R.A. Skelton, the head of the map room, and George D. Painter, the assistant keeper of printed books who was also an accomplished biographer of Proust and Caxton. If anyone could determine its pedigree it was these two. Skelton was a methodical specialist, the greatest authority on medieval cartography, and his imprimatur on the Vinland Map was regarded as crucial. However, he kept his opinion – indeed, the fact that he had viewed the map at all – secret for eight years.

How did Skelton and Painter first come to see it? At least three months before Enzo Ferrajoli had shown the map to Witten, he had first offered it to the leading London dealer Irving Davis. Davis borrowed it 'on approval', but, troubled by its origin, took it to the British Museum. Skelton and Painter then spent several days with it, and even traced it, a practice generally frowned upon.

Both Painter and Skelton believed the map to be genuine. Years later, Painter called it 'a major and authentic message from the middle ages, on a hitherto unknown moment in the history of world and American discovery. It is a true voice from the past, which still lives and need never be silent again.' But both Painter and Skelton knew it would be the most controversial opinion they had ever given, and Skelton was unwilling to stake his reputation on it and guarantee it absolutely. He may also have been concerned that the map was stolen. So the map went back to Davis, and Davis returned it to Ferrajoli.

Knowing nothing of the British Museum episode, Witten was forced to conclude that he was alone in his conviction that the map was 'good'. But in the winter of 1958, a stroke of luck convinced him that he had bought the map of the century. He received a call from a man called Tom Marston, the curator of medieval and Renaissance literature at the Yale University Library.

Marston said that he had some exciting and genuine new manuscripts to show him. Witten was initially reluctant to see them, thinking Marston (who was a friend and regular customer) intended to gloat. But he relented and one acquisition in particular caught his eye, a manuscript that contained a part of the *Speculum Historiale*, an encyclopedic history of the world composed by the Dominican friar Vincent de Beauvais. It was in a re-backed, badly-worn binding that was probably from the fifteenth century, and Witten thought that its cursive gothic script might offer some useful parallels with the lettering of the Vinland Map and the Tartar Relation. The word 'Speculum' also set bells ringing.

He borrowed it for the night, and compared the separate volumes. Intriguingly he found that the manuscripts shared the same watermark – a bull's or ox's head. 'A household ruler gave me the next tremendous shock,' he recalled in October 1989. 'The leaves of the two volumes were of exactly the same dimensions.' He also found that the columns of the documents had very similar handwriting with identical spacing. And then there was something else, the final piece of the puzzle: the worm holes matched up. The holes at the front of *Speculum Historiale* were in exactly the same points as on the Vinland Map, and the holes at the back lined up exactly with the *Tartar Relation*. Witten deduced that all the documents had once been bound together. 'My adrenaline began to flow,' he remembered. 'The reunited parts not only corroborated one another, but seemingly made it nearly incredible that any part could be a modern forgery.'

He called Marston later that night. Marston had been one of the first people Witten had shown the Vinland Map to after his return from Europe in 1957, and although he had been enthusiastic about it he too was concerend about the errant

Larry Witten (left) and the British Museum's expert, R.A. Skelton. Skelton believed the map was genuine, but thought twice about risking his reputation on it.

worm holes. But now, with the 'sandwich' of the *Speculum Historiale*, they were able to construct a plausible sequence of events. Enzo Ferrajoli had originally obtained both the Vinland Map/*Tartar Relation* and the *Speculum Historiale* (probably already separated), had not recognised them as a former single volume, and had sold the *Speculum* to the London dealer Irving Davis. Davis had placed it in one of his catalogues, from which Marston had chosen to buy it for less than £100.

Witten and Marston then ran through all the possible explanations that would suggest that the map – clearly the most valuable and important of the three elements – was a forgery. They speculated that a forger could have 'improved' the *Speculum* (the authenticity of which was never in doubt) by constructing the map on blank pieces of vellum in a random medieval manuscript (the *Tartar Relation*) and then binding them together. But they concluded that this lacked a plausible motive. They could also not understand why the forger would then separate the whole volume, the very thing that gave it authenticity. To ensure the documents would not be separated again, Marston donated the *Speculum Historiale* to Cora Witten. 'This was not a wholly quixotic gesture on my part,' he wrote later, 'for I hoped that this generosity would give the Yale Library some element of control over the disposition of the map in the event that Mrs Witten should decide to sell it.'

Witten and Marston then began a slow process of convincing others of their find. Not everyone was convinced, but one person was particularly captivated – the wealthy Yale alumnus and philanthropist Paul Mellon. Mellon also believed the Vinland Map to be genuine, and said he was prepared to donate it to Yale anonymously for further research and then, if thought bona fide, for permanent display. But first he had to own it: he offered Cora Witten $300,000, eighty-five times the sum her husband had paid, which she accepted.

But then it appeared the money would have to be paid back. In 1961, Enzo Ferrajoli was arrested and imprisoned by the Spanish police for stealing books from Zaragoza Cathedral. He spent eighteen months in jail before release on parole. Witten had always protested Ferrajoli's innocence of theft, claiming that all his transactions with the cathedral had been done with the canons' blessing. After several months' uncertainty he was relieved to find that the long list of items alleged to have been stolen did not contain the Vinland Map or associated manuscripts.

Over the next five years, under a cloak of sworn secrecy, a group of the world's leading cartographers explored every aspect of the map. Parchment, ink and binding experts were consulted throughout Europe. R.A. Skelton and George Painter of the British Museum flew to Yale to devote more time to these documents than any other single work they had ever considered, while several specialists at the university, including Tom Marston, did the same. Finally, with varying degrees of conviction, all arrived at the same conclusion: they would stake their reputation on the map being genuine. They prepared lengthy documents in support of their claims.

Alexander Orr Vietor, the Curator of Maps at Yale University Library, observed that 'in some great matters of history … the discovery of a single new document may significantly alter the accepted pattern; and its publication becomes an imperative responsibility.' In this case, Vietor said he was prepared to shoulder such a burden because he was dealing with objects 'of dramatic novelty'. He claimed that the Vinland Map 'contains the earliest known and indisputable cartographic representation of any part of the Americas', and concluded that all evidence, and all the experts, justified 'without reservation the genuineness of the manuscript.'

The map and manuscripts became public knowledge on 9 October 1965. Yale University Press published a 300-page book decoding the documents and analysing their likely origins, while the objects themselves went on display at the recently opened Beinecke Library, a striking new addition to the campus designed specifically to house Yale's exquisite collection of rare books and manuscripts. In this way the Vinland Map took its place beside a Gutenberg Bible.

The media explosion that followed was akin to the detonation of a small bomb. Laurence Witten recalled that 'it seemed that all Italians and Italian-Americans were incensed that anyone dared to assail the priority of Columbus's discovery of the New World, and the insult was compounded because Yale's announcement fell just before Columbus Day.' Witten was inundated with press inquiries, and reporters did what was an uncustomary practice in 1965 – they knocked on his door. Witten told them the bald outline of the details, although he was careful not to reveal where or precisely how he had obtained the documents.

Clearly, the onus was now on the disbelievers to prove their case. Familiar arguments regarding the material and provenance rallied back and forth for several years. An international conference was convened at the Smithsonian in 1966, and a year later it went for further analysis at the British Museum. But little changed until 1974, when the first scientific examination of the map turned everything on its head.

Advances in microscopy enabled the leading Chicago-based research company McCrone Associates to analyse the map's ink, and its findings were devastating. Walter McCrone and his wife Lucy removed 29 microparticles of ink from the map, and concluded that they contained between three and fifty per cent of titanium in the form of anatase, a pure titanium dioxide pigment that was only available commercially from about 1920. They also found a suggestion of subterfuge: a layer of black ink had apparently been placed over yellow-brown ink by a separate application and then largely scraped off in order to 'simulate the appearance of faded ink.' Walter McCrone claimed that the likelihood of the anatase he detected being found in medieval ink was on a par with 'the likelihood that Nelson's flagship at Trafalgar was a hovercraft.' Yale was forced to admit that the map 'may be a modern forgery'.

But Witten, George Painter and others refused to accept this analysis as final; they preferred to trust their instincts and historical reason. And in the years that followed, the Vinland Map appeared to be fighting back. In 1985, the map was analysed again. Science had marched on, and a new glorified X-ray machine at the Crocker Nuclear Laboratory at the University of California, Davis, would take the story further. The map went under the proton beam of a cyclotron, and the results cast doubt on the McCrone report, suggesting that the tiny particles of ink in their tests were unrepresentative of the ink of the entire map. The Davis scientists found only minute trace elements of titanium, and reported it absent from about half of the surface examined. Further, the cyclotron detected twenty other trace elements, many missed by McCrone, including copper, nickel, cobalt and lead, all natural substances common in medieval inks but almost never present in modern inks. The problematic titanium was also found to be present in the Gutenberg Bible, and in greater amounts than on the Vinland Map.

And so the map was back in play again. Another international conference was organised and Yale University Press announced an updated edition of the 1965 catalogue that accompanied the map's public unveiling. When this appeared in 1995, its editors hoped it would lead to the 'rehabilitation of one of history's most important cartographic finds', while George Painter, who had first examined the map almost forty years before at the British Museum, called it 'a true voice from the past, which still lives and need never be silent again.'

Other specialists, however, remained sceptical. William Reese, one of the world's leading experts on Americana, has handled the Vinland Map many times and believes that 'the odds are 80 percent that it's wrong and 20 percent that it's right'.* And in 2004, the map historian Kirsten Seaver tried not only to show it was a fake, but also suggested who had faked it. This was a consideration often ignored by previous investigations, and she placed the blame (or credit) on an Austrian Jesuit named Josef Fischer. Fischer was an expert in medieval cartography and the history of Nordic exploration, and Seaver suggests he had a deeper motive than financial gain (of which there appeared to be little). Rather, it was a sort of revenge: Fischer and fellow Jesuits had fallen foul of the Nazis in the mid-1930s, and when war broke out he was forced to

* William Reese curated a show at Yale's Beinecke Library called 'Creating America' in which he showed the Vinland Map alongside another controversial specimen, a map illustrating the supposed voyages of Nicolo and Antonio Zeno. The Zeno Brothers may have seen North America around 1380, although their exploits were only documented on a map made by a descendant in Venice more than 200 years later. This showed Greenland as part of the European mainland, but there was also an intriguing inclusion of 'Estoliland' and 'Drogho', which may have been Labrador and Newfoundland.

move from his work and home several times, finally settling at Wolfegg Castle in Baden-Württemberg in south-west Germany, where he may have drawn the map with the intention of taunting Nazi scholars. The Nazis may have approved of the map's promotion of Nordic superiority, but they would have been dismayed by the idea that the discovery of Vinland was driven by a desire to expand the Roman Catholic church. A review of these theories in the leading map journal *Imago Mundi* found them 'ingenious and compelling', but quite without proof.

These days, the Vinland Map and its attendant manuscripts continue to be potentially the most valuable component in the Beinecke Library catalogue (it was once valued for insurance purposes at $20m), and is described with delicious understatement as 'the subject of considerable debate'. But even if it is a fake (and we may never know for sure), its true and lasting value goes beyond its authenticity or fraudulence. Its value is in the narrative. The mystery of Vinland shows us the power of maps to fascinate, excite and provoke, to affect the course of history, to serve as the silent conduit to the compelling stories of where we've been and where we're going.

Chapter 6
Welcome to Amerigo

In the middle of the fifteenth century Jacobus Angelus, a Florentine scholar, was wandering around Constantinople's manuscript dealers, looking for early versions of Homer to translate from Greek into Latin. But instead he found something rather more valuable – a work that would once again change the way we looked at the world. Ptolemy's *Geographia* was suddenly back on the map.

Translated by Angelus, the first printed copy of Ptolemy's book appeared, without maps, in Vicenza in 1475. But it was an edition published in Bologna two years later that swiftly became the most sought after and influential publication of its day. The reason? Of its sixty-one leaves, twenty-six were engraved maps, making it the first atlas of the ancient world to be printed in the modern one.

There is evidence that Ptolemy's work had been circulating in the Arab world since the eighth century. But in fifteenth-century Italy it was a different story. *Geographia* was regarded as a revelation and, when visualised with woodblocks or copper-plate engravings and then coloured, a thing of beauty. The Ptolemy maps were also augmented for the first time with intricate cartouches, typeset place names and adorned with red-cheeked heavenly cherubs blowing gales from the edges. It was the rediscovery of the world in all its complex and strict alignment, and although the projection was to change (and the geography expand), the recreated Ptolomaic worldview at the height of the Italian Renaissance set a template that we still recognise when we look at a map today. It was, finally, definably where we live.

The rediscovery of *Geographia* signalled a golden age of map making. The new editions of the Ptolemy atlas – vividly instructional and genuinely exciting – established the novel concept of cartography as both art and science. They also triggered the first craze for collectors, with maps and globes becoming expressions of wealth and influence.

But why had it taken so long for maps – particularly world maps – to be considered important again in Europe? Perhaps it was serendipity, a timely alignment of the burgeoning printing industry, sturdier ships for travel and trade creating demand for updated maps, and a new banking and merchant class for finance. Intellectual reasons, too: a less fearful religious worldview created a quest for knowledge that for centuries had been considered irrelevant to a life of modest Christian duty.

Claudius Ptolemy – icon of Renaissance cartography – depicted on the 1507 Waldseemüller map.

Maps were no longer just things for cathedrals, shrines and palaces. Indeed, the Church's last great medieval cartographic hurrah had occurred with the completion of the Fra Mauro map in Venice in 1459. It is difficult to imagine two more contrasting cultural artefacts within such a short period (merely two decades apart) than Mauro's hand-coloured parchment and the modern printed updated editions of Ptolemy from the 1470s. It was as if the entire world had been modernised overnight by a combination of old mathematical geography and new technology.

But just as the Italian wealthy began to pore over their Ptolemys in crisp new printings, a thought probably dawned: they no longer actually lived in the ancient Greco-Roman world. Soon, Ptolemy editions were being supplemented by modern maps reflecting discoveries since 150 BC. The finest of these was printed in Ulm in 1482 by one Johannes Schnitzer of Armsheim. The first great German contribution to cartography, it featured five new woodblock maps and was the first printed world map to show Greenland. But there was a bigger issue to contend with. Europe stood at the dawn of the Age of Discovery, the age of great navigators eager to open up the world. To Bartolomeu Dias, John Cabot, Christopher Columbus, Vasco da Gama and Hernán Cortés, the components of Ptolemy's map, if not its temperament, would shortly appear distinctly limited. Not least when a new and unexpected continent drifted into view.

We cannot be sure precisely which maps accompanied Christopher Columbus on his four celebrated trans-Atlantic voyages between 1492 and 1504, but it would be fair to suggest he had a recent printing of Ptolemy, the *Travels of Marco Polo* and a letter of guidance from Paolo dal Pozzo Toscanelli, the

Florentine physician and astronomer who decades earlier had suggested to the King of Portugal that a journey to the riches of Asia might be attained far more easily by sailing westward rather than around the base of Africa. Columbus had presented this 'great idea' as his own, along with Toscanelli's greatly underestimated measurements of the globe, which explains not only how he persuaded the Spanish court to sponsor his voyages, but how he confused China and Japan with the Bahamas.

Columbus's sailings are familiar – as, too, his striking miscalculations. Following Toscanelli, he argued that the journey from Lisbon to Japan would be only 2,400 nautical miles rather than about 10,000, and the kingdom of Cathay would appear as a huge glittering bauble soon after. But we should remember that he was working from a system of 'dead reckoning', navigating by a nervous combination of compass and stars, making exact measurement impossible.*

When Columbus eventually set sail from the southern Spanish port of Palos de la Frontera on Friday 3 August 1492, he had spent a decade trying to convince European courts of the worth of his voyage. The Portuguese king had rejected his plans when Dias had rounded the Cape of Good Hope in 1488, believing the desired route to the east had already been settled. Queen Isabella and King Ferdinand required much persuading before finally consenting, against the advice of a Royal Council which argued that Columbus had got his sums wrong and his demands were too high. He was eventually allowed governance of any lands he discovered and a cut, passed down to his family, of the value of the natural treasures he found.

He sailed with three ships and about ninety men, and it wasn't long before the onset of disillusion. Japan failed to

* Latitude had been established in ancient Greece, but the correct measurement of longitude, a function of time, was only made possible in the late eighteenth century, when John Harrison's chronometer won a fabled competition.

appear as anticipated, and Columbus was forced to calm a mutiny when his crew reasoned that they had been duped. After varying winds and a journey far longer than expected (about 150 miles a day on some days, 25 on others), land was sighted on 11/12 October, and Columbus set foot on a small island in the Bahamas called Guanahani, which he renamed San Salvador (Holy Saviour). He assumed he had reached Asia and called the local Taino tribe 'Indians'.

He spent a few days visiting nearby islands and the north coast of Cuba, which he initially took to be China. He then rounded Hispaniola, so-named because it reminded Columbus of Spain, and after a storm wrecked one of his ships he set down thirty men at La Navidad, a Haitian bay. It was the first acknowledged European settlement in the Americas.

Columbus's longer second voyage drew him deeper into the Bahamas in 1494, where he still thought he was on the eastern shores of Asia. Then on 4th August 1498, on his third voyage, Columbus and his crew became the first Europeans to set foot on mainland South America, setting down at the Paria Peninsula of what is now Venezuela.

The first map to show Columbus's discoveries is less familiar than his own story. In 1500, his Spanish-Basque navigator Juan de la Cosa drew something that should have made him one of the most famous and enduring cartographers in history; that it didn't work out like that was due to the fact that his work was lost for more than three hundred years, only to be rediscovered in Paris in 1832.*

* The man who found it was the German explorer Alexander von Humboldt, who had many other claims to fame in the fields of geography and the emerging science of meteorology. In 1816 he devised the concept of isothermal lines on maps, indicating comparative atmospheric temperatures across the globe.

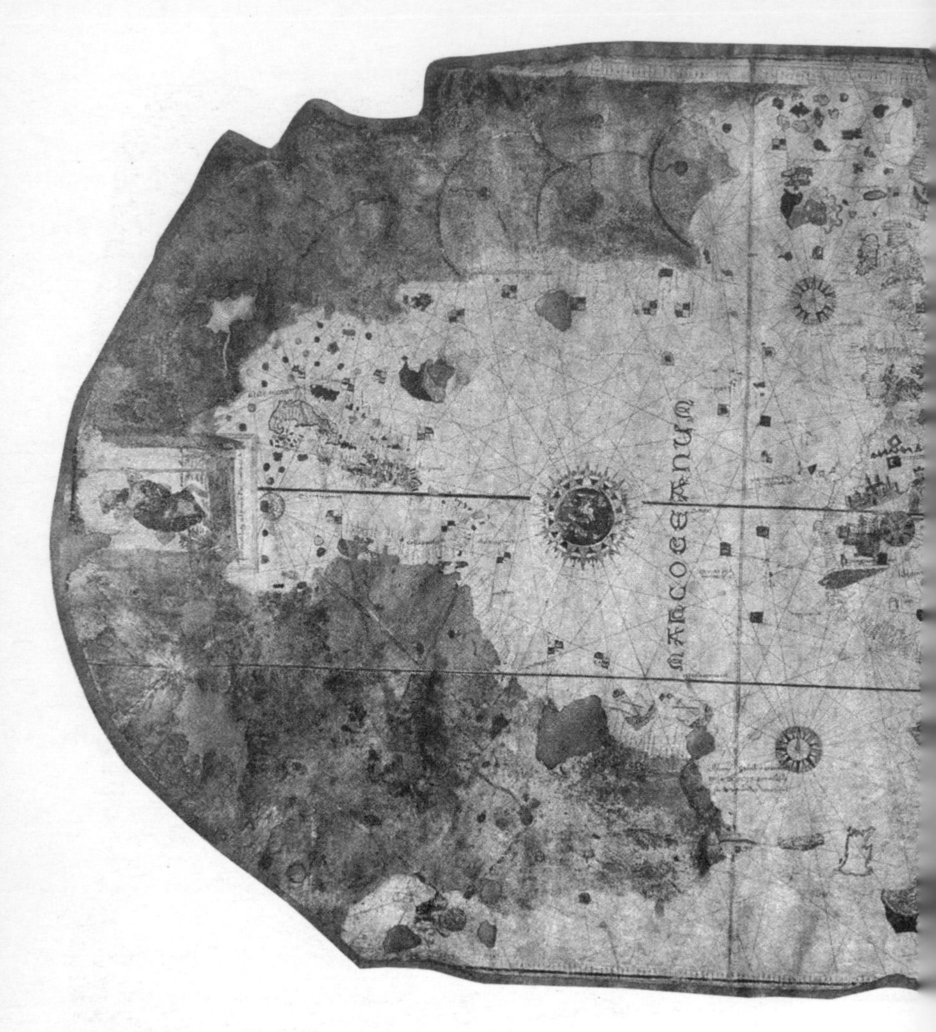

The beginnings of the New World as Juan de la Cosa remembers the voyages of Columbus.

Measuring 99cm by 177cm, dotted with coloured inks, de la Cosa's lost map combines the naive wonder of the New World with symbols from the Middle Ages. There are castles and enthroned monarchs in tents; three wise kings are trotting through Asia; the compass rose shows a nativity scene. It may now be found under glass in the naval museum at Madrid, where

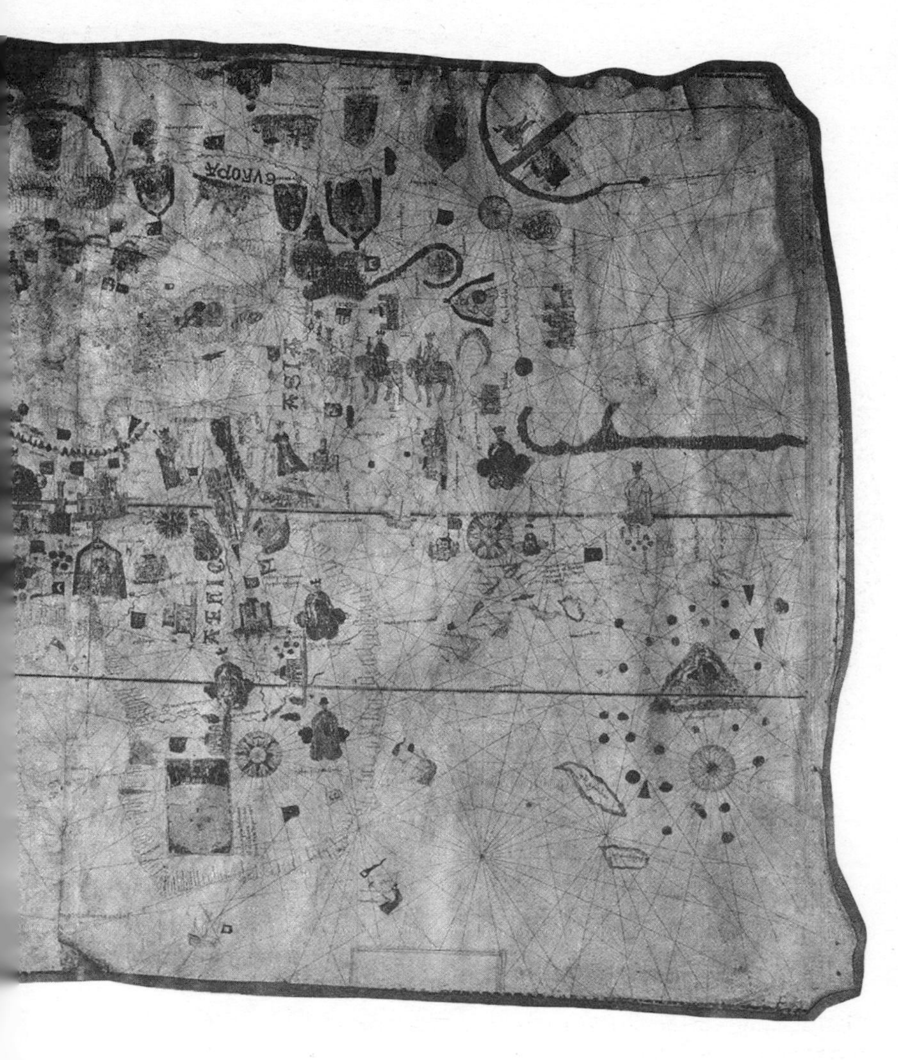

it tells one important story above all others – the solid land barrier blocking a smooth western route to the Orient.

It is assembled from several pieces of parchment, and you need to walk around it to view all the text and drawings. The old world looks reasonably accurate: Britain and Ireland are clearly recognisable; Africa has mountain ranges and an unusual absence of animals; while the description on the map of Vasco da Gama's arrival in India in 1498 makes it newly accessible. The New World is drawn from personal sea charts and memory, as Juan de la Cosa sailed with Columbus as his master pilot on the *Santa Maria* in 1492 and 1494, and subsequently on other Castilian voyages.

The map is part of a specific cartographic tradition – the portolan manuscript sea chart. Portolans are as old as the European mariner's magnetic compass, and for about two centuries the two depended on each other for the growth and safe passage of Mediterranean trade. From about 1300, navigators used maps of straight but criss-crossing lines (known as 'rhumb' lines) to plot their way through open water and along coasts, each line fanning out from up to thirty-two compass points (the name derives from the Italian *portolano*, relating to ports or harbours). The lines didn't represent direct routes in the way a road map can, but more of a safety net for increasingly adventurous sailors; they marked a way back to dry land the way a silken thread may once have guided mythical heroes.

Juan de la Cosa's portolan was one of the last of the breed, soon to be replaced by printed maps representing a more accurate projection of the world (the rhumb lines did not take the earth's curvature into account, and so would be accurate only over short distances). His map was unusual in portraying the whole world rather than a particular trading area or coastal route, and as such has several compasses, each with its own directional markings. But this was less of a map to be taken on a voyage than one announcing great news and important discoveries, not least the landfall off the North American coast of Labrador and Newfoundland by John Cabot, clearly defined on the map by flags and text: *Mar descubierta por los Ingleses*. ('Ownership' and 'discovery' are onerous words, of course, considering the islands in question are already occupied by an indigenous population.)

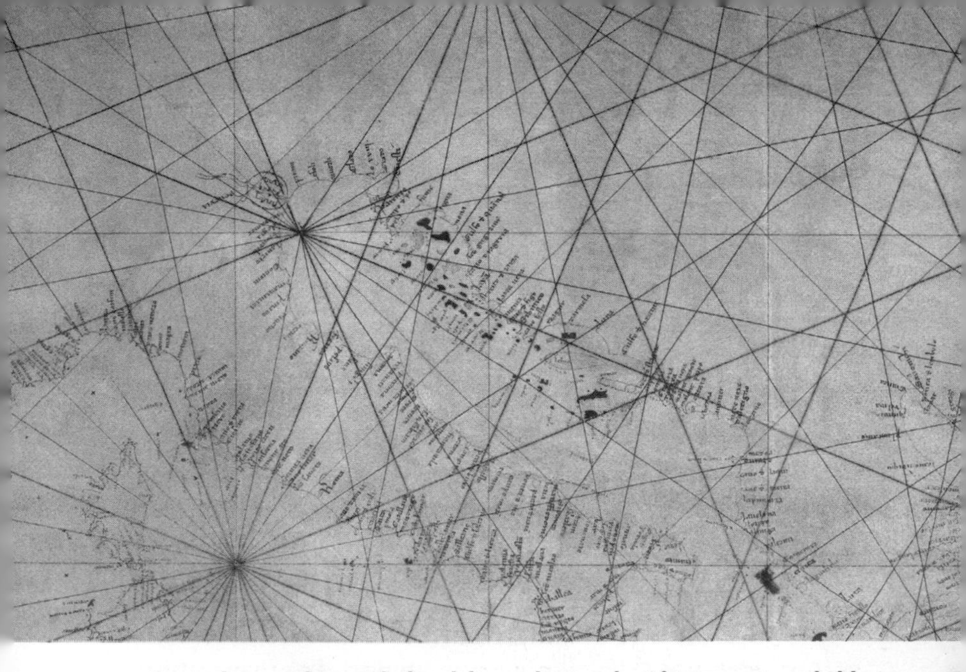

A portolan sea chart with rhumb lines plotting the African coast, probably drawn in Genoa around 1492.

The new finds in the Bahamas are shown on a slightly larger scale: the archipelago of Guanahani, the settlement of La Navidad, the town of Isabella on Haiti. Martinique and Guadeloupe are grouped together as The Cannibal Islands, a practice witnessed by both Columbus and de la Cosa. To the extreme west, at the map's narrowest point, is a trailing remnant of Christian map-iconography – Saint Christopher carrying a staff, symbolising Columbus bringing Christianity to new lands. Where the sea ends at the west of the map there is a large arc of green, presumably a bountiful but unidentified landmass; had de la Cosa identified it, he would have almost certainly mislabelled it Cathay (China).

Which is exactly what Giovanni Contarini did six years later, when his cone-shaped world map was the first to include the New World and the first to mis-identify it (the mis-identification became all the rage for a short while, as his map was reproduced in an edition of Ptolemy's *Geographia* published in Rome). Contarini placed Japan (Zipangu) between Cuba and Cathay, while another huge white unexplained landmass beneath it, larger than Africa, went by the name of Terra S Crucis (Land of the Holy Cross).

But a year after that, the world changed forever: the word 'America' appeared on a map for the first time. Too bad that the man after whom it was named, Amerigo Vespucci, didn't actually have much to do with the continent's 'discovery'.

In 2003, the Library of Congress completed the purchase of an artifact that it had been pursuing for a century. Known as the Waldseemüller map, after its principal draughtsman, it consists of twelve woodblocks, each showing a different section of the world, roughly 8ft by 4ft when pieced together. Perhaps a thousand were printed at the beginning of the sixteenth century, but only one known copy survives. After protracted negotiations with its German owners, a sum of $10 million – the highest sum ever paid for a single map – was agreed for its transfer to Washington DC.

Those who view the Waldseemüller map in its hushed and low-lit environment beneath glass in the capital's Thomas Jefferson Building swiftly come to regard it as money well spent, for it is certainly one of the most arresting and historically significant maps in existence. Like the Hereford Mappa Mundi, one may never tire of looking at it. Even if one may never fully understand it.

The map was made in 1507. Its creator knew it might baffle, and so unfamiliar were some of his revelations about the world that he inscribed a request for the patient indulgence of his viewing public; it was a request, in fact, not to be laughed at. The map included a new continent in the western hemisphere. But from a modern perspective the discovery of America is far from the strangest thing about it. The big conundrum is why he didn't call the new continent Columbus.

The facts of the map are limited. It was drawn up in the north-eastern French town of St Dié and its principal draughtsman was the German cleric Martin Waldseemüller, possibly with assistance from his colleagues at the Gymnasium Vosagense, an intellectual circle that met to discuss theology and geography. The map was part of the *Cosmographiae Introductio* by Waldseemüller and Matthias Ringmann, a publication which also included a much smaller world map cut into gores for a small globe, an introduction to geography and geometry, and an account of travels in the New World.

The knowledge portrayed on the map is far more detailed than anything that had preceded it. The map broadly follows one of Ptolemy's projections but it shows the latest coastal news from Africa and India. Waldseemüller drew upon many sources and recent maps, almost certainly including the globe that his fellow German Martin Behaim constructed in 1492, just a few weeks before Columbus first set sail. Behaim, however, would have been astonished by Waldseemüller's depiction of a large ocean stretching uninterrupted to the coast of Asia, evidently the Pacific. This was six years before Vasco Nunez de Balboa first described it, and fifteen years before Magellan's first circumnavigation of the world in 1522 confirmed that it was there.

How could Waldseemüller possibly have known of it? A cartographer's ghostly intuition? Or was there, perhaps, another map, containing news of other explorations, that has since been lost to us?

Waldseemüller's other great revelation, his new western hemisphere, was depicted on three vertical panels on the left side of his map. We vaguely recognise the shapes, and we may forgive his depiction of a clear waterway between North and South America, rather than an isthmus. But we are still struck by the lonely presence, in the bottom left-hand corner, of a word that the world was previously unfamiliar with: 'AMERICA'.

It appears over land we would now regard as central South America, and in a box of Latin text beneath it there is an explanation. The land is something 'of which the ancients make no mention' and its inclusion here is based on 'true and precise geographical knowledge'. Intriguingly, the text box is rather larger than the type that fills it, suggesting there may have been plans for a more detailed description that never materialised.

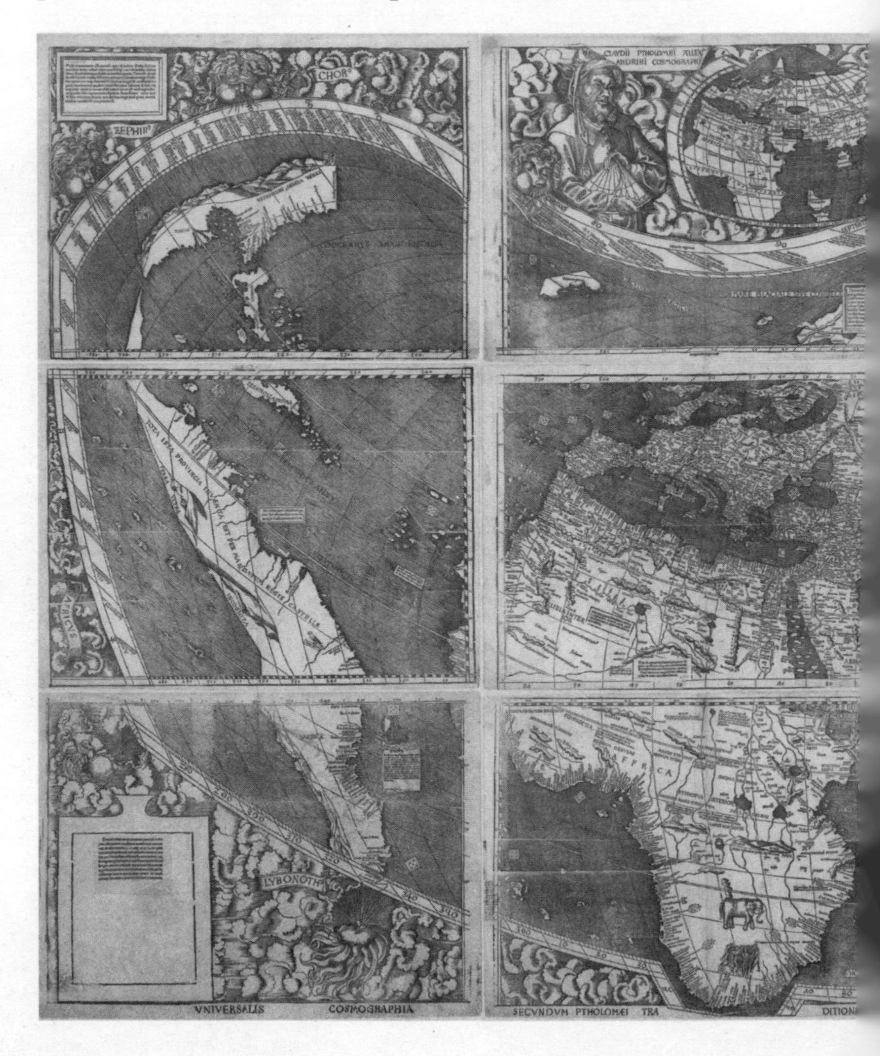

America was named after Amerigo Vespucci, a skilled but minor Florentine navigator with a background in finance; for a while he worked for a bank in Seville that provided some of the funds for the early voyages of Christopher

The twelve panels of the majestic Waldseemüller map – all the better for an elongated America (far left). The figures at the top are Ptolemy and Amerigo Vespucci.

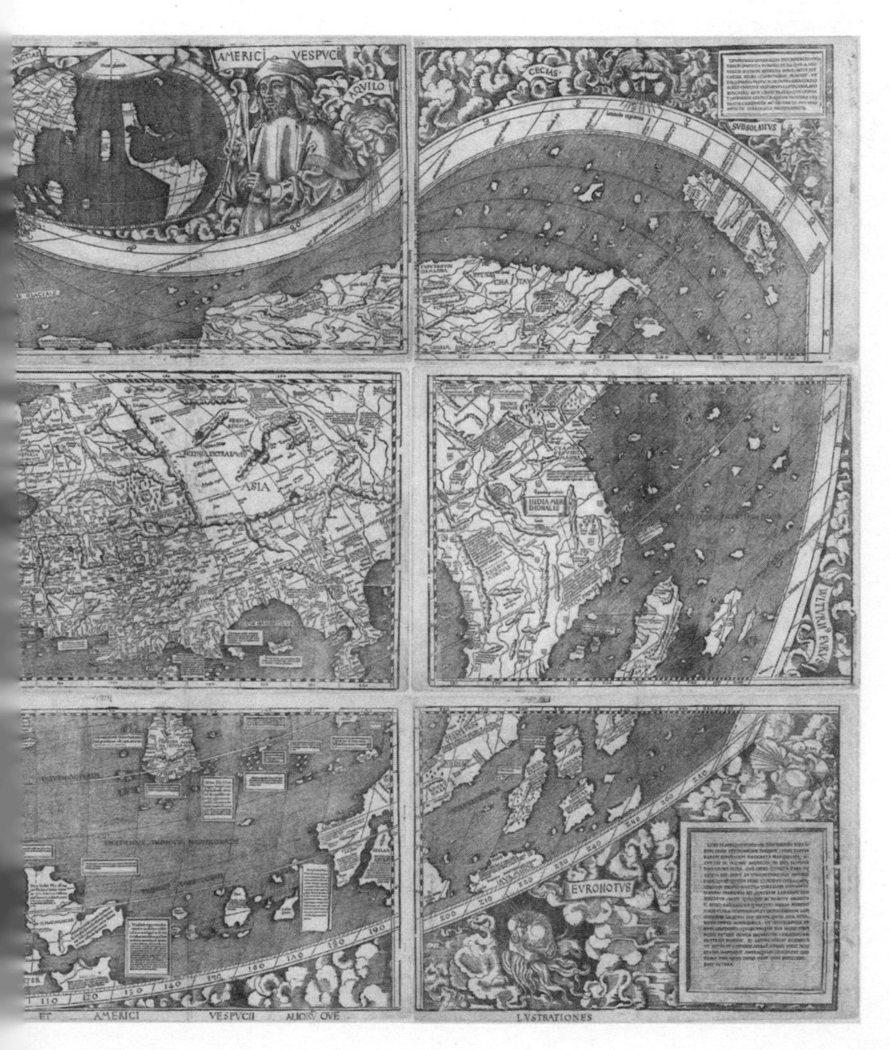

Columbus. Vespucci and Columbus became friends, and it is likely that Columbus fired Vespucci's passion for exploration. But only one of them set sail in 1497 to land on the coast of Venezuela.

Between 1495, when he loses his job at his bank, and 1499, when he is seen aboard a Spanish ship bound for the coast of South America, we have no record of Vespucci's whereabouts. On that trip, which set off more than a year after Columbus landed on the Paria Peninsula, he was under the command of the Spanish conquistador Alonso de Ojeda (the crew also included Juan de la Cosa), and it may have been Ojeda who recommended Vespucci for further voyages to the coast of Brazil. He ended his days in what may be called cartographic administration, providing sea charts for many Spanish voyages along the South American coasts. When he died aged 60 in 1512, leaving his impoverished widow to apply for financial assistance from the state, Amerigo Vespucci (or Americus Vesputius, as he was sometimes known) was probably unaware that yet another version of his name would render him immortal.

How did this come to be? The anomaly appears to have arisen from the wide circulation of two printed letters, at least one of which had come to the attention of Martin Waldseemüller in northern France by the time he drew his map in 1507. The first, the copy of a four-page letter apparently written by Vespucci, was published in Florence in 1503 and described a voyage to the coast of South America in the summer of 1501. The voyage was not as significant as its reporting: Vespucci's letter popularised the phrase 'New World', and it was the first to describe such an appealing and abundant coastline, 'a more temperate and pleasant climate than in any other region known to us'.

Vespucci's second letter, written in 1504 to a childhood friend who had risen to become the head of the Florentine government, described four voyages over thirty-two pages. The second and third trips took place between 1499 and 1502

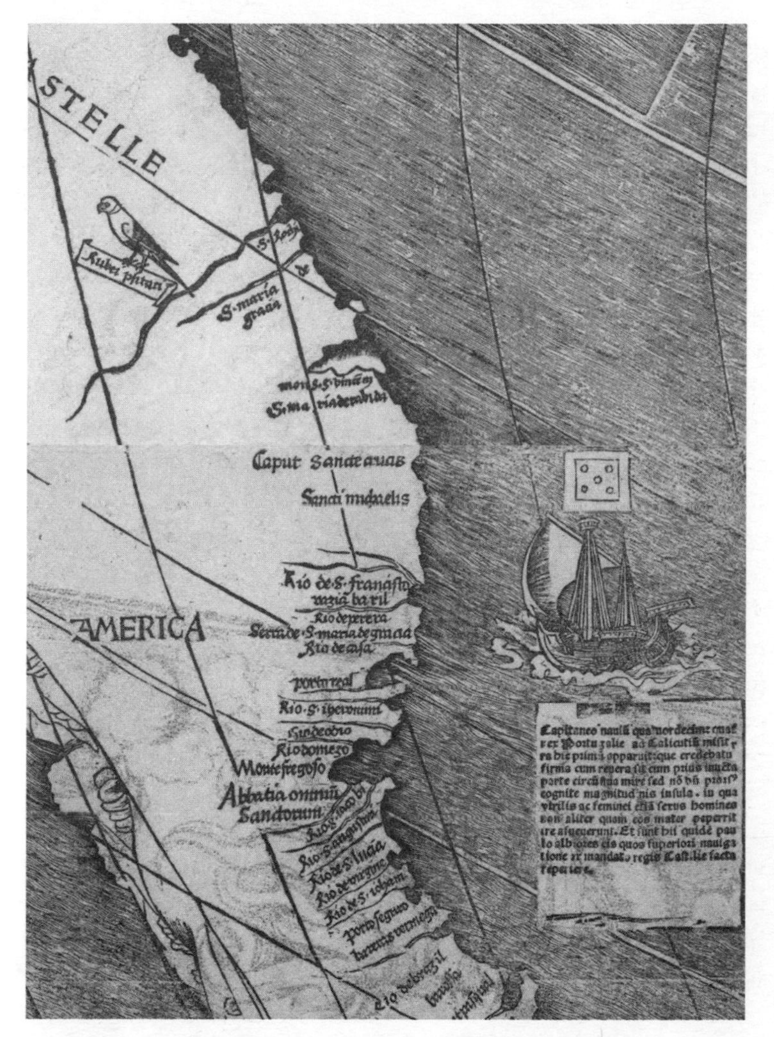

Waldseemüller's label: a new continent takes shape – and name.

and ventured to a similar stretch of coast described in his first letter. The fourth trip, from 1503 to 1504, is unlikely to have taken place at all, as the records show Vespucci was stationed on dry land in Spain. But it is the first trip that has caused most controversy. The letter claims that Vespucci landed on the South American continent a year before Columbus.

In his book *The Mismapping of America*, the forensic cartographic historian Seymour Schwartz has no qualms in calling this voyage 'a fraud' and one that 'totally misrepresented the fact that the voyage ... took place in 1499, one year after Columbus.' But we are uncertain as to who perpetrated the fraud – or downright forgery of the letters – or why.

How can we be sure they were forgeries? Schwartz has several planks of evidence. Between 1500 and (probably) 1504, Vespucci wrote three letters whose authenticity has never been in doubt, and in one he clearly describes landing at the Paria Peninsula a year after Columbus. Schwartz also asserts that, throughout his life, Alonso de Ojeda always supported the primacy claims of Columbus over Vespucci. Further evidence comes from a trial that took place in 1516 brought by the heirs of Columbus against the Treasury of Spain (though famous, Columbus did not leave a fortune). During the trial, none of the hundred witnesses disputed the fact that Columbus was the first to set foot on South American soil.

Amerigo Vespucci depicted on Waldseemüller's map, beside the new western hemisphere.

All of which makes Waldseemüller's misjudgement the more curious. Vespucci is honoured with a portrait at the top of the map, where he is shown opposite the only other figure, Ptolemy. There are two images of globes here, too: Ptolemy sits beside the older known eastern hemisphere, while Vespucci sits beside the new western one. Vespucci is also included in the map's title: *Universalis cosmographia secunda Ptholemei traditionem et Americi Vespucci aliorum que lustrationes* ('A drawing of the whole earth following the tradition of Ptolemy and the travels of Amerigo Vespucci and others). Christopher Columbus is certainly one of the 'others', but Waldseemüller justifies his naming of the country he 'discovered' in unapologetic terms within the written introduction to the map. 'Inasmuch as both Europe and Asia received their names from women, I see no reason why anyone should justly object to calling this part Amerige, i.e. the land of Amerigo, or America, after Amerigo, its discoverer, a man of great ability.'

A few years later there is some evidence that Waldseemüller regretted his choice. In 1513 he produced his first edition of Ptolemy's Atlas with new maps in Strasbourg, and on the page showing the New World, South America is labelled 'Terra Incognita'. But there is also an inscription that translates as 'The lands and adjacent islands were discovered by Columbus sent by authority of the King of Castile.' This time it was Vespucci who went unnamed. Then, three years later, when Waldseemüller published a new twelve-sheet world map called the Carta Marina, the two get equal billing. Both are mentioned in the text, although South America now has two new names that credit neither: 'TERRA NOVA' and 'TERRA PAPAGALLI' (The Land of Parrots).*

* There is, almost inevitably in a tale such as this, an intriguing contemporary aside. The Waldseemüller map was lost for hundreds of years, only to be rediscovered in 1901 in Wolfegg Castle in Southern Germany. At the very same castle some decades later Josef Fischer may have forged the Vinland Map.

But it was too late. The name America had already begun to appear on other maps, including influential mass-produced works by Peter Apian and Oronce Fine. And then forever more.

The misnaming of America has caused both alarm and amusement for five hundred years. In the seventeenth century, the important Scottish cartographer John Ogilby speculated that the prime reason Vespuccio took preference over Columbus 'by a lucky hit' was due to 'the gingle of his Name Americk with Africk'. Yet as a coda to the wayward naming of things, consider this story about the conquistador Hernán Cortés.

Cortés has a significant place in map history, having made the first printed map to show the Gulf of Mexico, the first dated map to name Florida, and the first plan of an American city, a place he named Temixtitan (on the site of the present Mexico City). But the map story that endures about Cortés is his naming of another place.

In 1519, about to set foot in Mexico, Cortés invited some natives to join him for a conversation aboard his ship, and asked them for the name of the place he was about to pillage for its gold. One man replied, 'Ma c'ubah than', which Cortés and his men heard as Yucatan, and named it thus on his map. Just over 450 years later, experts in Mayan dialects examined the tale (which may in any case be apocryphal) and found that 'Ma c'ubah than' actually means 'I do not understand you.'

Pocket Map
California as an Island

Before The Beach Boys, before Hollywood, before even the Gold Rush, California was known for being distinct from the rest of America. In fact, it was an island.

We are now sure – because we have seen it on maps – that California is firmly attached to Oregon, Arizona and Nevada. Even south of San Diego, when it eventually becomes the Mexican state of Baja California, it is firmly hitched to the mainland. But in 1622, something untoward happened. After eighty-one years officially attached to a huge landmass, California drifted free. It wasn't a radical act of political will, nor a single mistake (a slip of an engraver, perhaps), but a sustained act of cartographic misjudgement. Stranger still, the error continued to appear on maps long after navigators had tried to sail entirely around it and – with what must have been a sense of utter bafflement – failed.

The name California first appeared on a map in 1541. It was drawn as part of Mexico by Domingo del Castillo – a pilot on an expedition by Hernando de Álarcón – and it is shown as a peninsula and labelled. Its first appearance on a printed map occurred in 1562, when the Spanish pilot and instrument maker Diego Gutierrez again wrote its name at the tip of a peninsula, a very minor detail on a busy and very beautiful engraving of the New

World. The map, the largest then made of the region at 107cm by 104cm, may have been engraved after Gutierrez's death by Hieronymus Cock, an artist who clearly took great delight in imaginative trappings: huge ships and legends populate its seas, with Poseidon driving horses on a seaworthy chariot, a huge gorilla-type creature breaking the waves while it dines on a fish, and terrible goings-on in Brazil, where the natives are seen slaughtering human flesh, curing it from a tree, and then roasting it.

California subsequently appeared attached to the mainland for sixty years. And then off it floated into the Pacific, where it remained a cartographic island for more than two centuries.

Its first known insular appearance occurred in 1622, on an inset on a title page of a Spanish volume entitled *Historia General*. Two years later it was drifting free, bounded by the Mar Vermeio and Mar Del Zur on a Dutch map by Abraham Goos. But it received its most prominent currency on a London map of 1625 entitled 'The North Part of America'. This accompanied an article about the search for the Northwest Passage by the mathematician Henry Briggs. He supplemented the great untracked northerly spaces towards the Arctic with text describing the wonders of his map, 'Conteyning Newfoundland, new Eng/land, Virginia, Florida, new Spaine ... and upon ye West the large and goodly lland/ of California.' On the eastern seaboard both Plymouth and C Codd are placed in Massachusetts, but not yet Boston (and not yet Manhattan: the first mention on a printed map, by Joannes de Laet, occurred five years later, when it was named as Manhattes.)

The misconception persisted for decades. It was the seventeenth century's forerunner to a mistake on Wikipedia – doomed to be repeated in a thousand school essays until a bright spark noticed it and dared to make amends. Compiling a paper for the California Map Society in 1995, Glen McLaughlin and Nancy H. Mayo catalogued 249 separate maps (not including world maps)

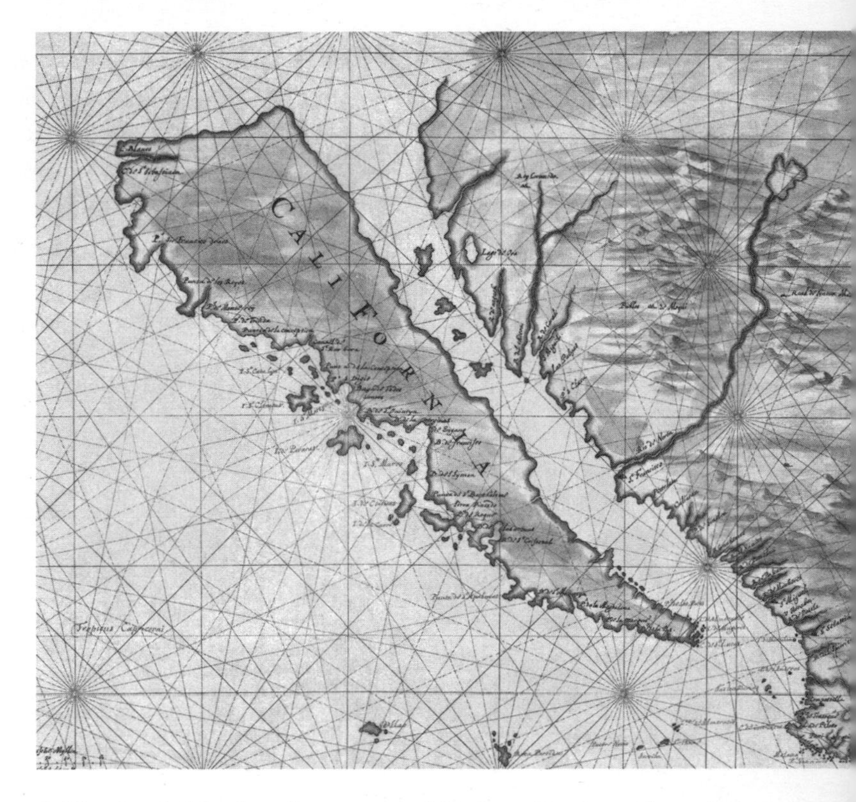

California – drifting happily away as an island on a Dutch map from 1650.

which cast the Golden State adrift. Their names carry bold assertions, with no wiggle room: 'A New and Most Exact map of America' claimed one, while another promised 'America drawn from the latest and best Observations.' Between 1650 and 1657, the French historian Nicolas Sanson published several maps which showed California as an island, and their translations into Dutch and German ensured that they superseded Briggs as the most influential mythmakers for half a century. But they also promoted newer, truer discoveries, including the first cartographic depiction of all five Great Lakes.

Even when new maps were published showing California attached to the mainland (the most significant accompanying the personal accounts of Jesuit Friar Eusebio Kino, in 1706), the island kept on appearing. In the end, though, it was killed off by a royal decree issued by Ferdinand VII of Spain in 1747, which denied the possibility of this Northwest Passage with the reasonably clear statement: 'California is not an Island.' Yet news travelled slowly. California appeared as an island on a map made in Japan as late as 1865.

And how did it all begin? The cartographical point zero has been tracked to a Carmelite friar named Antonio de la Acensión who sailed with Sebastian Vizcaino along the West Coast in 1602–3 and kept a journal. Two decades later he is believed to have mapped his trip on paper, which featured California as an island nation. The map was sent to Spain, but the ship on which it travelled was captured by the Dutch, and it ended its journey in Amsterdam. In 1622, Henry Briggs wrote of seeing this map of California in London. And shortly afterwards, the map drawn from the one 'taken by Hollanders' was set in copper and began its journey through the world.

Chapter 7
What's the Good of Mercator?

He had bought a large map representing the sea,
Without the least vestige of land:
And the crew were much pleased when they found it to be
A map they could all understand.

'What's the good of Mercator's North Poles and Equators,
Tropics, Zones, and Meridian Lines?'
So the Bellman would cry: and the crew would reply
'They are merely conventional signs!

'Other maps are such shapes, with their islands and capes!
But we've got our brave Captain to thank:'
(So the crew would protest) 'that he's bought us the best –
A perfect and absolute blank!'

Lewis Carroll, *The Hunting of the Snark*

Well, what *is* the good of Mercator's famous world map of 1569? It's riddled with distortions and full of countries many times larger than they really are. And yet, astonishingly, it's still essentially the map we use today. Countries have been added of course, and the shapes of coasts and borders have been corrected and politically adjusted, but the map that shaped the end of the Renaissance, saw in the Enlightenment and adorned Victorian classrooms remains the display of choice, right through to the latest Google Maps. It is the definitive icon of our world and to mess with it looks like terrorism. Not that people haven't tried.

We aren't looking at one map, of course, but a projection of the world – a template for all maps. Which is perhaps a little ironic for Gerardus Mercator, born in Flanders and working at the time in Duisburg, on the Rhine, was not himself a prodigious cartographer. When he laid out his famous world projection in 1569, at the age of fifty-seven, he had produced less than ten maps. But his new one was an undoubted wonder – mathematically meticulous and constructed with startling scale and ambition. It measured roughly 2 x 1.25 metres over eighteen printed sheets and must have stunned all who saw it.

The things that look wrong to us now – Greenland the size of Australia rather than a third of it, an Antarctic continent that bumps raggedly and indefinitely along the base – were not the strangest things then, for exact proportionate sizes were not yet known and the polar regions were but dismal myth. The strangest thing to his contemporaries was that Mercator, a man who had never been to sea (and would never go), would so effectively help the mariner plot a true course across the oceans after so many centuries of intuitive guesswork. The military would also have cause to be grateful to him: he helped them more accurately fire their cannons.

The Mercator map's main attribute was technical: it provided a solution to a puzzle that had been troubling mapmakers since the world was recognised as a sphere, which is to say back to Aristotle. The problem was: how does one represent the curved surface of the globe on a flat chart? The strict and well-established grid of latitude and longitude was all very well for theoretical coordinates, but the navigator pursuing a constant course sailed on an endless curve.

Mercator had already displayed this curving course on his globes through his rings or 'rhumb' lines, and now he wanted to convert them to a map, and enable any navigator to swiftly locate his position and find his way to any destination.

Mercator had struggled with the problem for a while. You can too: take a nice furry tennis ball, draw a few shapes representing countries on it and slice it in two. Then make some more nicks on the cut sides and flatten it out. The countries will bulge up in the middle, and in order for the tennis ball map to lie flat the middle must be shrunk and the edges expanded. Now try to do this accurately, so that sailors bring their cargo home. Mercator's quest was to find a way of doing so by mathematical formula.

In 1546 he wrote to a friend that the same journey by sea between two places would often be described in ships' logs with very different latitudes. The maps were simply misleading: 'I saw that all nautical charts ... would not serve their purpose.' He wasn't the first one to see this, but the problem only really presented itself in the sixteenth century with the refinement of the compass and the classic voyages of discovery that tacked their way across new oceans. In the space of a few decades there were numerous new and often cranky projections of the world: the Azimuthal and the Azimuthal Equidistant, the Orthographic, Gnomic, Stereographic, Cordiform, Pseudocordiform, Globular, Trapezoidal and Oval.

Almost all of these projections depended on the graticule system of latitude and longitude, and most marked the Tropics and the Equator. Not all of them were aimed at seafarers; some were better suited to celestial or polar mapping, while others were illustrative and impressionistic. Leonardo da Vinci and Albrecht Durer had also had artistic shots at the problem.

Inevitably Ptolemy had already tackled the issue first – twice. He named one projection 'inferior and easier' and one 'superior and more troublesome'. The former, his by now classical grid system, was naturally limited: in the first projection, for

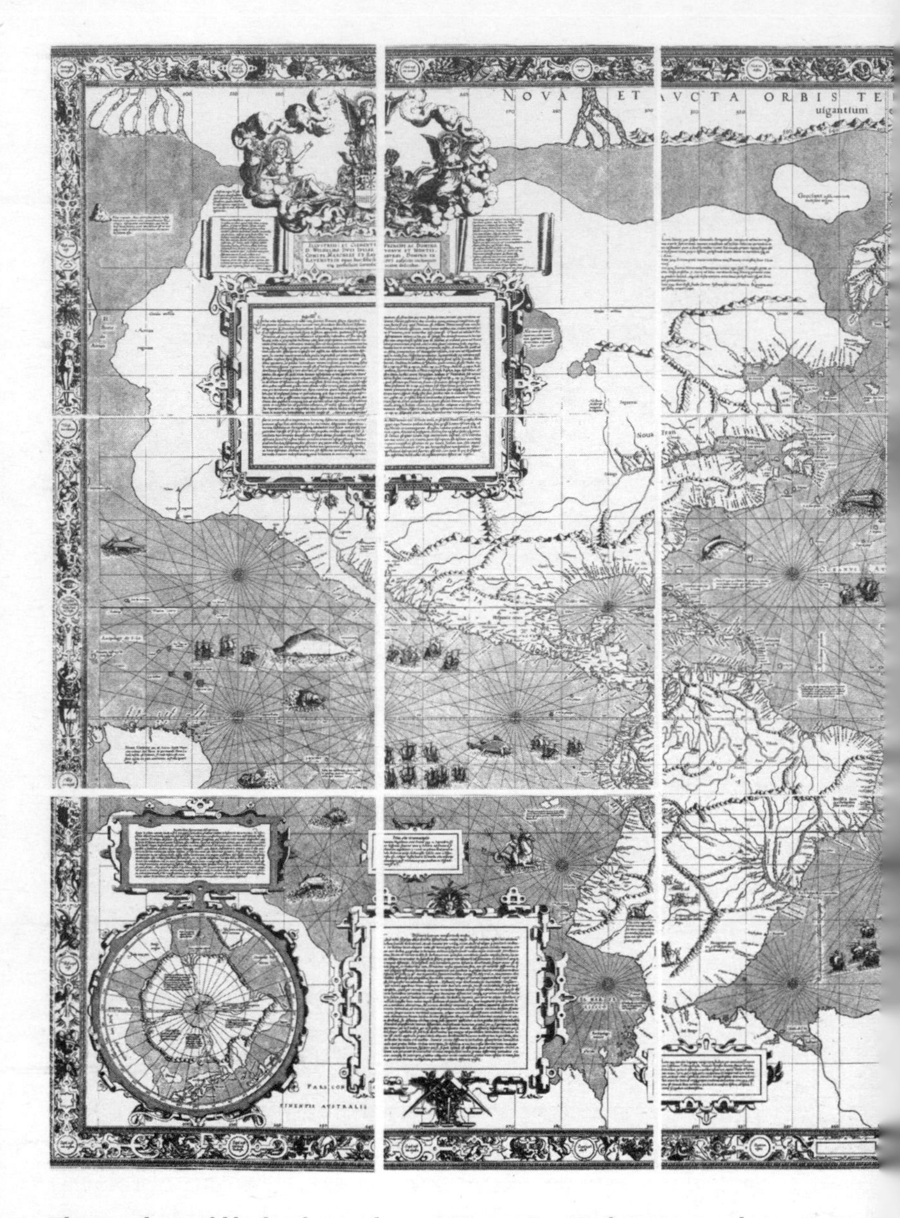

The way the world looks, then and now: Mercator's 1569 classic, printed over eighteen sheets and demanding a double page of any book.

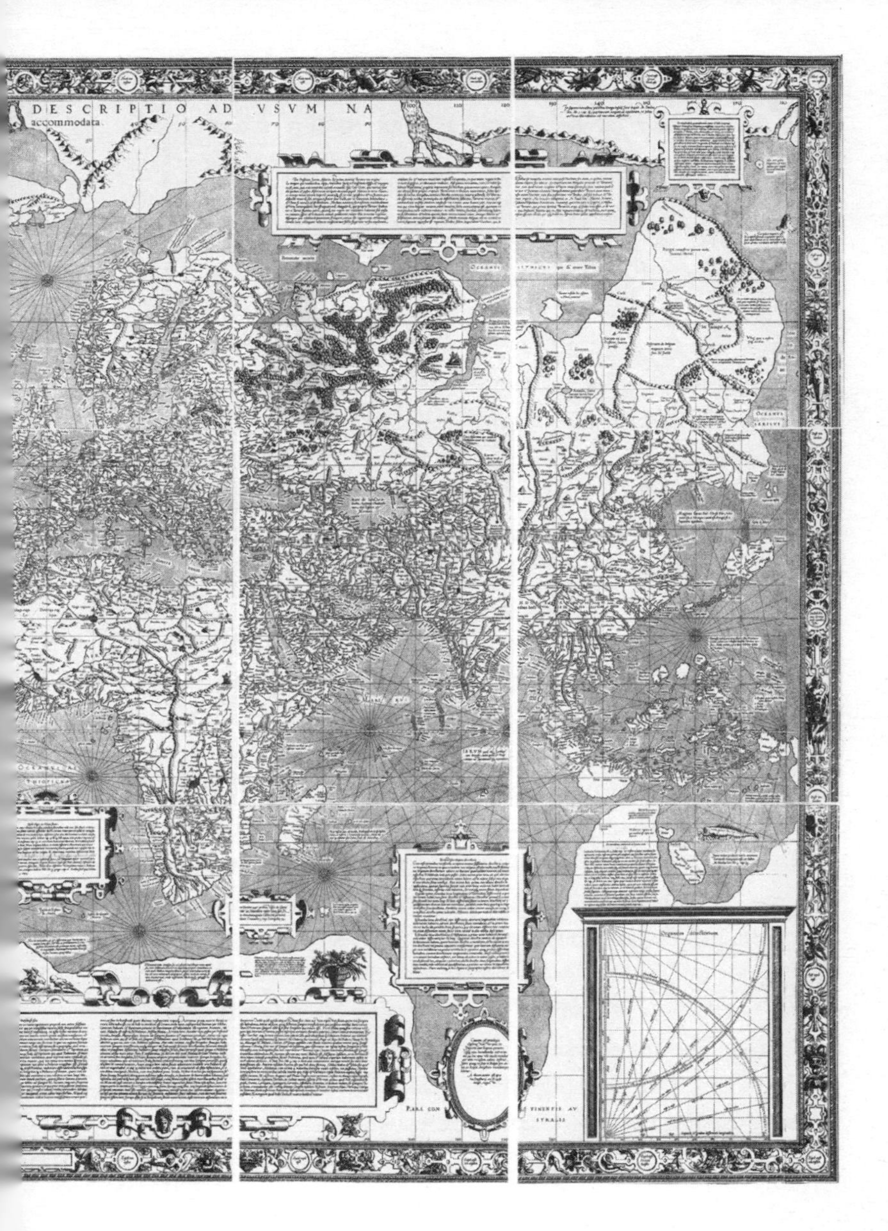
DESCRIPTIO AD VSVM NA
accommodata

example, his latitude began at his equator (pitched at 16·25° south to 63° north), while longitude, extending to a mere 180 degrees of the sphere, had a zero meridian beginning at the Blessed Isles, a land now regarded as either the Canary Islands or Cape Verde. Nonetheless, given the limitations posed by an inadequate supply of coordinates, the area covered by his projection is a remarkably good approximation of the true relationship between countries that we recognise today.

Mercator's map drew heavily on Ptolemy's gazetteer and refined it with recent discoveries, notably the outline of North America, which was fully realised, indeed almost plump. But his enduring breakthrough was his new Conformal projection, the method by which he manoeuvred his latitudinal rings to keep all the angles straight (the lines of latitude became further apart as they moved from the Equator). Mariners would thus be able to navigate across the map in straight lines, in keeping with the desired direction of their flickering compass.

Mercator used the blank space on the unexplored interior of North America and his empty oceans to justify his new device to all who might find his projection unfamiliar. He explained that he intended 'to spread on a plane the surface of the sphere in such a way that the positions of places shall correspond on all sides with each other both in so far as true direction and distance are concerned, and as concerns correct longitudes and latitudes.' In so doing, Mercator had created a grid which, in the words of his recent biographer Nicholas Crane, 'would prove as timeless as the planetary theory of Copernicus. In seeking the essence of spatial truth, he had become the father of modern mapmaking.'

❋ ❋ ❋

What has happened to Mercator's projection since? It has inevitably been modified and improved.

This process began almost as soon as his world map was first published (most notably by Edward Wright, Edmund Halley and Johann Heinrich Lambert), and has continued up to Google – which, extraordinarily, found Mercator's neat and symmetrical rectangles perfectly suited to the pixelated tiles that make up a digital map.

The projection's resilience is even more remarkable when one considers the forces that have raged against it for the last four hundred and fifty years. In 1745 a Frenchman named César-François Cassini de Thury suggested using a cylindrical projection, sometimes shown as two hemispheres placed on top of each other with their centres at the poles. This showed a true scale along its central meridian and all places at right angles to it, but a varying level of distortion elsewhere. A more radical transformation was proposed by the Scottish astronomer James Gall at a meeting in Glasgow in 1855. Gall highlighted the essential fault with the Mercator projection – the shapes of the land masses were vaguely right, but their sizes were wrong. Applying his new 'stereographic cylindrical' theory first to the constellations and then to earth, he found a way of flattening the earth to a more compact scale, while also decreasing some of Mercator's distortions (although introducing others).

Without due acknowledgement, many of the attributes of Gall's work were picked up by the German Arno Peters in the mid-1970s and turned into a hot political quarrel that has still not entirely subsided. The argument was relatively simple: because of its high-latitude distortions, the Mercator map over-emphasised the size and significance of the developed world at the expense of the under-developed (which tended to be closer to the Equator). Peters' cylindrical projection (now generally known as the Gall-Peters projection) was therefore put forward as both an anatomically and politically correct alternative, and even though its claims were not novel (and it was often compared to a washing line on which countries had been hung out to dry), its alternative to the 'cartographic imperialism' and 'Euro-centric ethnic bias' of Mercator's map took on a voguish momentum.

The screenwriter Aaron Sorkin summarised it caustically in an episode of *The West Wing* in 2001, a scene in which the Press Secretary C.J. Cregg and Deputy Chief of Staff Josh Lyman attend a briefing by members of the fictional Organisation of Cartographers for Social Equality. The OCSE were pushing for the President to 'aggressively' support legislation that would make it mandatory for every school to use Peters rather than Mercator. 'Are you saying the map is wrong?' Cregg asks. 'Oh dear yes,' the OCSE representative replies as the slideshow behind him displays same-size images of Africa and Greenland. 'Would it blow your mind if I told you that in reality Africa is fourteen times larger?'

Another OCSE member then explains that Mercator's Europe is drawn considerably larger than South America, whereas in reality South America's 6.9m square miles is almost double Europe's 3.8m square miles. Then there is Germany. Germany appears in the middle of the map, whereas in fact it should be in the northernmost quarter. 'Wait,' Josh Lyman

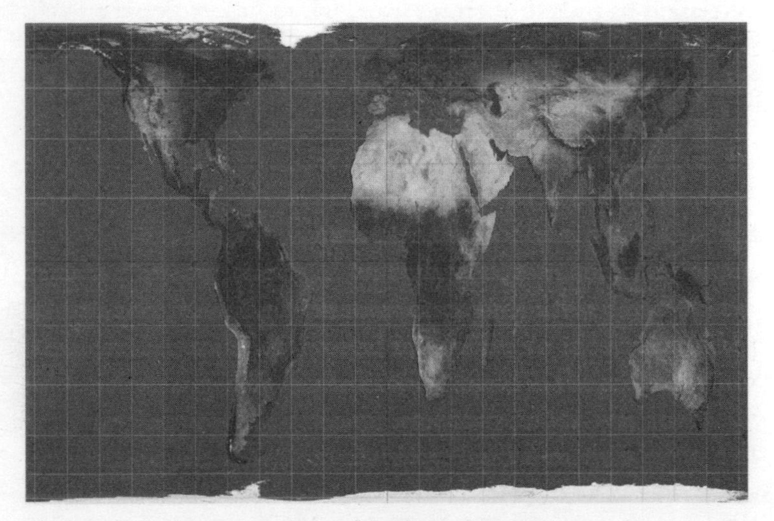

The Peters-Gall Projection: a 'washing line' of countries

says. 'Relative size is one thing, but you're telling me that Germany isn't where we think it is?'

'Nothing's where you think it is,' the chief OCSE man says. He then clicks up the Peters Projection and the OCSE propose that the world map should be flipped so that the northern hemisphere is put on the bottom. A new slide shows what it will look like.

'Yeah, but you can't do that,' C.J. Cregg reasons ... 'Because it's freaking me out.'

❊ ❊ ❊

Peters died a year after the episode was broadcast, his projection ridiculed as much for the smug superiority of its proponents as for itself. In fact, the principal objections often centred on the fact that the supporters of Peters exaggerated both their claims and their outrage, perpetuating the myth that two thirds of Mercator's map is dedicated to the northern hemisphere and only one third to the southern. And the Gall-Peters has its own distortions (particularly severe between 35° north and 35° south, and between 65° and the poles), rendering some African countries and Indonesia twice as long north-south as they really are. The Royal Geographical Society's quarterly journal began its review of Arno Peters' book *The New Cartography* (1983) thus: 'Having read this book many times in German and English, I still marvel that the author, any author, could write such nonsense.'

Other projections have also found favour, including one produced by the American cartographer Arthur Robinson which combined elements of Mercator and Gall-Peters, and was adopted by the US map-making company Rand McNally. It first appeared at the height of the Cold War in the early 1960s, but argued against the perceived menace of the USSR by greatly reducing its surface size.

There are so many possible projections, each with their own particular political agenda and limitations, that there is a way of measuring their spatial prejudices as a chart, known as the Tissot Indicatrix of Distortion. This could take, say, the Winkel Tripel Projection of 1921 (yes, this is a real thing) and overlay it with a pattern of stretched circles to show the degree of corruption over any one area (a perfect circle showing true unity, an oval stretched north-south reflecting a north-south distortion).

Will one projection emerge victorious? It's already happened. Mercator's map casts its shadow over the digital world just as it did in the world of those navigators opening up new trade routes half a millennium ago, and the possibilities for its future manipulation are therefore limitless. It is the projection used not only by Google Maps, but also (with a spherical interpretation) by its rivals, Microsoft's Bing and OpenStreetMap. Even in the virtual age it is the ocean-crossing of least resistance. Any alternative would have to be imposed centrally by a court bigger than the United Nations, whose logo, incidentally, is a projection of a globe centred on the Arctic Circle and wreathed in olive branches that first appeared twelve years after Mercator's.

Yes indeed: the Postel Azimuthal Equidistant of 1581 still has its influential supporters.

Pocket Map
Keeping It Quiet: Drake's Silver Voyage

When in 1580 Francis Drake returned triumphant from his unintentional circumnavigation of the world, Elizabeth I declared two things: her delight at the fact that his cargo had enabled her to pay off the national debt (she knighted him the following year); and her wish that Drake's route to the world's untapped riches remain secret by staying off the maps. Nervous of their necks if they disobeyed, the nation's cartographers upheld her decree, at least on paper. When one of them finally broke cover after nine years, the map that emerged was intricate, accurate, and struck as a solid silver medallion to be worn around the neck.

There are nine known copies of the Silver Map of Drake's Voyage, two of them in the British Museum, one in the Library of Congress. Eight of the medallions are almost identical, with a diameter of 69mm, and a small tang at the top, which can be pierced to take a chain. But only the Library of Congress medallion has a tiny oval addition on one side with details of its date, maker, and origination: *Michael Mercator, 1589, London.*

Michael Mercator was a grandson of Gerardus Mercator, and the world displayed on his Silver Map was drawn from various

Dutch, Flemish and English sources. Significantly, it was based on his grandfather's famous projection. But precisely how he obtained the details of Drake's route (which he shows with a dotted line) is unclear. Several accounts of Drake's great voyage were published shortly before the Silver Map was cast, most notably by the English geographer Richard Hakluyt, but any new discovery that appeared on a map was certainly not credited to Drake before 1589, the year after the defeat of the Armada.

How hard must it have been for Drake and his crew to keep the details of their circumnavigation concealed for nine years? Columbus had been under no such restraint in 1492, and nor was

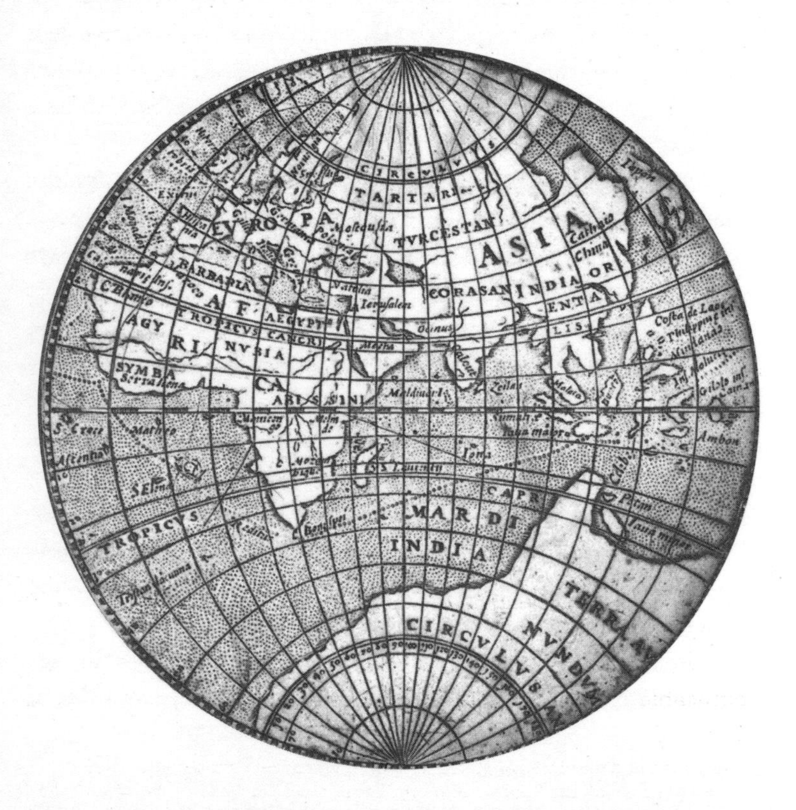

The world around your neck: Drake's silver medal, enlarged to show the impressive detail of his ragged circumnavigation.

Juan Sebastian Elcano, who completed Magellan's circumnavigation in 1522. Drake had only his riches, valued at some £1000, to console him.

Inevitably, secrecy spawns speculation. And no one speculated more than the two most famous map-makers of the age, Gerardus Mercator and Abraham Ortelius. Drake reached Plymouth at the end of September 1580, and only ten weeks later Mercator wrote to 'Master Ortelius, the best of friends' that 'I am persuaded that there can be no reason for so carefully concealing the course followed during this voyage, nor for putting out differing accounts of the route taken and the areas visited, other than that they must have found very wealthy regions never yet discovered by Europeans ...' But he got it wrong: he assumed that Drake's expedition 'pretend[ed] they secured through plunder' their 'huge treasure in silver and precious stones'. But, in fact, that was precisely what had happened.

Mercator and Ortelius were intrigued not only by Drake's haul but also by rumours of two sightings that, if true, would once again transform the look of the world. And these rumours were true: Drake and his men had landed in the upper parts of California (which he named Nova Albion), and sailed past the islands of Tierra del Fuego, which were thought previously to be part of the giant unmapped southern continent of Terra Australis.*

That all this appeared credited to Drake for the first time on a fancy piece of jewellery designed to be hung around the neck of privileged Elizabethans was remarkable in itself, but the medallion offered even more: it is without doubt the smallest map to document so much geographical significance on one side, and conceal so much ruthless piratical history on the other.

The map contains 110 place names. Europe includes the recognisable landmarks of Hibernia, Scotia, Moscouia and Gallia,

* Magellan had previously noted and named Tierra del Fuego on his circumnavigation, as had the Spanish explorer Francisco de Hoces in 1525/26, but the news failed to change the work of most cartographers, who were still convinced of some giant southern landmass attached to South America.

while Africa boasts Aegypt, Maroco, Mozambique and Serra Lione. China and Japan are present without further detail. On the western side, North America shows both Nova Albion and Californea, while South America features Panama, Lima, Chili and Peru. Frisland still sits mythically in the Atlantic, while there are also the enticing Pacific possibilities of Cazones (Santa Domingo), I. d. los Reyes (possibly Christmas Island) and Infortunates Insules (conceivably Easter Island).

Drake's route through this world was shown as a dotted line, already an established technique in the sixteenth century (Magellan and his crew, who predated Drake's circumnavigation by sixty years, had their journey dotted on several globes and maps). In addition to the line, eight inscriptions provide an unexpectedly large amount of additional information about his voyage, including departure and arrival dates, the passage through the Strait of Magellan and the discovery of New Albion. But the map cannot show everything.

The historian Miller Christy published a study of the Silver Map in 1900 and it became clear as he traced Drake's route on the medallion that it relied almost entirely on opportunism. All sailors are reliant on fair winds, clear skies and suitable weather, but in the sixteenth century other obstacles were just as likely to throw out what passed for planning. Navigational instruments were unreliable; other explorers from other dominions wanted the same things you did and thus had to be engaged; and the maps beyond Europe were both incomplete and wrong. Drake's passage was affected by all these factors. He left Plymouth on 13 December 1577 with five ships, gave Spain a wide berth to avoid detection, and struck the north-west coast of Africa a fortnight later. He reached the Cape Verde Islands, sailed down the coast of Brazil, and entered the Strait of Magellan in August; the medallion shows his newly named Elizabeth Island. But he was then forced south against his will by a two-month storm, which enabled him to recognise Tierra del Fuego as an archipelago and, beneath it, the latterly named Drake Passage, the strip of ocean connecting the Atlantic and Pacific (Drake did not sail through this notoriously stomach-churning passage himself).

The rest of Drake's route depicted on the map was equally eventful. Two ships in his convoy foundered in storms (two others had already been broken up after crossing the Atlantic); then sailing northward up the coast of South America he encountered his biggest hauls of silver from the Spanish, and, just as valuable, some of their maps. His hull full and glistening (the Silver Map may have been made from his haul), Drake feared revenge if he returned home by doubling back, so he kept going, towards what he hoped would be the famed Northwest Passage from the Atlantic to the Pacific around the north of North America. Instead he sailed into the area described ominously only two years before by Sir Martin Frobisher as 'The Mistaken Strait' (renamed Hudson Strait in 1609), and thereafter resolved with his crew to return home from San Francisco, through the Indian Ocean and round the Cape of Good Hope. Miller Christy contends that 'it is probable that he had never, up to this point, contemplated a circumnavigation of the Globe.'

Chapter 8
The World in a Book

In the spring of 1595, five months after his death at the age of eighty-two, Gerardus Mercator introduced a new word into the European dictionary: *atlas*. His inspiration wasn't the figure we're familiar with – the muscle-bound Titan holding up the heavens on his shoulders – but a rather more learned, bearded fellow, a mathematician and philosopher draped in fuschia robes as he measures a basketball-sized celestial globe with a pair of compasses. That at least is how he appears at the beginning of *Mercator's Atlas*, alongside a 36,000-word dissertation on the Creation, several poems in Latin, and 107 maps.

It was the magnificent culmination of a life's passion. You could buy it at the Frankfurt Book Fair that year, and if you didn't put your back out getting it home (it was five volumes bound as one) you could marvel at what were consistently the most accurate and complete country maps available, dexterously hand-coloured, with the world elegantly flattened in his novel world-changing projection.

The *Atlas* was the work of someone who prided himself not on the ornate but on the painstaking. We have seen that Mercator was not hugely prolific, and his maps and globes were intended for a discerning market rather than the masses targeted by his commercial rivals. His son Rumold and his grandson Michael, who completed the atlas after Mercator's stroke in 1590 and then saw it through the press and binding, shared a similar dedication to their work, travelling to London from their base in the Rhineland to obtain the latest discoveries and coordinates.

The *Atlas* was dedicated to Queen Elizabeth and the praise lavished on the British Isles was profound. The island was blessed with 'all the goods of heaven and earth ... neither the rigours of winter are too great ... nor is the summer's heat ... Indeed, Britain is the work of joyous nature; nature seems to have created her like another world outside the world.' (It was quite a contrast to the Ancient Greek view of Britain as wretched and perennially wet). The *Atlas* also contained other unrecognisable novelties, such as a circular map of the North Pole, shown as a rocky island divided by four rivers, and the fictional island of Frisland, at that time a popular apparition near Iceland.

Despite everything, the *Atlas* did not sell well.* To some it was not ornate enough, but others were just happy with what was already available. For while Mercator was the first to give us a name to describe a bound collection of maps of the same dimensions, he didn't create the concept. That had already happened in northern Italy.

📖 📖 📖

The early collection of twenty-seven ancient Ptolemy maps printed in Bologna from 1477 could justifiably be called the first of the breed, with Martin Waldseemüller and two

* But the Mercator Atlas did become popular fifteen years later. Following the death of Mercator's son, the Dutch cartographer Jodocus Hondius bought the family's copperplate engravings, added almost forty maps of his own (including new interpretations of Africa and America), and the new volume went through twenty-nine popular editions and translations in as many years.

colleagues making the first modern atlas in 1513 by combining the Ptolemy maps with twenty contemporary regional woodcuts of their own. This contained one of the earliest examples of colour printing, and the first map in an atlas entirely devoted to America (it is titled *Tabula Terre Nove*, and has an unusual textual reference to Columbus as a Genoese explorer sailing under orders from the King of Castile).

Atlas lends his name to the dictionary via this appearance in Mercator's masterful work from 1595.

It was in Venice that the atlas became a craze. In the 1560s mapsellers had the idea of allowing customers to build their own atlas from the stock on display. If you didn't like the Spanish maps on offer, you simply didn't put them in your book. But if you were intrigued with the emerging face of South America you could choose two or three (perhaps conflicting) impressions. Most buyers would select one single-sheet copy of the latest work of the leading cartographers – Giacomo Gastaldi was strong on Africa and Arabia, whilst you might choose Paolo Forlani for South America and George Lily for the British Isles. These would then be folded and bound between covers of your choice, a unique and discerning collection, the cartographic iPod of its day.

This bespoke service was also popular in Rome, where one publisher, Antonio Lafreri, lent his name to the practice and produced the finest known example, a two-volume compilation known as the Lafreri-Doria Atlas. This contained 186 printed and manuscript maps, and in 2005 was sold at Sotheby's in London for a princely £1,464,000.

The Doria Atlas was bound in about 1570, the same year that saw the publication of the first atlas we would recognise as such – a book of uniform size and style, containing maps primarily drawn or compiled by the same hand. The *Theatrum Orbis Terrarum* of Abraham Ortelius was a huge and instant success, despite the fact that it was the most expensive book ever produced. Its title ('Theatre of the World' – the word Atlas was still twenty-five years away) was both apt and dramatic, for in its various editions over forty-two years it ran to 228 different plates, ranging from local maps of Palestine, Transylvania and the island of Ischia to the latest impressions of America, China and Russia. The *Theatrum* also included a collection of historical and mythical maps: the Kingdom of Alexander the Great, the Roman Empire, the voyage of Jason and the Argonauts in search of the Golden Fleece.

The books were produced at the press of Christopher Plantin and their colours were rich and saturated, the lettering (in Latin) elaborately cursive. The cartouches (a map's decorative emblems) burst with vivid additional information – the natural history of a region, a town plan, or a genealogical tree. Ortelius was also a generous publisher: by including an index of map-makers he had relied upon for source material he created an invaluable checklist for future historians.

The *Theatrum* sold 7,300 copies in thirty-one editions, and at least 900 survive. To leaf through one today is to get a (misguided) sense of a world as a fully realised enterprise, an ordered place from which cartographic dalliances with geographical guesswork and overbearing religion had been banished in favour of science and reason. The Age of Exploration was not quite over, but Ortelius' great work already looks unimprovable, and certainly must have seemed so to its eager purchasers.

📖 📖 📖

The fact that the *Theatrum* was published not in Italy but in Antwerp marked the beginning of a major shift in cartographic power. The path of the popular atlas – from Italy to the Rhineland and Belgium, and later to the Netherlands, France and Great Britain – provides an accurate weathervane of the course of the golden age of cartography. The factors that determined this movement were predictable: the rise and decline in economic strength brought about by trade and naval power. This in turn reflected the ability and willingness of monarchies to commission new explorations, and the availability and prosperity of skilled draughtsmen, paper-makers, printers and bookbinders.

The New World shapes up nicely in Ortelius's 'Theatre of the World', an atlas featuring 228 exquisite and detailed plates.

MARIS PACIFICI,
(quod vulgò Mar del Zur)
cum regionibus circumiacentibus, insulisque in eodem passim sparsis, novissima descriptio.
C. Hermoso
Cabo blanco
Cabo del Engaño
Tierra de las palmas
Baia hermosa
R. grande
R. hermoso
Braco
Mar Ver-mejo.
Cali-fornia.
Ponto de buena Speransa
Ancon de S. Andres
S. Augustin
Laguna del rasto
R. de S. Francesco.
Noua Hispania.
Laguna de caliterra
Islas hermosas
Islas de los Cedros
Islas de los diamantes
Messico.
P. de Pedro y S. Paulo.
Florida.
MARE
SIVE
Cuba
Y. de S. Thomas
C. de California
Las Marias
Isleos coltinos
P. di navidad.
Colima
Quesibito
Guatulco
Coconusco
Stapa
IVCATAN
La anublada
Rocca partida.
Iamaica
Nombre de Dios
Cartagena
Nicaragua
Blanco
Y. de Cocos
Malpelle
C. de forta.
Rio de S. Ioan.
Mangrales.
Rio de S. Tiago.
P. de Gue-vara.
Y. de Galopagos
Quito
24o
25o
26o
27o
28o
29o
30o
Circulus Aequinoctialis.
Isola de la plata.
Badia de caraque.
Charapanton
de puna.
R. de S. Iago.
S. Clara.
R. de tambes
R. de S. Miguel.
Isola de lobos
Paita.
C. de laguia
P. de Salinas.
Peru
Isolas de lobos
Limosin
Cividad de los reiz
Pachacamia
Garico.
Laguna
Isolas de cuervos.
Lamasco
Machate.
R. de Monte
Isleo de arecisse
R. Decumas
R. de arecipo
Iambopale
S. Petri
C. de Fortuna
R. de buena madre
Cunhamu
C. Blanco.
C. dela isla
OD VVLGO
NOMINANT
DEL
ZVR.
Prima ego velivolis ambivi cursibus Orbem,
Magellane novo te duce ducta freto.
Ambivi, meritoq; vocor VICTORIA: sunt mi
Vela, alæ; precium, gloria; pugna, mare.
RALIS,
ICA, NON-
Cum privilegijs Imp. & Reg. Maiestatum,
nec non Cancellariæ Brabantiæ, ad decennium.

But there was another factor too – raw and rare talent, a combination of inspiration and tutored craftsmanship, for surveying, plotting, drawing, engraving, compiling, illuminating and colouring – that cannot be wholly explained by financial or other economic perspectives. The vision of seeing the world anew, and the ability to express it, was what set Waldseemüller, Mercator and Ortelius apart.

It is too simplistic to plot the shifting cartographic dominance of one European country over another in terms of decades. But certainly there are trends: Germany's role in the revival of Ptolemy was crucial in the late fifteenth century (there were important printings in Ulm and Cologne), while Martin Waldseemüller and Martin Behaim both produced maps and globes with exciting new discoveries and ways of showing them. Italy's vibrant printing trade undoubtedly also benefitted map-making in the same period. But it was the Low Countries that turned cartography into a new commercial artform. In the late-sixteenth and seventeenth centuries, a new generation of map-makers transformed an arcane, intellectual and exclusive activity into a booming industry.

This work wasn't confined to atlases, but it was an atlas published in Amsterdam that best demonstrated what maps had become. The *Blaeu Atlas Maior* was quite simply the most beautiful, elaborate, expensive, heaviest and stunning work of cartography that the world had ever seen. And everything that followed it – right up to the present day – seems a bit of an anti-climax in comparison.

📖 📖 📖

The Blaeu dynasty would dominate European map-making for a century, aided by its own vast printing press, believed to be the largest in the world. But its beginnings are cloudy. We still do not know for certain when its cartographic founder,

Willem Blaeu, was born, nor precisely where. Alkmaar (Northern Holland) in 1571 seems to be the most likely combination, with Blaeu moving to Amsterdam in his youth, when he was still known as Willem Janszoon (he joined at least four other Willem Janszoons in the city, which may explain the Blaeu – 'Blue' – addition to his name). He began working for his father's herring company, developed an interest in mathematics and instrumentation, and was apprenticed to the astronomer Tycho Brahe. He then became a bookseller and maritime cartographer, forging an early reputation for his pilot charts, which soon became the principal navigational tools for Dutch shipping.

Willem gradually moved into terrestrial maps to expand his income, though his skill lay in compilation rather than draughtsmanship. He would commission copperplates from local cartographers, but more often he would buy already engraved plates from other printers in Europe and enhance them with a few additions and lavish colours. But what appears to have accelerated both the scale of his production and his ambition was a feverish battle with a local competitor called Johannes Janssonius. For decades, these rivals fought for an increasingly lucrative market, and together they dominated the Dutch map trade.

Both Janssonius and Blaeu had back-up – the former worked with Henricus Hondius, while Blaeu was assisted by his two sons Joan and Cornelis – and their focus soon narrowed to one particular specialty, the atlas. Intriguingly and honourably, they didn't steal each other's maps but tried to outdo each other on sumptuousness and range. It was with this in mind that in 1634 Willem Blaeu announced plans for his first 'large map-book', a compendium of 210 maps, only to find that Janssonius was preparing one of 320 maps. And so the stakes were raised and Blaeu promised an even grander project. When he died in 1638, his son Joan took up the challenge, and the *Atlas Maior*, produced between 1659 and 1672, was the almost ruinous result. Lavish was only the half of it.

It was published in eleven volumes in folio size (52.7 x 32.1 cm). Some countries (Germany, England, the Netherlands, Italy and America) got a volume each, while Spain and Portugal shared a volume with Africa, and Greece shared a binding with Sweden, Russia and Poland. The first Latin edition contained 594 maps but there was much else too, the apotheosis of geography as history. There were 21 frontispieces (containing paintings of the Creation, optical and measuring instruments, globes and compasses, portraits of Ptolemy and Greek gods). And there were 3,368 pages of text, in which Blaeu and his colleagues explained a country's history and customs in a tone both exhaustive and reductive. Germany was 'very rich thanks to its commerce, its mines of gold, silver and other metals, and to its corn, cattle and other products.' Scotland was praised for 'the excellence of the minds that it produces.' The description of China betrays a reliance on far-off fable. 'Found in this province [of Peking] are long-haired totally-white cats with floppy ears, as prized as little Maltese dogs, and which the ladies adore.'

Blaeu's maps of America contain the first cartographic appearance of 'New Amsterdam' as the capital of 'New Netherland', but the name would soon be out of date: it became New York City in 1664. America as a whole is still a distant land, and both Greenland and Iceland are considered to be 'the northernmost part' of it. 'Half of America stretches to the west,' Blaeu wrote. 'This part is entirely unknown in its interior…'

The maps themselves are lusciously Baroque. A cartouche would be considered squandered if it wasn't draped in cherubs or heraldic arms or unicorns. Each sea would be either augmented with trade winds and navigational directions, or else full of galleons, serpents and menacing fish. The wording on the maps is set either with movable type or hand-drawn, while the names of surrounding seas are often drawn in an extravagantly curling script resembling fishing lines. The coastlines are dramatically thick with shading, while mountain ranges, generally of uniform height, look like obstinate rashes.

The Dutch scholar Peter van de Krogt has calculated that the Blaeu printing works made about 1,550 copies of the atlas in a thirteen-year run, some 1,830,000 sheets. And the cost to the customer reflected the outlay: the uncoloured editions were priced at between 330 and 390 guilders, while the coloured editions cost 430 to 460 guilders depending on the translation and number of maps. At today's values, this would price a coloured edition at approximately £25,000 or $40,000. What else could you get for this sort of money in the mid-seventeenth century? You could buy ten slaves at 40 guilders each. And for 60 guilders in 1626 you could have bought the island of Manhattan from its native Indians.

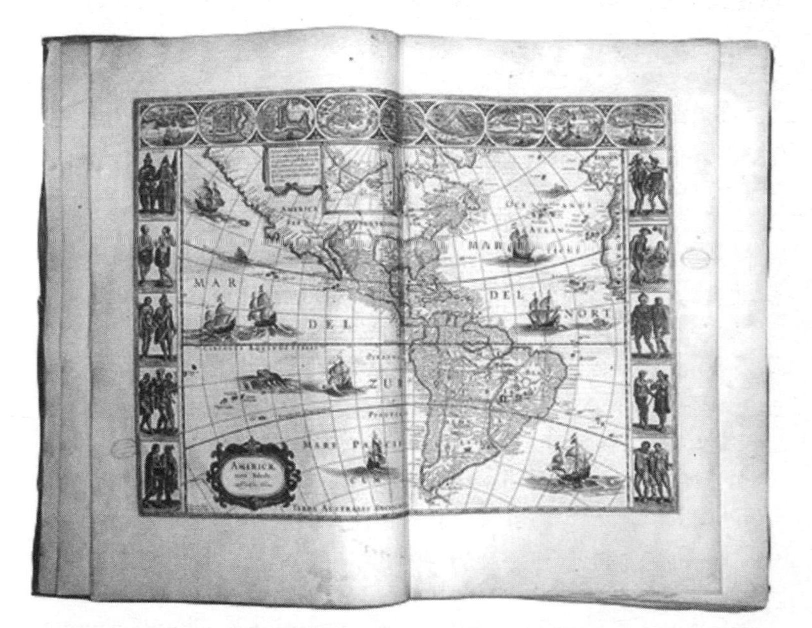

Baroque and berserk: the most beautiful atlas ever – the Blaeu *Atlas Maior*, published between 1659 and 1672.

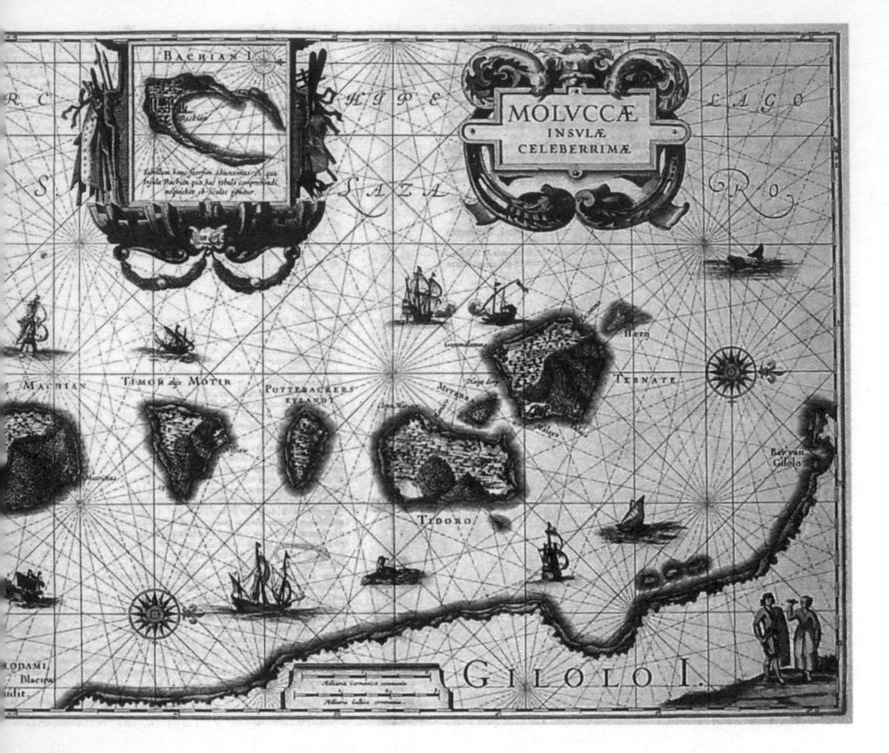

Like teeth adrift in Indonesian waters: the Moluccas, also known as the Spice Islands, as crafted in the appendix to Blaeu's *Atlas Maior*.

That Blaeu was proud of his achievement is clear from his address to his 'gentle reader' at the beginning of the atlas. 'Geography has paved the way not only for the happiness and comfort of humanity but for its glory,' he writes. 'Were kingdoms not separated by rivers, mountains, straits, isthmuses and oceans, empires would have no confines nor wars a conclusion.' He might have added 'nor wars a cause or purpose,' but he went on to outline the joy felt by so many who have picked up a map in all centuries: our ability to 'set eyes on far-off places without so much as leaving home.' He praised Ptolemy, Ortelius, Mercator, and England's William Camden (whose descriptions of the British Isles he copied), and he ends with a plea to the reader to forgive his mistakes ('easily made when describing a place one has never seen') and to send him maps of their own making. It was a humble address considering the bombastic product that followed it.

Astonishingly, Blaeu would help to create one further and even greater atlas: the Klencke Atlas. This was a one-off, produced in 1660 by Joan Klencke and a group of Dutch merchants as a gift for King Charles II of England at the restoration of the monarchy. At 1.78m high by 1.05m wide, it was the largest atlas in the world, and featured countries and continents from the Blaeu and Hondius dynasties. It's been at the British Library for almost two centuries, and in the *Guinness Book of World Records* since its records began.

Like the Klencke wonder, Blaeu's atlas was not a collection of maps to take on a voyage; it was a rich man's plaything, the sort of thing that in a different age would appear alongside Ferraris in glossy magazines. The fact that Blaeu's atlas was not always the most up-to-date was not of prime concern (his maps of England, for instance, were more than 30 years old). For cartography had entered yet another phase – a period of flamboyancy and ornamentation, where the luxuriousness and sheer weight of an atlas was judged to be more important than its practical quality and accuracy. It would take a century – and the emergence of the French 'scientific' school of mapmaking – to reverse this trend.

It is not known whether Joan Blaeu recouped his costs, but in the end the gods may have done his final accounting. As new editions of the *Atlas Maior* were being planned in 1672, a huge fire swept through the Blaeu workshop and destroyed many of the copper plates required for further printings. Joan Blaeu died the following year at the age of seventy-six. No one would produce such a tragically magnificent book again. But many would try, and to glimpse the most famous we need to leap forward a couple of centuries.

On Wednesday 2 January 1895, a brief advertisement appeared in the London *Times* classifieds: it announced that in early April a new atlas would be published containing 117 pages of new maps with an index of more than 125,000 place names. As was the fashion, the atlas would be issued in serial form, fifteen weekly parts at one shilling each. A month later, another advert carried more information and a harder sell:

> 'To newspaper readers of the present day a good Atlas is absolutely indispensable. In order to enable the public to obtain the full advantage of the information contained in the paper, it has been decided to offer to them, at a price within the reach of all, an Atlas of the very highest quality, which will form an inseparable and indispensable companion to the newspaper.'

And thus was *The Times Atlas* born. The 117 pages would be 17 x 11 inches and display 173 maps in colour. Eleven pages would be devoted to Africa, while there would also be extensive coverage of the 'Indian Question', the war between China and Japan, and all the latest developments in polar exploration. The advertisements even carried a review from the *Manchester Guardian* ('Superior to English atlases at 10 guineas. We have no hesitation in saying that the publication of this atlas will mark an epoch in English geographical teaching and study'). When the atlas was published in bound form, the 125,000 place names had increased to 130,000, and there was an explanation on the Contents page. 'Owing to the fact that the maps ... embody additions and corrections which have been made, in some cases, only a few days before going to press, it has been found impossible to incorporate the whole of the names of places to be found in the present edition in the main Index.' So a supplementary index was attached, which contained thousands more. A great many were in Africa and South America.

Six years later there was an updated edition, and the trend was edging towards the luxury seventeenth-century Dutch items: for 24 shillings you could purchase the atlas in 'extra cloth'; for 30 shillings you could get it in 'half Morocco' (a bit of calf-skin that might put readers in mind of the Mappa Mundi); and for 50 shillings you could obtain the utterly leather 'Edition De Luxe'.

The original *Times Atlas* was an impressive and popular product, but at the time it didn't stand greatly apart from its competition – in particular rival tomes from Philips and the map shop Stanfords. Each of these offered small details that the others didn't, had claims on comprehensiveness, and an eye on posterity. But half a century on, *The Times Atlas* emerged in a new version that made all others look like yesterday's news. The 1955 edition appeared in five volumes, making it massively more comprehensive than before and with publicity

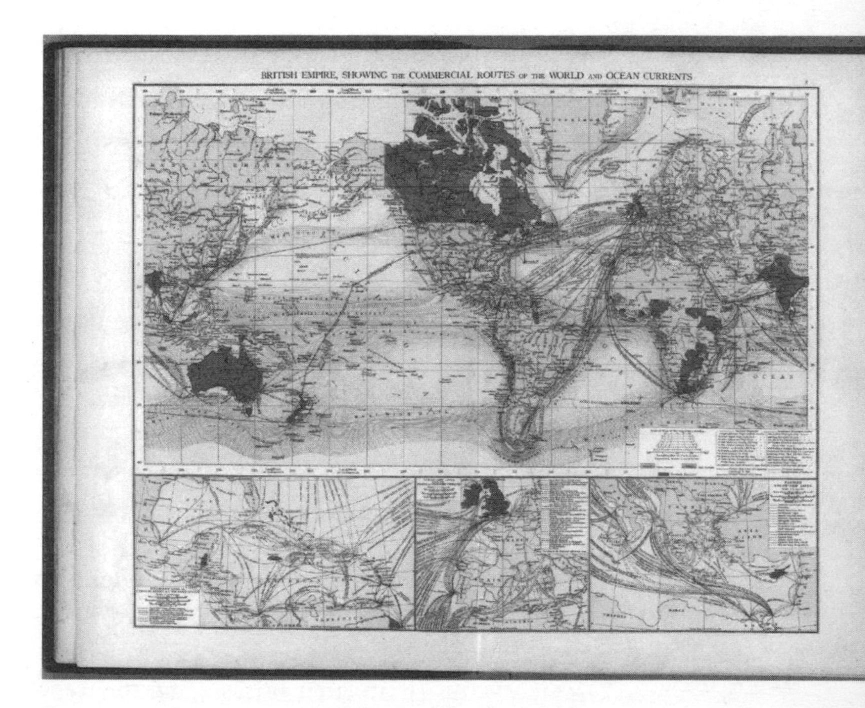

The Times Atlas – a decent size for the Empire in 1895.

materials determined to make librarians put in an order at once (for £22). The volumes prided themselves on their new political boundaries and the fine gradation of ink tints to denote altitude. The maps signalled a shift in geography, too. The first *Times* atlases were based on German maps, but they had subsequently switched to the Scottish cartographic firm of Bartholomew, and the new atlas was Scottish to the core.

📖 📖 📖

It is curious how rival map-makers come up with major innovations at similar times. Like Janssonius and Blaeu, *The New Times Atlas* had a significant rival – this time across the Atlantic.

The newcomer was the *World Geo-Graphic Atlas*, published in 1953 in Chicago, which is regarded by many as the most beautiful and original modern atlas ever made. Edited by Herbert Bayer, a former master at the Bauhaus school who fled to New York in the 1930s, the book was a daring early exercise in information graphics. Bayer and his designers believed that it wasn't enough to just print and bind current maps – they wanted to explain what the maps showed and how the world was changing; the reality, in other words, of what it was really like to live in the transformative years after World War II. (Bayer worked for a while at the leading advertising firm J. Walter Thompson; he was one of the archetypal Mad Men.)

The book was published in a small private edition by the Container Corporation of America and was never commercially available, although many of the world's major reference libraries recognised its value and managed to obtain a copy. Almost sixty years after publication it remains a stunning thing to behold, as fresh and groundbreaking as the Helvetica typeface of the same decade; both reflect a clear-headed modernism. The atlas is sub-titled *A Composite of Man's Environment*, with the standard world maps abetted by sections on economics, geology, geography, demography, astrology and climatology. But what really sets it apart is the use of diagrammatic and pictorial images (some 2,200 illustrations, charts and symbols in all – and pre-computer) to demonstrate the way we live. The slightly awkward title (*Geo-Graphic*) was designed to highlight the use of pictograms, charts and other infographics.

Some of the diagrams are just fun, such as the 'arrow map' of the United States showing the best East-West route through every US State from Maine to Washington State. Others point up ecological concerns; even in 1953 it was clear that the world was running out of the natural resources required to feed a rapidly growing population, let alone meet our demands for mineral fuels. As Bayer made clear in his preface, the atlas was designed with a consciously pacifist and environmental 'whole-earth' direction, taking root several

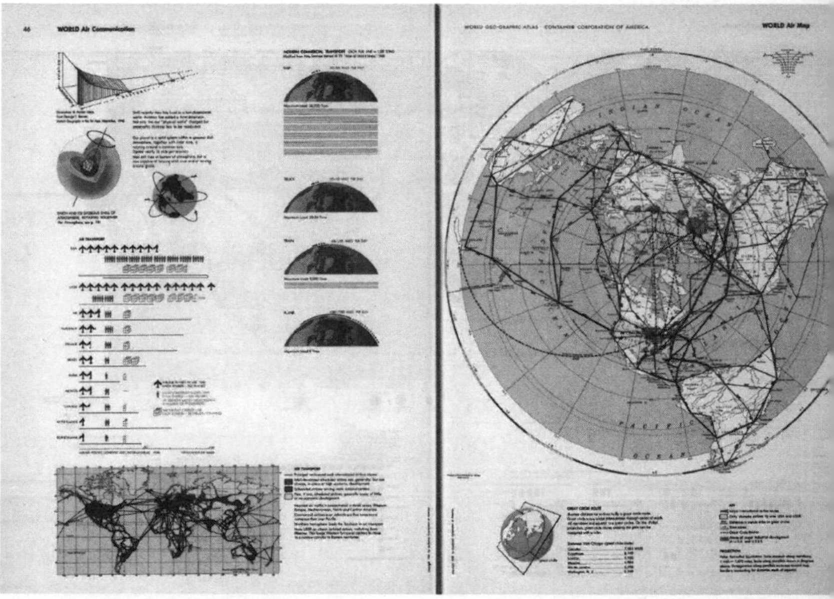

The world meets the modernist: Bayer's magnificent and influential *Geo-Graphic Atlas* from 1953.

years before the approach became a liberal foundation-stone of the 1960s. 'Political inferences have been avoided whenever possible,' Bayer wrote, 'because a global concept of this earth, its people and life sources necessarily rejects implications of power, strategy, force and suppression.' But it remained provocative throughout, illustrating such things as the US migration routes of Indian tribes and Norse peoples.

The atlas was also bang up to date: an addenda included the news that there was a new 'highest point reached by mountain climbers: 29,002 feet,' marking the ascent of Everest in May 1953. The atlas achieves an immediacy and relevance that the grander, more traditional atlases never could, and it is inspirational to all who see it.

📖 📖 📖

Two people who probably did fall under the *Geo-Graphic* spell were Michael Kidron and Ronald Segal, the radical duo behind *The State of the World Atlas*. This first appeared in 1981 as a self-styled volume of 'cartojournalism' published by the radical socialist Pluto Press, and became a bestseller. It was as much an anti-capitalist manifesto as it was geography, and it encouraged the reader to view the world in a new light, the same way one's conventional idea of a portrait might change after seeing a Picasso.

Every double-page spread was essentially the same – a condensed map of the world portrayed either with each country drawn in its recognisable shape or as a rectangular block – but overlaid with a message of inequality. These ranged from a survey of countries' trade union membership to a map of state censorship, in which countries were divided up into varying degrees of (il)liberalism. The maps help to turn agitprop into witty graphic art with spreads entitled 'This Little Piggy' (to denote workers who toil unproductively) and 'Dirt's Cheap'

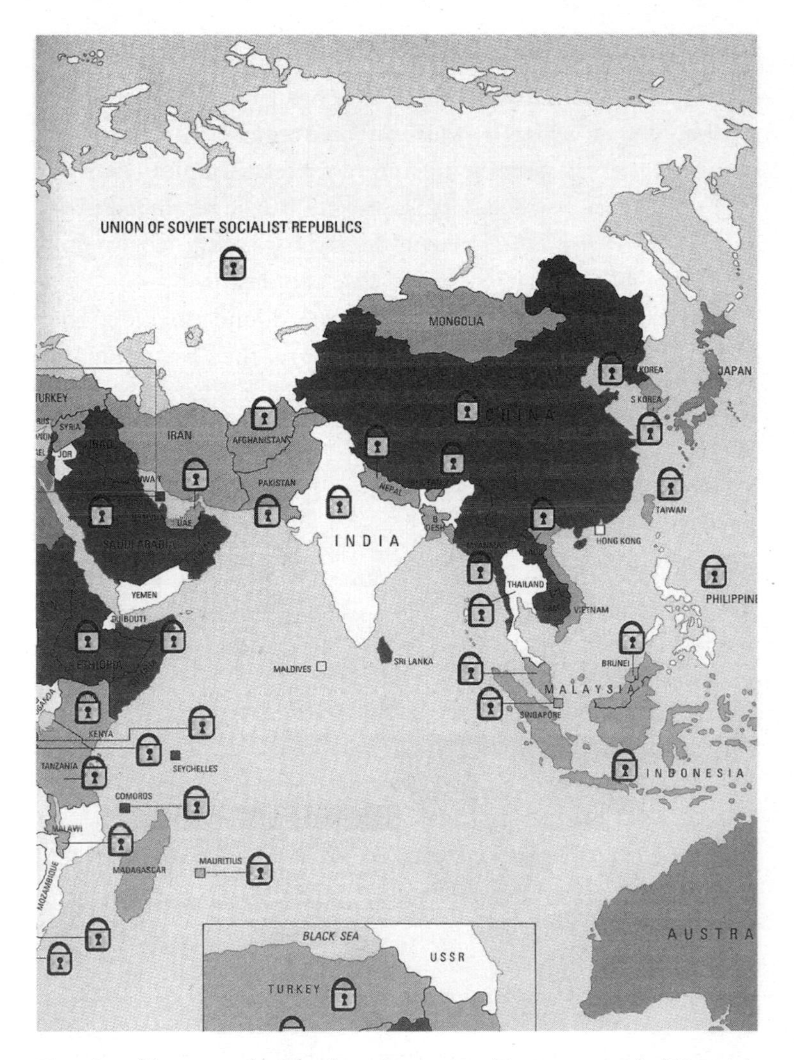

Not a lot of free speech around: a state censorship map – entitled 'See, Hear, Speak No Evil' – from the 1990 *State of the World Atlas*.

(to denote polluted air). In the fourth edition (a decade after the first), the atlas contained many symbols that had yet to be recognised by the Ordnance Survey: padlocks, men in suits holding champagne glasses, armed militia and very thin people with begging bowls. By the eighth edition from 2008, the topics had expanded to terrorism, obesity, sex tourism, gay rights and children's rights.

The State of the World Atlas was followed by *The War Atlas*, which had fewer visual jokes and more retreating armies, and by a succession of politically themed books: *The Atlas of Food*, *The Atlas of Water*, *The Tobacco Atlas*, all displaying what would otherwise be deadening tables in a devastating way. They remain in print and online.

📖 📖 📖

And whatever happened to the hernia-inducing really big atlas, the one that reached its peak with Joan Blaeu and the walk-in Klencke atlas in the 1660s? It's back, more hospitalising than ever. In 2009, a company called Millennium House published something called *Earth* – a 580-page, 24 x 18.5 inch monster. According to its promotional blurb it takes 'cartography and

The island of serious paper cuts: the *Earth Platinum Edition* unveiled page by heavy page.

publishing to a new stratosphere,' a huge mixture of maps, pictures and six-foot gatefolds, each individually numbered 'by our calligrapher in Hong Kong.'

It was, almost inevitably, bound in leather and hand-tooled and hand-gilded, 'a legacy for future generations.' There were two editions, the first Royal Blue version with a 2,000 copy print run costing £2,400, and the Imperial Gold edition of 1,000 copies with a price upon application. But before you could apply for the price, Millennium House brought out its *Earth Platinum Edition*, which made the others look like a postage stamp. This was the biggest book ever made – bigger, at 1.8m x 1.4m, than the British Library's Klencke. And better because it was for sale, albeit at $100,000. The print run was thirty-one.

The atlas needed its own plane to fly it to its wealthy owners, as well as six people to lift it. The people from the *Guinness Book of World Records* duly confirmed that it was indeed something special, and at the beginning of 2012 it officially became the biggest, most expensive, most user unfriendly atlas of the world in the world.

Pocket Map
Lions, Eagles and Gerrymanders

The *Blaeu Atlas Maior* took the atlas to new heights of clarity and comprehensiveness – but one thing it didn't forsake were the animals, which had roamed freely on maps for centuries, usually adorning a border, or a blank expanse of land or ocean, and occasionally taking over completely.

In the Low Countries, the carto-animal *par excellence* was Leo Belgicus, a lion that arrived in 1583 and just refused to leave. There is a reason Leo endured – he just fits. He was introduced in Cologne by an Austrian cartographer and nobleman, Michael Aitsinger, when Belgium and the Netherlands were both part of the Spanish Empire and almost every province in the Netherlands featured the lion on its coat of arms. There weren't a great deal of other map 'jokes' at that point and Leo was an instant hit in Low Country homes, the *Keep Calm and Carry On* of his day.

The original map first appeared as a folding panel in a book, and then went through many editions and twists. In Aitsinger's original, Leo faced right, tongue panting, his upper jaw by Transylvania, his left paw named Luxembourg. Great Britain

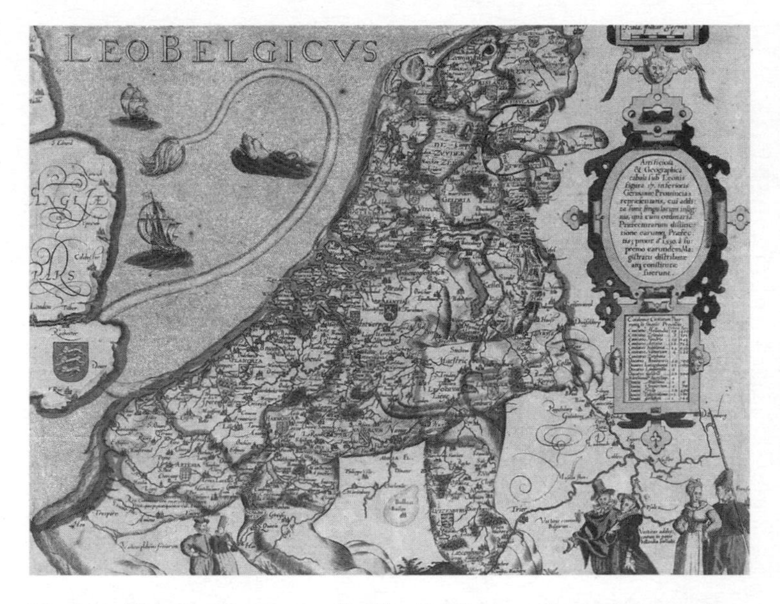

The friendliest lion in cartography: Leo Belgicus semi-rampant over the Low Countries in 1617.

received a political seeing to from the tail, which swished over Norwich, Ipswich, Colchester and London. By the time the Amsterdam engraver Claes Janszoon Visscher had a go in 1609, Leo was squatter, less mobile and fierce, his lower jaw dominated by the Zuyder Zee, with a background not of the British Isles but of Dutch traders, coats of arms and views of Antwerpen, Bruxel and Amsterdam. But when Janszoon made his paper lion in 1611, he was facing the other way and was far longer, with the Zuyder Zee now at its rump.

The body of the lion varied over time as borders and rulers changed. After the Treaty of Munster of 1648 curtailed the Eighty Years' War and recognised a Dutch Republic separated from the southern Spanish Netherlands, Visscher redrew his lion yet again. He was turned around once more, bedraggled and exhausted, and seemingly far less populated, not least because he now represented the new independent Netherlands alone, and was rebranded *Leo Hollandicus*.

Popular examples of Leo continued until the beginning of the nineteenth century, at which point engravers and collectors presumably wearied of the gag. Perfect timing, then, for the dramatic, short-lived and slightly compromised zoomorphic emergence of the American Eagle. In 1833 an engraver named Isaac W. Moore stretched the eagle over a map of the rapidly transmogrifying United States, his work published in Philadelphia in Joseph Churchman's geography book *Rudiments of National Knowledge, Presented to the Youth of United States, and to Enquiring Foreigners*. It is a very rare map (little change from $20,000), measures 42 x 53 cm, and it came about by a trick of the light.

Churchman explained that he was looking at a wall map of the US when the dim light in the room cast its shadow in such a way as to suggest the shape of an eagle. He was ready to dismiss it when he realised that such an image might increase the 'facility with which [geography] lessons may be impressed and retained upon the youthful memory.'

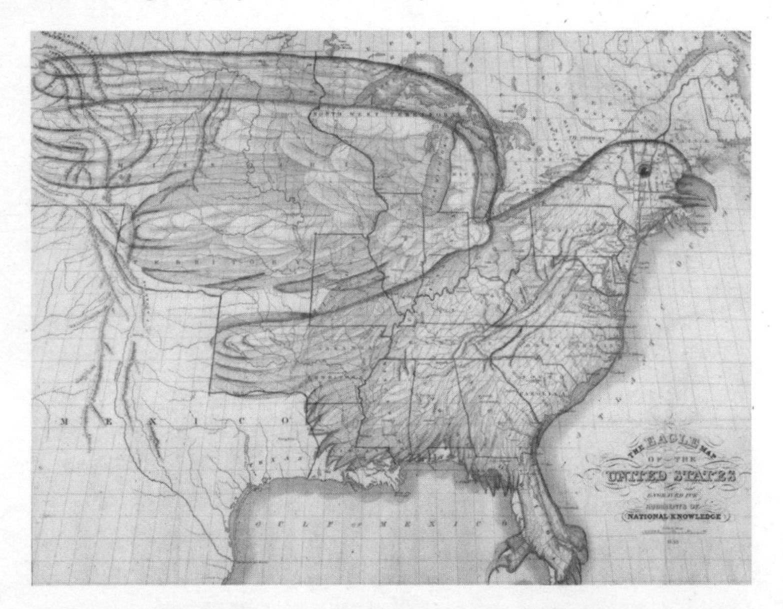

A dead parrot sketched over the United States, though it's supposed to be more of an Eagle.

The resultant carto-bird, portrayed in smudgy brown over sharply defined red states and borders, tries very hard to keep its subject in check: its feet and talons look good extending down through Florida and towards Cuba, while its breast effectively covers the eastern seaboard. But its head, with an eye in Vermont, is not big enough to cover Maine (for which the author apologises); its tail feathers extend not far beyond Arkansas; while its wings smother an ill-defined 'Missouri Territory'. Within sixteen years Californian statehood rendered the bird inoperable.

There is another awkward anomaly: the eagle looks more like a parrot, and the author who made him has a reason for this. The eagle is usually portrayed prey-like, he explains, eager for swooping and flesh-ripping. 'Here, on the contrary, having possession of the whole country, and no enemy to contend with, it is designed to appear as the placid representative of national liberty, and national independence; with an aspect of beneficent mildness, and in an attitude of peace.'

So where in the world to look for carto-aggression? Towards Russia, clearly, which didn't have an eagle or a lion on its vast land but an octopus, the animal you use on a map if you want to denote greed, suction and unremitting tentacular ambition. The octopus is cartographically versatile, for it is really eight animals in one. Its globular reach is unmatched by anything else on land or sea – in fact, it is the only sea creature (unless one counts the amphibious dragon) which seems unusually happy on land, even in Siberia, even without its normal dietary supply of whelks, clams and other molluscs. That is because it is eating everything else.

On the famous *Serio-Comic War Map for the Year 1877*, drawn by Frederick Walrond Rose, the message is both powerful and sinister, one of the most lucid expressions of menace in the entire map drawer. An obese Russian octopus spreads its thick tentacles round the neck of Persia, Turkey and Poland. Germany is portrayed as the Kaiser, England as a colonising businessman

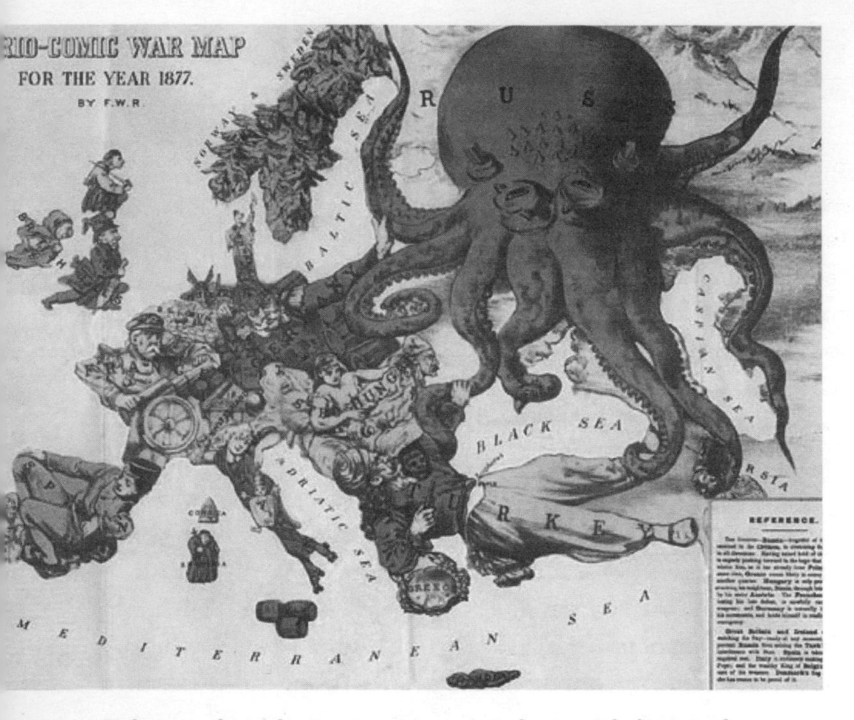

So long, suckers: the Russian octopus gets heavy with the rest of Europe in 1877.

with a moneybag labelled India, Transvaal, Suez. A rapier-wielding kilted Scotsman stands on England's shoulders, a sleeping Spain has its back to the rest of Europe, France is a general with a telescope, Italy is a roller-blading child toying with a wooden figure of the Pope, Turkey is a swarthy gun-toting pirate, and Holland is a gentle land of windmills. The stereotyping is now almost jailworthy.

Rose's map is an image one can't easily put away, and it is little wonder that octopuses have had their character besmirched on many maps since. A decade later, an American cartoonist portrayed the British Empire's ceaseless colonialism in the shape of John Bull smirking in choppy waters. He is more than an octopus: his eleven hands rest on Jamaica, Australia, India, Malta and the rest, while his arms tuck Ireland and Heligoland close to his body. Some of the possessions seem solid enough; some such as Egypt seem already to be drifting away.

In 1890, the United States was in the grip of what the newspapers called The Lottery Octopus, another gift to cartographers as a skinny snake-like thing with its body in Louisiana and its tentacles all over the states from Maine to Washington. The lottery had begun in New York in the late 1860s, with its tickets travelling to cities all over the country by train, making a fortune for its corrupt owners in the process. When legislation to renew its charter came up in 1892 it was granted three more years to overcome widespread church-led opposition, and, failing so to do, was killed in 1895 and erased from the map.

But the award for most influential animal on a map goes to the salamander – an amphibian that gave the English language a new word that was both verb and noun. Its tale begins in February 1812, when the supporters of the ninth Governor of Massachusetts, a man named Elbridge Gerry, decided it might benefit his Democratic-Republican Party to reconfigure the electoral boundaries in Essex South County, north of Boston. The plan was simple: sacrifice a few Senate seats by packing a few districts with opposition Federalist voters, while gaining a Republican majority in many more.

So far, so predictable; it wasn't a new political ploy, and Gerry's opponents soon became aware of the chicanery. (Gerry himself, a distinguished diplomat who had signed the Declaration of Independence, helped found the Library of Congress and would one day serve as Vice President, was not himself the prime instigator of this 'redistricting'.) And then, the story goes, there was a dinner party. Over beef, the resemblance of the reshaped districts to a salamander became clear: a creature curved from left to right, with Chelsea as its behind, Danvers and Andover as its prime torso, and Salsbury as its head. And inevitably the dinner party, with several newspapermen in attendance, produced the immortal line: 'That's not a salamander, that's a Gerrymander!'

The amphibian that lost the election: the Gerrymander encircles Boston in 1812.

The following month, there it was: a respected Boston miniaturist and cartoonist named Elkanah Tisdale reshaded the map to strengthen his point, added claws, wings and viperous jaw, and struck the point home. Elbridge Gerry lost his seat, and the map may have been partly responsible.*

* But we have also lost something. The "G" in "Gerry" was hard; we have long since learnt to mispronounce it.

Chapter 9
Mapping a Cittee (without forder troble)

We will come, in a few chapters' time, to one of the most useful and used maps of all time – the *London A to Z* – and the legend of its creation. But great and useful maps of cities were not invented in the twentieth century. For that distinction, we need to look back to 1593, when John Norden published *A Guide for Cuntrey men in the famous Cittee of LONDON, by the helpe of wich plot they shall be able to know how farr it is to any street. As allso to go unto the same, without forder troble.*

Norden's map stretched from Islington in the north down to St Katherine Docks near the Tower of London, and was engraved with great attention to the arrangement of churches and other public buildings, with trees denoting open land and the coats of arms of city livery companies (Grocers, Drapers, Fishmongers) framing the survey in two vertical panels. Key areas included Grayes Inn, Creple-gate, Lambeth mersh and More feyldes, while the banks of the Thames had only one bridge at Southwark but a great many other landmarks: Black friers, Broken wharfe, Three cranes, Olde swann, Bellyns gate.

The map's other significant feature was that it was designed by a Dutchman, Pieter van den Keere.

The places are phonetic exercises to us now, but the map also had one feature of a truly modern street map: letters and numbers were placed at strategic points, and identified in a table at the bottom. It could justifiably claim to be the first A-Z: *a* marks Bushops gate streete, *c* is Allhallowes in the wall, *k* is Holborne Conduct, and *z* is Cornehill. In a printing from 1653 the index has been greatly expanded to include ninety-five other locations, from Grub streete to Nightfryday streete (passing Faster lane and Pie Corner on the way).

You could have bought Norden's map from Peter Stent at ye Whitehose in Giltspur street neere Newgate. Stent was one of six prominent print and mapsellers in the capital in 1660, but he was soon to face fierce competition. The number had tripled by 1690, a burst of activity that reflected two things: London's new trading prosperity along the Thames, and a craze for printing and collecting maps.

Most of these new maps, covering every part of the known world, were not bound for explorations; they were instead records of them. And most were not intended as symbols of power or influence. They were the first signs that people – or at least London's merchant classes (Samuel Pepys among them) – found maps fascinating. In their newly affordable form, maps were educational, decorative, imaginative and journalistic. And they reflected the opening up of the world.

Between 1668 and 1719, the *London Gazette*, the first official newspaper of record, published more than four hundred advertisements from London mapsellers. Their tone varied

The first London A-Z? Norden and van den Keere's map of London in 1593.

The Way
More feyldes
Moregate
Creple gate
West Smythfeyld
Aldersgate
Newgate
Ludgate
Bryde wel
Black friers
Baynardes castle
Paules wharfe
Broken wharfe
Queene hythe
Three cranes
The Stilliarde
Shrewesburye howse
Olde swann
A M Y S
Banckes syde
S. Marye Overyes
The Beare howse
The play howse
South

dent at ye Whitehorse in
te neere Newgate

streete.
howse.
Conduct.
ry lane.

n. Holbourn.
o. Grayes Inn lane.
p. S. Androwes.
q. Newgate.
r. S. Iones.

t. Cheap syde.
u. Bucklers burye.
w. Brodstreete.
x. The stockes.
y. The Exchannge.

2. Colmanstree
3. Bassings hal
4. Hounsditche
5. Leadon hall
6. Grations str

from plain to frantic – some of them seemed to suggest they were selling not paper representations of new discoveries but the actual land itself. Long before auction catalogues and the opening of Stanfords map shop in the Strand, they provide the first London snapshot of commercial cartography.

What did they show? 'There is now Extant a new Mapp of the Estates of the Crown of POLAND,' began one notice in November 1672. 'Containing all the Dutchies and Provinces of that Kingdome; as Prussia, Cujavia, Mazovia, Russia-nigra, Lithuania, Volhinia, Podolia and the Ukraine. Shewing all the principal Cities, Towns and Fortifications, wherein may be seen the Advance and Progress of the Turkish Armies.' It was offered by three vendors: John Seller, Hydrographer to the King, at his shop in Exchange Alley, Robert Morden at the Atlas in Cornhill, and Arthur Tooker, 'overagainst [ie opposite] Salisbury House in the Strand.'

Other shops offered maps of the Netherlands, France and Germany, new plans of North America with special attention paid to English plantations, along with curiosities (John Seller had a map of the Moon) and newly drafted sea charts. In March 1673, James Atkinson, a mathematical instrument maker, offered a map of the Magellan Straits 'shewing all the depths of Water and Anchorage, Shoulds, and places of danger', available from the east side of St Savories Dock.

In 1714, a new celestial and astronomical map was promised by the London cartographer and engraver John Senex from the Globe in Salisbury Court. This would include 'Mr Professor [Edmund] Halley's Description of the Shadow of the Moon over England in the total eclipse of the Sun'. Due on 22 April, 'the sudden Darkness will make the Stars visible about the Sun, the like Eclipse not for 500 Years been seen in the Southern Parts of Great Britain ... The Map shews every part of England over which the total Darkness will pass.'

And then there were the maps of London. These tended to emphasise newness, and no wonder – they marked the complete rebuilding of the city after the Great Fire of 1666. Indeed the fire marked an entirely unprecedented burst of civic cartography. It was evident to anyone who survived the flames that Norden's map was simply no longer relevant, and it helped that the ruling monarch, Charles II, was himself a map enthusiast.

In 1675 a new London map was sold by Robert Green at the Rose and Crown in Budge Row with particular attention to Westminster's 'Lanes, Allies and Courts, with other Remarks, as they now are' (at the same time, Green also had 'A Map of Pensilvania by William Pen Esq'). In 1697, Robert Morden announced a map of London measuring 8ft by 6ft, divided into wards and parishes 'with all the new Buildings and Improvements of these late years'. The major selling point (and at forty shillings it was the most expensive map of London on offer – most cost one shilling) seemed not to be its size, however; it was the fact that the area covered had been 'actually Surveyed'.

Ten years later, a London map offered further treasures. A visit to the Bishop's Head in St Paul's Churchyard would be rewarded with an eight section, two-volume publication that boasted 'a more particular Description thereof than has hitherto been publish'd of any City in the World.' This contained not only 'all the Streets, Squares, Lanes, Courts etc' but also their distances from Charing Cross, St Paul's and The Tower. There was also a list of all the prisons, statues, churches, hospitals, workhouses, fountains, conduits, public baths and 'Bagnio's' (which may mean either a bathhouse, a brothel, or both at once).

Among all these advertisements and all these maps, there was one printer-cartographer whose name was mentioned 'above the title', in the way a Hollywood star is used to sell a film. It was a name to be trusted, not least because it came with royal approval. The name was John Ogilby.

Ogilby commanded such respect that in May 1668 he announced a licensed lottery – in which the winners would obtain a stake in an exciting new project yet to be announced. It was five years before he put them out of their misery, revealing the project was called *Britannia* and would comprise an extravagant multi-volume survey of England and Wales, featuring county maps, views of English cities and a topographical description 'of the whole Kingdom.' New investors, whom Ogilby called 'Adventurers', were called upon to repair to Garaway's Coffee House near the Royal Exchange, where they 'may put in their Money upon the Author' – and if they paid enough their name would appear on a cartouche of one of the maps. While they waited, his adventurers could have immediately satisfied themselves with the second volume of *Ogilby's English Atlas*, which consisted of his maps of America. Or they could admire his maps and topographical works on China, Japan and Africa, many of them embellished with engravings by Wenceslaus Hollar.

Alongside John Speed, John Ogilby did more than anyone in England to set cartography on a respectful, practical and commercially popular footing, an achievement made even more remarkable because he had come peculiarly late to the trade. In fact, no one in his profession could rival the variety, misfortune or tireless reinvention of his earlier lives.

The Ogilby saga reads like a tale from the circus. He was born near Dundee in 1600, but by the age of six he was in London and his father was in jail for bad debts. His first love was dancing. He became apprenticed to a dancing master in Gray's Inn Lane and was soon performing at London's grandest balls. But a particularly complex manoeuvre performed for James I resulted in a broken leg and permanent lameness, after which he turned his attention to teaching and theatre management in Ireland.

A rough financial patch followed, followed by a sea crossing to England during which he was almost shipwrecked. His recovery was aided by one of his previous dancing clients, Lord Strafford, who also recognised his literary talents, and supported Ogilby as a translator of Virgil, Homer and Aesop's Fables, all of which sold well. He believed he had escaped another personal disaster in 1665 when he fled the Great Plague by moving to Kingston-on-Thames, only to learn a few months later that his house and books had been burnt in the Great Fire. And only then did he move into maps.

Ogilby had previously won favour from the royal court for his lavish account of Charles II's coronation, and he was appointed a 'sworn viewer' (surveyor) of the reconstruction of London. It had become an imprisonable offence to depict the damage caused by the fire, but sketches and plans for a new city were widely encouraged, and several mapmakers set to work to assist civic authorities. Ogilby's map was by far the

John Ogilby presenting his Subscription List for *Britannia* to Charles II and his queen, Catherine of Braganza.

most ambitious, promising mapping 'curiously and accurately performed beyond whatever has yet been attempted for any city of the Universe.' For his pains, he was given a fifteen-year copyright, one of the first times a cartographer had gained protection for his work.

Ogilby didn't labour alone. His chief surveyor was the mathematician and astronomer William Leybourn, whose task it was to walk the new streets, plotting every building and garden, before returning to Whitefriars to enter his day's findings on the map. It was truly exhausting work. Writing in 1674, Leybourn wearily noted that he hoped 'with God's assistance in a few months time to compleat it.' But it would take two more years.

The map, at a scale of 100 feet to an inch (1:1200), was first sold in January 1677, and its publication was a major event, comparable with John Bunyan's *The Pilgrim's Progress* the following year. It was printed on twenty sheets, ideally to be backed and joined by linen, giving an overall size of 8ft 5in by 4ft 7in. In geographical extent, too, it was ambitious,

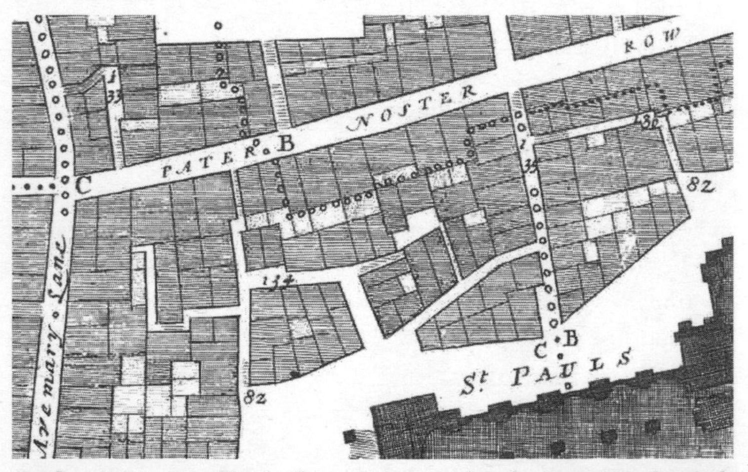

London 1677, mapped by Ogilby and Morgan after the Great Fire, 'beyond whatever has been attempted for any city of the Universe'. A detail from the map showing St Paul's. The numbers refer to the most detailed index yet to accompany a London map (or indeed any single map of the time).

extending from Gray's Inn and Lincoln's Inn in the west to Whitechapel in the east, and from Upper Moorfields in the north to the bank of the Thames. It was entitled 'A Large and Accurate Map of the City of London Ichnographically Describing all the Streets, Lanes, Alleys, Courts, Yards, Churches, Halls and Houses, &c Actually Surveyed and Delineated.' (An Ichnographic map was typically one that showed a realistic plan rather than a bird's eye drawing.)

A fair amount of credit for the new map should go to Ogilby's step-grandson, William Morgan, who succeeded him as royal cosmographer and improved the map in subsequent editions. Their work, the most complete and historically significant survey of the capital to date, set a new benchmark in accurate mathematical cartography of cities. The plotting of individual houses and their backyards had not previously been publicly available, and although the level of detail was still far from the standard we would come to expect from Ordnance Survey, Ogilby's map was perhaps the first to perform the one duty we have come to expect from all city maps since – it enabled visitors to find their way around.

Running one's finger across its dense hatching and graphite tension today, one still experiences excitement. The streets are wider, the Fleet river is dredged and serviceable again, the city just looks cleaner. In fact, it looks like the sort of model metropolis that architects of today might offer a new client, with agreeable mixed-use housing and green spaces. It is a place of boundless opportunity, all clarity and defined rectangular boundaries. The pastures, free of cattle and ordure, appear just the spot for the Sunday stroll; Billings Gate Dock stands quiet by London Bridge, awaiting cargo. We know from bawdy Restoration theatre that London was exploding with slums, squalor and assault at this time, but the only clue is from the lengthy index that Ogilby and Morgan supplied in a separate booklet: Hooker's Court, The Fiery Pillar, Scummer Alley, Dagger Alley, Pickaxe Alley, Dark Entry, Slaughter Yard.

Sadly, Gropecunt Lane, a popular late-night venue in many British cities in the century before Ogilby was born, does not appear.

And not everything is quite right: St Paul's, shown as an outline, was probably based on one of Wren's early sketches rather than the final plans (it is highly likely that Ogilby and Wren knew each other from the Covent Garden coffee houses; it was where the future of the post-destruction city took on its most collaborative and practical shape). And the depiction of the Thames Quay as an attractive river frontage comparable with other European cities was wishful planning and was never realised (or at least, not until the Docklands boom of the 1980s).

But the way this landmark map looked spread out on a table – the impression it gave – was its greatest achievement. It was accurate, and it set a high watermark in civic pride. It reflected what Ogilby had observed when work on the map began, that the swift transformation of London after the fire was a 'Stupendious Miracle!' He saw how the 'Raising from a Confused Heap of Ruines' had occurred 'sooner than some believ'd they could remove the Rubbish.' One can see why Charles II and his courtiers were so supportive of his work: with its broader streets, with its hungry Thames and new docks, the map announced to the world that London was open for business again.

And if you were tired of London, then John Ogilby could also help you leave it. His great survey of the capital was only a part of his grander *Britannia* project, an undertaking that almost bankrupted him. His ambitions – to produce a huge atlas of England that would show each town and county in great detail – had to be scaled down when the enormity of the task became clear. But what remained of the scheme was far more original, and it was something that would prove his most famous, beautiful and covetable legacy: the strip map.

As with his map of London, Ogilby wished to improve the lot of the traveller; or at the very least he saw profit in enabling his wealthy patrons to travel the country's perilous roads in the most direct and hospitable way. His solution was the earliest form of popular road atlas, a collection of one hundred lavishly designed route maps, engraved on copperplate and printed on heavy paper, intended both for practical coach travel and domestic adornment.

At their simplest, they would guide the traveller from London to Abingdon in Oxfordshire, and then (on another map) from Abingdon to Monmouth in Wales. Yet another London strip set off north-easterly towards Cambridgeshire, taking in Waltham, Hoddesdon, Ware, Royston and Huntingdon before ending at Stilton. Plate 6 took you to Tuxford in Nottinghamshire, Plate 7 from Tuxford to York, Plate 8 from York to Chester-le-Street in County Durham, and Plate 9 from here to Berwick, just south of the Scottish border.

Fourteen of the strips began in the capital. You would set off at point A, and know that en route you'd encounter various staging posts, marshes, rivers, inns, churches, coal pits, arable fields and all manner of what Ogilby called 'scenographical ornaments'. They prepared the traveller, coachman and prospective highwayman as never before. It was now possible to read the distances and calculate where to stop for a meal or a night robbery.

The maps were extraordinarily accurate. A surveyor had actually gone out with a measuring wheel and walked each route, accompanied by a colleague on horseback carrying supplies. Ogilby insisted on high levels of factual information on each map. He standardised the measurement of the mile,

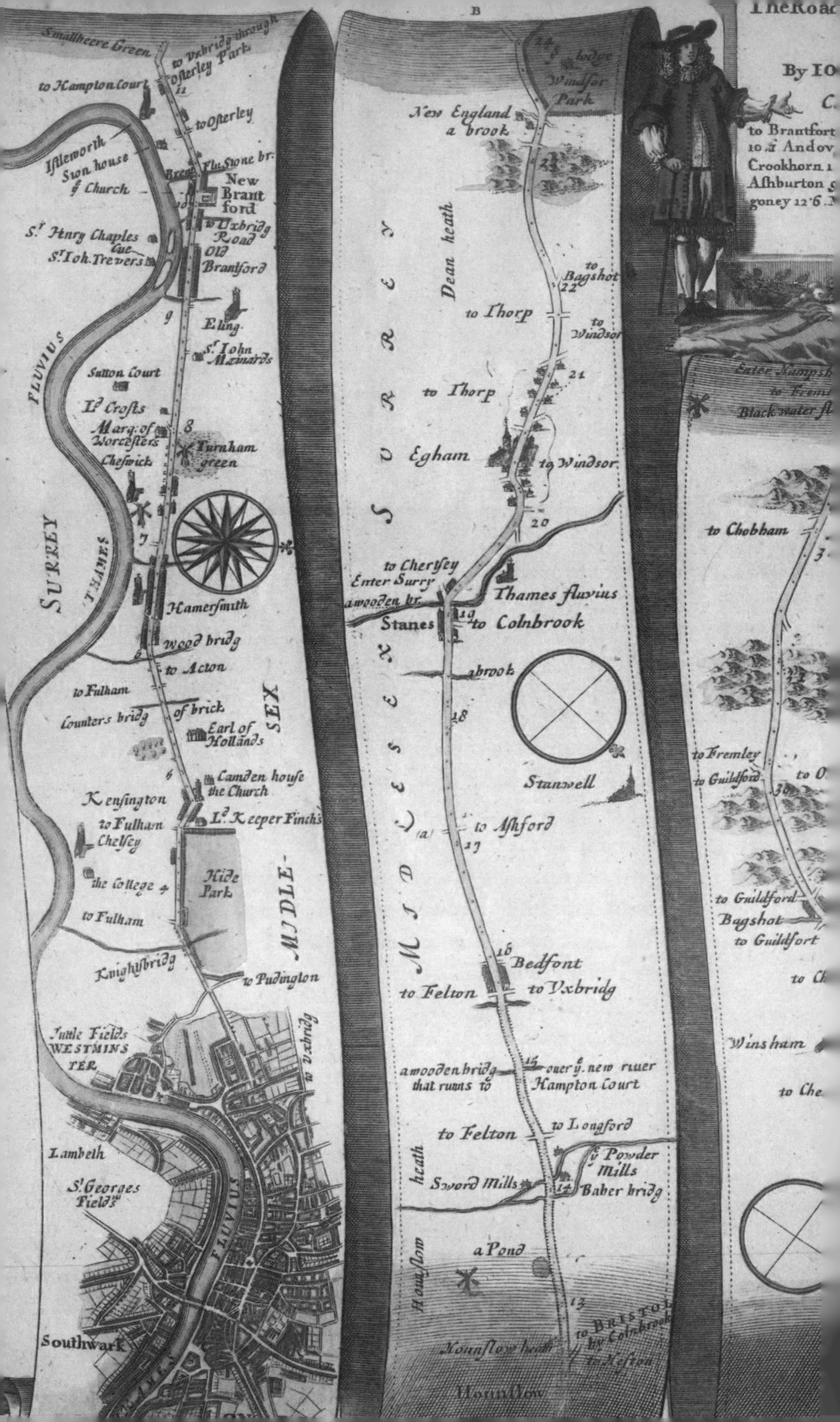

B
The Road
By IO
to Brantfort
Crookhorn
Ashburton
Smallbeere Green
to Vxbridg through Osterley Park
to Hampton Court
11
to Osterley
Isleworth
Syon house
& Church
Brent
Flu. Stone br.
New Brantford
to Vxbridg Road
Old Brantford
St. Henry Chaples
St. Ioh. Trevers
9
Eling.
St. John Meinards
Sutton Court
Ld. Crofts
Marq: of Worcesters
Cheswick
8
Turnham green
7
PLUVIUS
THAMES
SURREY
Hamersmith
wood bridg
6
to Acton
to Fulham
Counters bridg
of brick
Earl of Hollands
SEX
Camden house
the Church
Kensington
to Fulham
Chelsey
Ld. Keeper Finch's
the College
4
Hide Park
to Fulham
MIDDLE-
Knightsbridg
to Pudington
to Vxbridg
Tuttle Fields
WESTMINSTER
Lambeth
St. Georges Fields
FLUVIUS
Southwark
Windsor Park
New England
a brook
Dean heath
to Bagshot
22
to Thorp
to Windsor
21
to Thorp
SURREY
Egham
to Windsor
20
to Chertsey
Enter Surry
a wooden br.
Thames fluvius
Stanes
19
to Colnbrook
a brook
18
MIDDLESEX
Stanwell
(a)
to Ashford
17
16
Bedfont
to Felton
to Vxbridg
a wooden bridg
that runns to
15
over ye new river
Hampton Court
to Felton
to Longford
ye Powder Mills
Sword Mills
14
Baber bridg
Hounslow heath
a Pond
13
to BRISTOL by Colnbrook
Hounslow heath
Hounslow
Black water
to Chobham
to Fremley
to Guildford
30
to Guildford
Bagshot
to Guildfort
Winsham

setting it at 1,760 yards rather than the shorter Roman mile of 1,617 yards or the longer 'old English' mile of 2,428 yards. Each mile was clearly marked on the strip (at a scale of one inch to the mile). The maps would also show steepness and degrees of arduousness in the form of a pyramid of hills, each pointing in the direction of gradient.

One of Ogilby's biggest cheerleaders in this project was the great Restoration polymath Robert Hooke, the 'curator of experiments' at the Royal Society who had also conducted surveys of London after the Great Fire. In many of his previous projects Ogilby was content to repackage existing maps, adding only the flourish of a new cartouche or border; there were already valuable topographic and archaeological studies of Britain by William Camden and John Leland, and they would have been easy to reproduce and elaborate. But Hooke encouraged him to make something entirely novel.

The strip maps were more than useful. They looked stunning and they were fun – not unlike a spotters' game to keep children diverted on an interminable car journey. Here was a bridge, and not long to go now before there's a windmill, and only three miles before we're at the Old Red Lion. To modern eyes, each of the strips resembles a spinning barrel on a fruit machine, with most of the symbols recurring frequently (compasses, clumps of trees, churches), while others (a custom house, a castle) appear only as special attractions. They employed an intricately shaded trompe l'oeil effect, wherein each map looked as if it was written on a slim paper scroll, with the imaginary excess of the paper pleated at the back. Ogilby knew he was onto something. When the strips were complete, he suggested it would be 'bold to challenge the Universe for a Parallel,' for 'nothing of this nature requiring so vast a Charge and such infinite Labour and Disquisition was ever yet Attempted or even Thought of …'

A long pleasant scroll to the West Country: a detail from John Ogilby's route from London to Cornwall.

Expensive editions would be hand-coloured, which made them desirable objects to hang on a wall, one reason why so many of the atlases were broken up and maps removed for local interest; the other reason was that travellers would undertake one specific route and saw no reason to take the rest of the country with them as they went. Within a year there would be smaller, cheaper editions, including 'Ogilby Improv'd' and 'The Pocket Book of Roads', and Ogilby complained to no avail that he was being pirated by printers all over London 'who Have Rob'd my Book.'

It turned out to be the least of his problems. In 1676, shortly after his strip maps began escorting the English to places they had never been before, Ogilby died, at the age of seventy-six. He was buried in the vault in the 'Printer's Church' of St Bride's on Fleet Street, not long after Wren had completely rebuilt it after the Great Fire. And there he lay until 29 December 1940, when he was blown to bits as the next great reconfiguration of London was made necessary by the Luftwaffe.

Chapter 10
Six Increasingly Coordinated Tales of the Ordnance Survey

On 3rd and 4th December 1790, the first ever map auction took place at Christie's in London's Pall Mall. The lots included a map of London at the time of Queen Elizabeth, many privately-commissioned surveys of the English counties, recent coastal maps of North America, twenty-odd maps of Scotland (including several unfinished manuscript proofs), and seventy-eight sheets of the Cassini family's map of France. The sale also included a valuable selection of books: a good smattering of travel writing, including recent volumes on the interior of Africa and Captain Cook's last voyage towards the South Pole, and an extensive range of specialist books about mathematics and surveying, many of them published in France the decade before, including Cassini de Thury's *Description Géométrique de la France* and Cagnoli's *Traite de Trigonometrie*.

And then there was what the auction catalogue called 'A Capital Collection' of engineering and measuring instruments, which was also a great rarity for an auction house at this time (Mr Christie, as the sale catalogue had him, had only established his business in the 1760s). The lots suggested a life of curiosity and adventure: there were many quadrants, sextants, compasses and barometers, there were theodolites and telescopes, 'a four eye glass perspective' and a four-foot achromatic telescope made by the optics dynasty of John and Peter Dollond (150 years before they met Aitchison; Admiral Nelson also owned a Dolland). Then there were delineators, thermometers, plotting scales, a portable camera obscura, a Gunter's chain and something listed as 'a small electrical machine', whose purpose was unknown.

Every tool but the Gunter's Chain: theodolite and waywiser at work in the eighteenth century.

But who once owned these things?

They were the property of General William Roy, who had died five months before, and were the tools of the trade of the man who effectively invented the Ordnance Survey. The project he inspired mapped not only the entire contents of Great Britain – every orchard, each bracket, every saltmarsh and saltings, at a scale and thoroughness hitherto thought impossible – but also reshaped British understanding and appreciation of its landscape, its property boundaries, its urban and rural planning, engineering, archaeology, district and tax laws. The mapping of Great Britain was, initially, a project of more than sixty years, and it resulted in, effectively, a cartographic Domesday Book. It almost defined Britain and – just before the birth of the railways, and long before the BBC and the National Health – it became the envy of the world.

Does it continue to be indispensable? Stuck out on a Peak District promontory in all weathers you may still think there is nothing like it on God's blue-green planet. And you would be right.

The great Trigonometrical Survey of the Board of Ordnance began officially in June 1791. Prior to this date it is sometimes imagined that the country was hardly mapped properly at all, a blur of ploughed field and wayside inns broken by early stirrings of industrial sprawl. In fact, the opposite was the case – though compared to the rigour of the OS, it was cartographic anarchy.

By the dawn of the eighteenth century, the London survey and country strip maps made by John Ogilby were long out of date – and the county maps of Christopher Saxton and John Speed even more so. But Britain had become obsessed by mapping. From the 1750s on, a great many people were out at all hours with their Gunter's chains and their eyepieces, creating maps for commercial or land interests, or assessing tax liabilities ('cadastral' maps). Prominent (and accurate) county surveys were also conducted by expert plotters such as Carrington Bowles, Robert Sayer and John Cary. But each of these maps was particular to the demands of its patrons. Many of the maps were symbols of influence, coverage was patchy, and there was no agreement over what was included or ignored. Nor was there any standardisation of scale or symbols (though a church was denoted by a spire, and one horizontal bar signified a parish church, two an abbey, three an archbishopric).

The shift to national mapping was driven by one principal concern: military defence. In Scotland, the Jacobite Uprising of 1745 alerted troops loyal to the British government of the need for accurate mapping of Highland terrain far beyond the previous concentration on castles and other fortresses. And thus, under the leadership of a lieutenant-colonel named David Watson, General William Roy began a new survey of Scotland from 1747 to 1755, creating maps on a scale of 1000 yards to an inch, something he later considered 'a magnificent military sketch [rather] than a very accurate map of a country.' Much of it, indeed, was conducted from observations on horseback, and although there had been considerable improvements in measuring instruments (not least telescopes and theodolites made in England and Germany), it was still far from the rigorous mapping that Roy would advocate in the years to come.

His inspiration for such systematised mapping came from France. Beginning in 1733, the great national French survey took twelve years to produce a set of maps which, if not the most decorative to have emerged from Europe in this period (Prussia under Frederick the Great may stake a claim to that), was an exemplar that would soon have an impact as far away as India and America. The cost (labour, administration and printing) was almost certainly the greatest ever for a national mapping project, but its value exceeded it.

The 182-sheet *Carte de France* (78 sheets of which would sell at Roy's auction) was first published in 1745. It was the work of the Cassini family, backed by Louis XV, and majestic in scale. The maps could be pasted together to form a collage with a width and length of some 11.5 metres (they were also available as a bound atlas). The survey played an important role in establishing a fiscal framework for the new French state, and – improved by four generations of Cassinis up until 1815 – provided something of a topographic constitution during the country's endless European wars. It would be easy now to regard such a thing as an obvious step in a nation's cartographic progress, but at the time it was a geographic miracle. As the American cartographic historian Matthew Edney puts

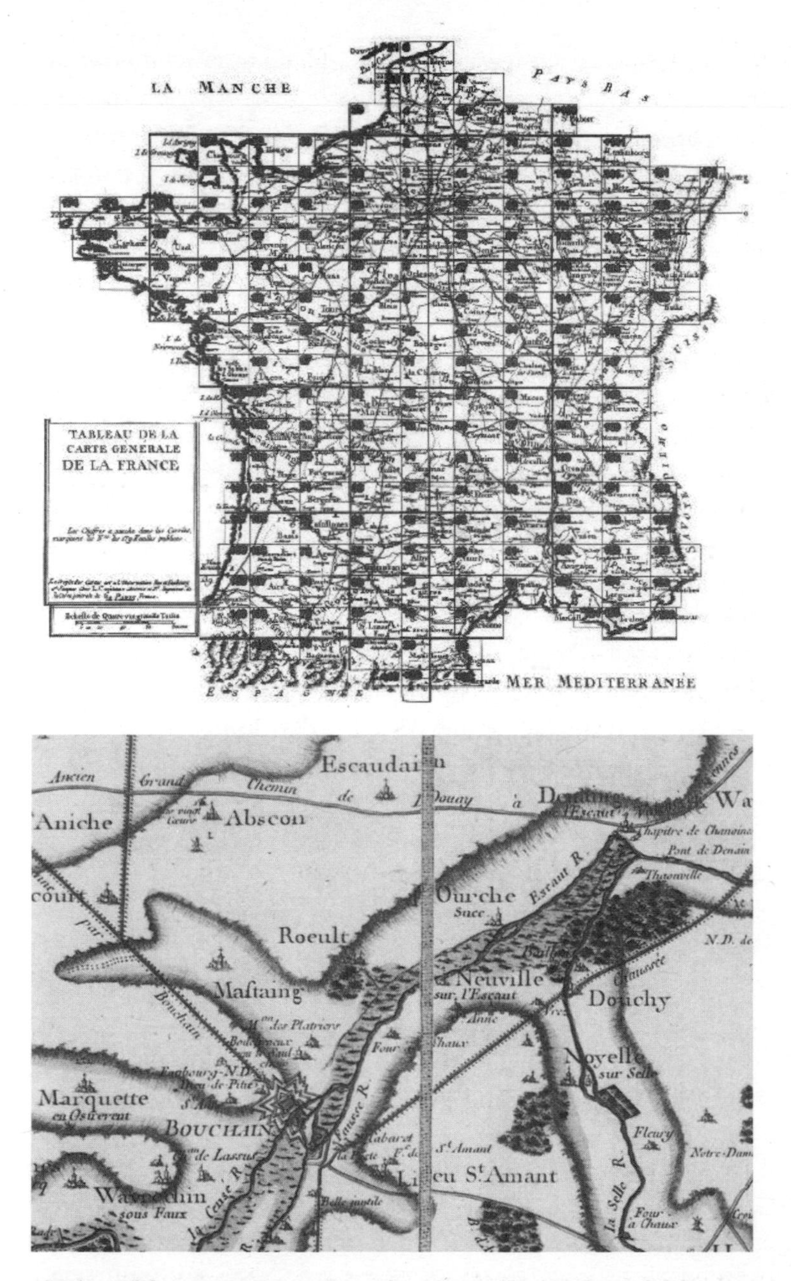

The first of the great map surveys – the Cassini family's *Carte de France*, created between 1733 and 1815.

it, 'Such high-quality surveys were the period's equivalent of atomic science.'

Certainly, the Ordnance Survey owes its French relative an immeasurable debt. The *Carte de France* was based upon an enhanced system of triangulation, the system of calculation that determines distance by establishing two angles measured from a fixed baseline. The method was popularised in the sixteenth century by the Dutch cartographer Gemma Frisius, although it is a theory that would not have been unfamiliar to Pythagoras.

The British Ordnance Survey would rely on triangulation for all its work until GPS took over almost two hundred years later. William Roy first championed it in 1763, when, in his role as Deputy Quartermaster-General, he proposed a triangulation of the whole country at one inch to the mile. He made a similar proposal two years later when he became an inspector of coasts for the Board of Ordnance, a branch of the military based at the Tower of London that was concerned with armament supply and other logistics (including maps) to the army and navy. Roy argued for more detailed maps to protect the south coast, although he also wrote about how such a plan could be extended with a meridian line 'thro' the whole extent of the island, marked by Obelisks ... like that thro' France.' His plans were dismissed on the grounds of time, labour and cost.

Roy finally got his chance, however, in 1784, when he was commissioned by George III and the Royal Society to conduct a triangulation to link the observatories in Greenwich and Paris (the initial impetus came from the Cassini family). It took three years to cross the Channel from Dover, the project delayed by what appears to have been the procrastination of the chief London instrument maker Jesse Ramsden. It was Ramsden's three-foot-high 'Great Theodolite' that was key to the accuracy of the work, but according to the historian R.A. Skelton, Roy became so irritated by Ramsden's deliberating

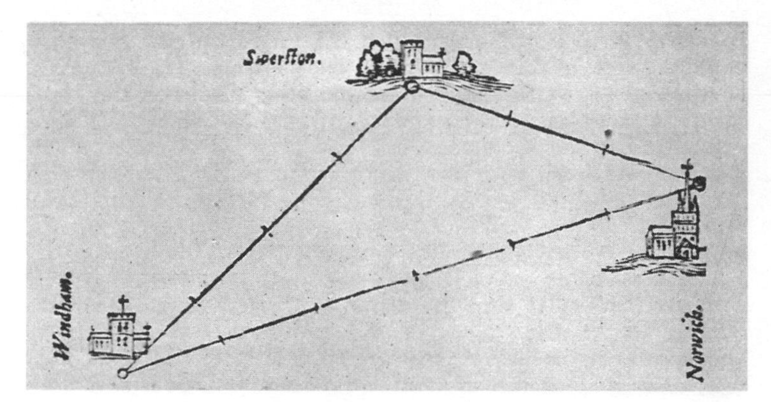

An early illustration of triangulation from the 1550s

over measurements that he damned him in his official report – expletives that were removed before printing.

Notwithstanding, the work was deemed a success, and a triumph for triangulation. Roy had again written of his ambitions for a national survey in two reports for the Royal Society, the last composed a few weeks before he died in 1790. Around this time he also wrote to one of his principal supporters, the Duke of Richmond, the Master General of the Ordnance, and it was the Duke who commissioned what we now recognise as the Ordnance Survey the following year.

The fact that French troops were gathering strength in Europe certainly helped his case for surveying the coasts. But in Roy's words, the plan was to use 'the great triangles' to survey not only vulnerable areas or 'Forests, Woods, Heaths, Commons or Marshes,' but also 'in the enclosed parts … all the Hedges, and other Boundaries of Fields'. Roy believed that this could not be achieved on a scale of less than two inches to a mile, although they could be reduced when printed to one inch to a mile 'for the Island In general.'

The first part of the island to be measured was the base at Hounslow Heath that General Roy had used for the Anglo-French work (a location near what was to become Heathrow Airport). This then extended to Surrey, West Sussex,

Hampshire, the Isle of Wight and Kent. As with so many great ideas, it would be fair to assume that all who pored over it in the weeks and months that followed must have wondered why no one had thought of making such a thing before.

The Ordnance Survey arrived in Ireland in the autumn of 1824, and immediately ran into a problem: impenetrable fog. In England, mists that were present at dawn were usually gone by mid-morning, but this was not the case in Donegal, Mayo and Derry. The Irish survey began as a systematic attempt to reform the Irish taxation system by measuring the boundaries of some 60,000 rural 'townlands', but it soon developed

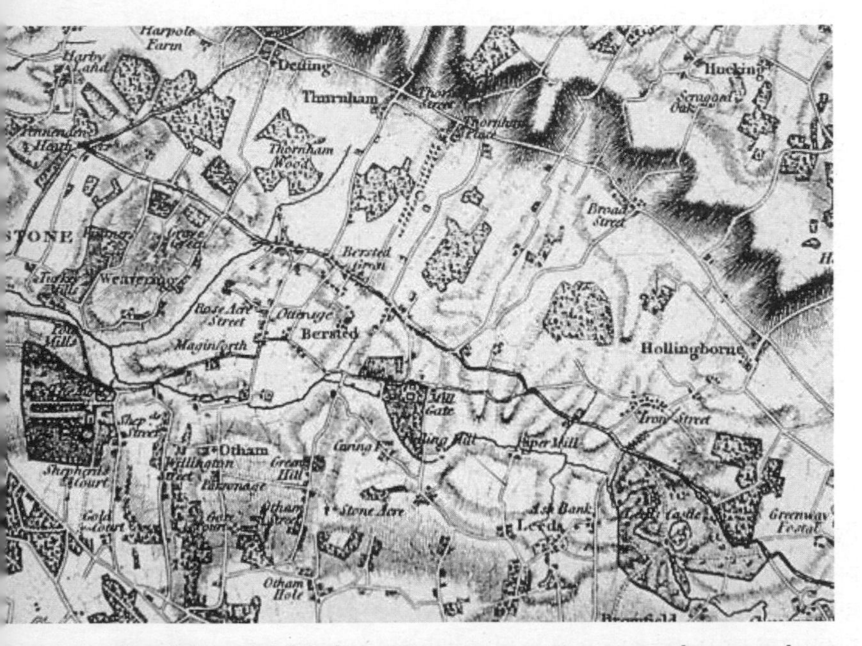

One of the original Ordnance Survey maps – Kent, mapped at one inch to the mile, published in 1801

into a scheme that mapped the country from north to south at six inches to the mile. An astonishing amount of men were employed in this exercise – more than two thousand at its peak in the 1830s – but there was one man who made their employment feasible, their output prodigious and their visibility possible, and like William Roy he was a Scot.

Thomas Drummond was born in Edinburgh and attended the university, but his formative education took place at the Royal Military Academy at Woolwich. He was proficient at mathematics and engineering, and Thomas Colby, director of the Ordnance Survey, enlisted him as a lieutenant on the mapping of the English countryside, and a few years later as his deputy in Ireland. Drummond was a practical man, and from his base in Dublin he began improving the technology of mapping at a rate that brought him to the attention of the Royal Society, who asked him to demonstrate his inventions before Michael Faraday, an event that Drummond recalled as the proudest of his life.

Much of the minutely detailed Irish survey was conducted with chains, but the goal of a 'full face portrait of the land' required a full arsenal of new sights. Drummond had made small improvements to the barometer, the photometer and an optical device known as the aethroscope, but it was his portable enhancement of the heliostat that garnered particular attention, a mirror that deflected the sun's rays towards a particular distant target. But what could be done after sunset, or on those frequent days when there was sleet, haze, smog and murk? Drummond found that the solution to the pea-souper was the 'pea-light', a small pellet of lime (calcium oxide) that, when burned with an oxy-hydrogen flame, produced a light hugely more powerful than either a burning torch, the popular Argand oil lamp or nascent gaslight. The intensity greatly increased the distance possible between trig points; in bad light or snow a signal could now be seen up to a hundred miles away.

And in this way did 'limelight' enter the vocabulary. It wasn't Drummond's invention (the first oxy-hydrogen blowtorch was developed by the Cornish scientist Goldsworthy Gurney a few years before), but he gave it its first great application and, in harnessing its use in lanterns, significantly refined its safety and glow-time. When the 'Drummond Light' was used with the 'Drummond Baseline' (a measuring technique involving metal bars), a highly accurate survey of the whole of Ireland was carried out in just twenty-one years. Many decades later, when they re-measured the baseline which had been set up in the 1820s, they found that it was accurate to within one inch in eight miles.

And across the Irish Sea, London's theatre goers also had cause to be grateful to Drummond. The first use of limelight onstage is believed to have been at the Theatre Royal, Covent Garden, in 1837, for the musical farce *Peeping Tom of Coventry*, when limelight complemented gaslight and was used primarily to focus on the star performer or a dramatic epiphany. Dan Leno, Marie Lloyd and Little Titch all flourished in the limelight, and the fading of the light at the turn of the century – when electricity took over – ran parallel with the slow demise of the music hall.

Great British colonial surveys extended far beyond Ireland. In fact, the greatest of them all extended to the highest point on earth. At the beginning of 1856, the tallest mountain in the world was known as many things, including Deodhunga, Bhairavathan, Chomolungma and Peak XV. At the end of 1856 it was known as Mount Everest.

It was a strange choice of name. George Everest, by all accounts a domineering and ruthlessly exacting man, was the Surveyor General of India from 1830 to 1843. Yet he had almost certainly never seen the mountain, and he suggested that the locals would have trouble pronouncing it (as do we: he called himself Eev-rest rather than Ever-rest). But the imperial British were doing what they did rather well in the middle of the nineteenth century – putting their names on places on the map over which they had no dominion. Despite local objections, the name stuck, a small but telling by-product of the arrival in India of the new science of surveying from the mother country.

The Great Trigonometrical Survey of India – conceived in 1799, commenced in 1802, but only officially titled in 1818 – was a close cousin of the Ordnance Survey, and was almost

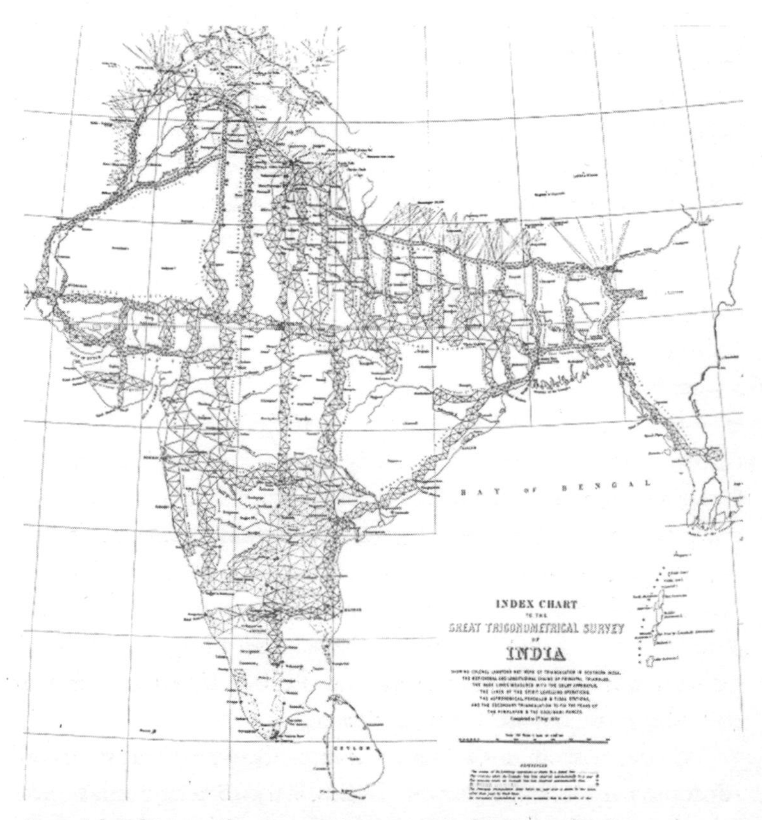

Imperial ambition – the Great Trigonometrical Survey of India

as transformative as its British counterpart. It marked the switch from broad Indian route mapping (the descriptive, landmark-based type useful for a traveller or trader) to the strict mathematical technique based on triangulation (the sort better suited to military planning, establishing a standard cartographical grid over which other maps could be matched or compared). It consolidated British dominance wherever a theodolite and trig point was placed, and the East India Company, the survey's initial sponsor, took full advantage in claiming new territory under the guise of scientific progress. The survey relied on the importation of heavy British measuring equipment for its success, and on British-trained surveyors (William Lambton, the Great Survey's first superintendent, was directly inspired by William Roy, while George Everest, who succeeded Lambton in the post, spent time with the Ordnance Survey in Ireland). The whole project, which lasted some sixty years, may also be considered particularly British in another sense, a ripping yarn worthy of a Pythonesque parody in which surveyors were ravaged by heatstroke, malaria and tigers as they struggled to cartographically tame the extremes of climate and jungle.

Beyond politically advantageous mapping, the Great Survey also achieved a genuinely scientific geographic breakthrough, establishing, at more than 1600 miles, the longest measurement of the earth's surface. Before we could view it from space, this Great Meridional Arc, running south-to-north from the southernmost tip of Cape Comorin (now known as Kanyakumari) in Tamil Nadu province to the Himalayas and the edge of Nepal, provided the greatest glimpse of our planet's curvature (an ambition first realised by Eratosthenes and his gnomon). The Arc produced the skeleton that followed – a great chain of maps stretching east and west from its central backbone.

By contrast, the measurement of Everest was a moment of supreme cartographic hubris, as much to do with imperialist boasting as with mapping (for those concerned with

Indian stamps commemorating two heroes of the Great Survey – Nain Singh and Radhanath Sikdar. Singh explored the Himalayas and mapped much of Tibet, while Sikdar calculated the height of Peak XV (later renamed Mount Everest), and determined that it was the highest mountain in the world.

triangulation it was the biggest triangle of all).* Its precise height was calculated by the brilliant Bengali mathematician Radhanath Sikdar as 29,000 feet, but it was announced at 29,002 feet lest it be considered a rough estimate (the height was the mean figure reached after computing the results of six separate survey stations, measured from between 108 and 118 miles from Everest's summit). Its exact measurement was a matter of great pride for Colonel Andrew Waugh, the man who succeeded George Everest as Surveyor General (and named it after him). And its accuracy should now be regarded as a source of some pride for nineteenth-century mapping. Although the precise height of the mountain varies over time according to shifting tectonic plates and variations in snow covering (and debates over whether one should measure the snow cap at all), the widely accepted 21st century figure of 29,029 ft (8,848m) is an increase considered to be neither here nor there, even by those trying to reach its summit.

* The British mapping of India provided some irrefutable evidence of the power of maps, not least to conceal as much as they reveal. A map of Calcutta, for example, produced in 1842 for the ominously named Society for the Diffusion of Useful Knowledge, showed public buildings such as banks and police stations, but no mention of temples or mosques. As Ian J. Barrow has pointed out in his history of mapping in India, 'apart from the depiction of Indians as porters or peasants, there is little indication in the maps that Calcutta was an Indian city inhabited by Indians.'

Back where it began, the Ordnance Survey was itself rapidly becoming a part of the British landscape. By the First World War its symbols were as recognisable as roadsigns, the maps sold in their millions, a new edition on a new scale was a significant event, and concertinas of sheets nestled close to gloves, scarves and flasks in every carriage, reading room and boot-room. A new written and visual language had evolved around them, alongside an almost universally accepted set of rules as to how we should map a rapidly changing territory. The maps were gently modifying the British sense of place, and defining a culture far beyond the space they miniaturised.

And once you had the mapping itch, you had to keep scratching it. Maps didn't sit still and stay happy – you had to amend, bend, excise and redraw according to such inconvenient things as population explosions and demolition. Oddly for a country whose national borders rarely underwent the mayhem of war that beset others in mainland Europe, Britain seemed to possess an innate need to map. Not merely for practical and professional purposes, but because it seemed like a birth-right.

For the Ordnance Survey, this meant a continual stream of revisions and new maps on a variety of scales. The job was never done, and would never be done. No sooner had an OS surveyor put their feet up to toast the completion of a new series, than the work would already be out of date. It was a wonder how they kept their sanity, for surely mapping was a terrible job:

each year the 'silences' on maps – those spaces where there was nothing to see or report – became rarer.

Although our choice of OS map scales has now shrunk, there once was a series for every use, pocket and location. There was the 1:2376 scale in the West Riding of Yorkshire of 1842, the 1:1250 scale first seen in Shoeburyness, Essex, in 1859, the 1:126,720 that became popular in the early 1900s, and the 1:10,560 'regular edition' town maps with the 1:100,000 county maps, both from the 1960s. But now things have largely settled down, so that we have the National Grid Series, (marketed as the Explorer or 'Orange' series at 1:25,000, or 2½ inches to a mile); and the Landranger or 'Pink' series at 1:50,000 or 1¼ inches to a mile.

You can work your way through a clump of OS maps today and be struck how they construct, map by map, a matchless social history of a vanishing country. They log not just industrial and technological progress, but domestic sanitation, trends in travel and architecture, leisure habits and linguistic quirks. And they do this without knowing, an imperceptible march of years, like the thinning of hair.

OS maps – indeed, all serious maps – are governed by rules of inclusion and exclusion, and by a highly commendable British rulebook that raised the bar for crabby and incontestable officiousness. The rules are numerous, varied and rigid, the result of both field practice and endless minuted committees at its headquarters in Southampton. In the 1990s, the cartographic historian Richard Oliver began compiling a list of regulations gathered from two centuries of internal style manuals to its surveyors, most of them previously unpublished, as to how OS should record what it observed. They are necessary, joyless, mundane and utterly compelling.

(These paraphrased examples are drawn from the so-called *Red Book* of 1963, but they have modern equivalents, not so very different.)

Allotments: Permanent ones are shown, but minor detail such as sheds are not.
Bracken: Should be mapped as distinct vegetation category.
Playgrounds: Gymnastic apparatus in public grounds was authorised to be shown on 1:2500 mapping in 1894. But play apparatus, e.g. swings and roundabouts, is not shown.
Outhouses (Lavatories): Are shown if permanent, and if large enough to show without exaggeration. (*Instructions to Draughtsmen and Plan Examiners, 1906*)
Taps: Taps on public drinking fountains will be shown.
Churches: Subject to the approval of the church authorities, it is customary to publish 'St John's Church' when it means 'St John the Evangelist', but it is also customary to keep 'St John the Baptist's Church' in full.
Trees: Are quite important but not crucial. In avenues or rows trees will be shown except where the symbol would obscure more important detail.
Heath: Is nowadays defined as when the vegetation is 'heather or bilberry'.
Letter Boxes: Are mapped, except when built into post offices.
Public Houses: Are licensed to sell intoxicants; they do not have overnight accommodation for travellers. (If they do, they are inns.)
Legal Value of OS Mapping: Indisputable. Two court judgements in 1939 and 1957 ruled that anything appearing on an OS map is prima facie evidence of its existence on the ground; if it's on the map, it's in the world.

And then there were the space-saving, often perplexing abbreviations, used on OS maps and hardly anywhere else. *San* was Sanitorium, *SM* was Sloping Masonry, *St* was Stone, *ST* was

Spring Tides, *St* was Stable and *Sta* was Station. *W* could mean Walk, Wall, Water, Watershed, Way, Weir, Well, West, Wharf or Wood. Best of luck with that.

But that's all in the past. Aren't paper maps finished for all but the incurably nostalgic? OS perceived the dawn of digital cartography in the early-1970s, when it began transferring its data to magnetic tape and looking forward to a time when a hiker could send a postal order for a map tailor-made to their next fortnight in the foggy Dales. Too bad they didn't foresee the £50 GPS handheld, nor the map-less clowns who yomp up Ben Nevis at teatime with a fading single bar on their iPhones.

But what if paper maps have a future? What if we've seen the small-screen limitations of GPS and are requiring the broader picture once again? Are we nostalgic for Britain on a scale of 1 to 25,000? What if young people were to put down their devices for a minute and feel the need to get muddy and soggy again with just a compass and a plastic map sheath around their necks? Is it possible, as OS would like to believe, that we may one day return to the fold? And if so, where would we go to learn about using these things?

The Ordnance Survey, more than two centuries old, runs map-reading classes in indoor environments, predominantly in outdoor sports and camping shops. But in May 2011 one of these occurred at grid reference SP313271 – Jaffe & Neale's Bookshop and Café in Chipping Norton, Oxfordshire.

We were given plastic compasses tied to large boards with maps on them to place on our knees. OS staff Richard Ward and Simon Rose announced that usually the session consists of them doing all the talking, but they had just that week taken possession of a set of instructional videos featuring the television naturalist Simon King, and so we sat around watching

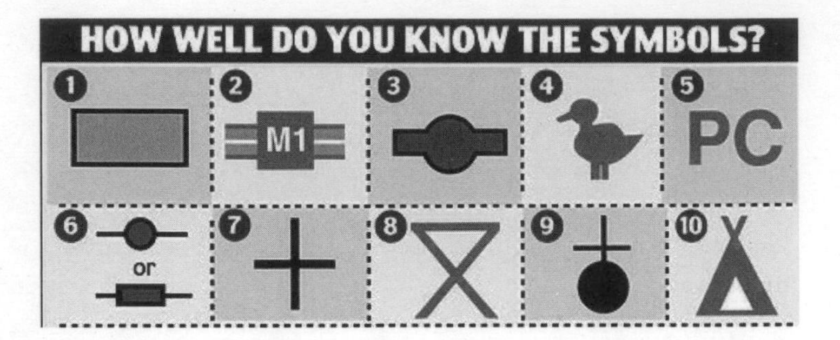

In 2007 a car insurance survey found fifteen million UK drivers couldn't identify basic OS roadmap symbols. These were the test questions. The answers are: [1] mud, [2] motorway, [3] bus or coach station, [4] nature reserve, [5] toilets, [6] train station, [7] place of worship, [8] picnic site, [9] place of worship with spire, [10] campsite. Women scored slightly higher than men, but overall 55% couldn't recognise a toilet, 83% a motorway, and an impressive 91% would be likely to get stuck in the mud.

those until it was time to find our way for ourselves. It was all propaganda, of course, but of the gentle sort, the sort that made you want to get out of the bookshop and start walking. King said that he enjoyed the way that his phone made things easier, 'but nothing replaces a paper map. Only by spreading a sheet out and looking at how the different features on the ground relate to each other can you get a clear idea of the landscape ... I just love them!'

He introduced the two main series, the *Explorer* and the *Landranger*, and in the next film he talked about grid references, and the one after that it was contours (where the contour lines are close together the area will be steep, but further apart they'll be more level). And then it was compass bearings and finding grid north as opposed to magnetic north (magnetic north is constantly shifting depending on where you are in the world and the strength of magnetic force; we had to shift the compass housing a few degrees anti-clockwise to orientate the map correctly). After each clip we had exercises. We had to locate symbols on a map and say what they meant: a pub and (one of those OS words that hasn't been used since 1791) a bunkhouse.

Like all paper empires facing ruin by satellite, the Ordnance Survey is doing what it can to keep up with the digital present. Its first maps were produced digitally in 1972, and four years later its Director General, B. St G. Irwin, a man whose very name suggests OS symbols, told a meeting of the Royal Geographical Society that one in every six of its large-scale maps was produced with the aid of computer coding. Irwin estimated that it would be 'towards the end of the twenty-first century' before this progress resulted in the complete digitisation of the OS database.

This has, of course, happened more quickly than planned. OS now offers a print-at-home subscription service, wherein users can select map coordinates of their choosing, plot a personal route for a nice day out, and then print it out or download it to their mobile phone. Its website offers an impressive range of handheld GPS devices and a guide to geocaching (a GPS-based treasure hunt), as well as socks, hydration packs, insect repellent and blister packs. There is also a range of free large-scale maps to download, largely as a result of pressure from OpenStreetMap, the organisation that has argued that UK taxpayers are otherwise effectively paying for the OS twice.

Towards the end of the session in the Cotswold bookshop there was a final video to watch – what to include in your rucksack. A head-torch is useful, perhaps lip salve, and don't forget some warmer clothing lest the weather turns against you. 'And, most important, your map. Never be without your map and your compass. Not only does it show you where you are, but you get so much more out of the walking experience.' When it was over, and Simon King told us to honour the country code and 'leave only footprints, take only memories,' one of the workshop leaders said, 'Now you're ready to find your way!' as if it was an evangelical awakening. Most of us then bought maps, paper maps, and promised ourselves we'd open them very soon.

Pocket Map

A Nineteenth-Century Murder Map

Mary Ashford is not yet a major part of cartographic history, but perhaps she should be, poor thing. At 8am on Tuesday 27 May 1817 her body was found in a watery pit in a field in Erdington, on the outskirts of Birmingham. She may have fallen in, or she may have been murdered, and the mystery of how this twenty-year-old woman failed to return home after a local dance became one of those great and terrible stories that enthralled the public and filled the papers for months. The trial brought about a change in English law; the scandal brought about what was probably the world's first commercially produced forensic murder map.

Ashford had attended the annual dance at a public house with a girlfriend, and attracted the attention of a man called Abraham Thornton, who had previously boasted of his reputation as a lothario. The two danced together, and afterwards walked back over open fields. Ashford was last seen alive at 4am; at 6.30am a local mill worker discovered her blood-spotted shoes and bonnet.

Thornton was immediately charged with her murder. At the trial ten weeks later two maps were used to describe what may have happened that night, one drawn up for the prosecution by local surveyor William Fowler, and one drawn for the defence by the Birmingham surveyor Henry Jacobs. Each side drew arrows on their maps to instruct the jury on Ashford's and Thornton's possible movements after the dance, and once-insignificant places such as Penn's Mill Lane and Clover Field became known on breakfast tables throughout the country. Thornton had an alibi and a good counsel, and the jury took only six minutes to acquit him.

It was an unpopular verdict. Mary Ashford's brother William, incensed by the outcome but cheered by the strength of public opinion against Thornton, managed to arrange a retrial at the King's Bench in Westminster the following November, and the news pleased not only the newspapers but pamphleteers and cartographers too. There was money to be made. The most successful map, in several editions, was made by a local teacher who inspired his pupils and his colleague George Morecroft to rise at dawn to construct a survey of the crime scene and surrounding area. The teacher, by the name of R. Hill, explained that he had not been impressed with the 'rude plan' produced in the *Midland Chronicle* that was 'apparently done without measurement ... a very imperfect representation.' Hill, a keen geographer, had previously sketched new maps of Spain and Portugal for his school, though no examples survive. He had also read a popular account of the birth of the Ordnance Survey by its director, Major-General William Mudge, and, 'finding it more interesting than any novel', taught himself trigonometry and other modern surveying techniques.

His map, a wood engraving printed in Birmingham, measures 38 x 48.5cm. Hill cast his boundaries wider than the maps in the newspapers 'so far as to include the place of the alleged alibi,' and added a cross-sectional drawing and an engraving of the pit where Mary Ashford's body was found (shown for what it really was, a rather bucolic tree-lined pond). The main map is augmented by helpful directions: 'The road which Thornton says he

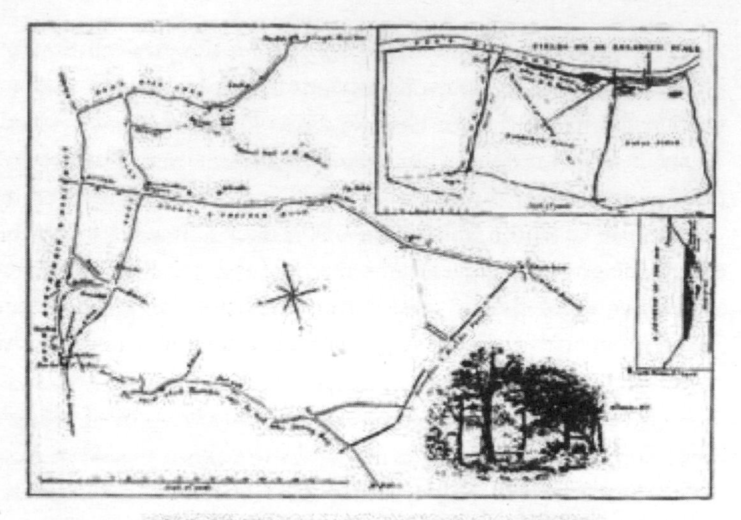

The fateful last journey of Mary Ashford in a map by the mysterious 'R Hill'.

took after leaving Mary'; 'Supposed track of the Murderer.' And then there is a bit of topographical sensationalism. In an inset of the fields where the murder was alleged to have occurred, Clover Field is renamed *Fatal Field*.

The map was accompanied by detailed explanatory text and a juicy title in a combination of plain and ornate Old English type: 'Map of the roads, near to the spot where Mary Ashford was murdered'. The text provided a basic outline of the case and the disputed claims of Abraham Thornton's route after accompanying Mary Ashford from the dance, composed in the peculiarly unemotional style of a policeman's evidence ('It may perhaps assist the inspector of the map to be reminded of some of the particulars ...'; 'the height of the water as here presented, is as near as possible [to] what it was when the Murder took place.') Hill's map was a great success, earning him and his class £15 profit despite widespread plagiarism.

No map, however, could hope to capture the courtroom drama to come. Thornton not only continued to plead his innocence, but invoked the archaic statute of trial by combat – challenging William Ashford to a duel by putting on a gauntlet and throwing another down by Ashford's feet. Ashford declined to pick it up. Thornton was then tried for a third time the following April (no one could really get enough of this case), when he was again acquitted, the judge declaring that he could indeed induce his 'wager of battle' as a valid defence, although a new statute would soon abolish it.

Thornton then fled to America, where he is believed to have died in his late-sixties. The lengthy inscription on Mary Ashford's gravestone in a Sutton Coldfield churchyard makes sober and moralistic reading, describing her terrible fate 'having incautiously repaired to a scene of amusement without proper protection.'

And what became of our principal map-maker? He did all right for himself. He left teaching and the Midlands for the civil service and a house in Hampstead, northwest London; he was promoted to the Treasury; and in 1840, twenty-three years after the murder of Mary Ashford, he transformed the world's postal system with the Universal Penny Post and the Penny Black stamp. The murder map is the only evidence we have of Sir Rowland Hill's cartographic career.

Chapter 11
The Legendary Mountains of Kong

In 1798, an English cartographer called James Rennell did something so audaciously memorable, so uniquely unpredictable, that no one in the map world has been able to match it since. He invented a mountain range. Not just any range, either: a central belt that stretched through thousands of miles of West Africa – mountains of 'stupendous height' that would prove an impassable mental barrier to Livingstone, Stanley, and any other European explorer with ambitions to penetrate one of the most lucrative blanks on the map.

The Mountains of Kong, named after a once-prosperous trading region in what is now Côte d'Ivoire and Burkina Faso, are one of the great phantoms in the history of cartography, and not just because of the ridiculous novelty of their length, extending west to east from modern-day Nigeria to Sierra Leone. The Mountains of Kong were also extraordinary because of their longevity. Once on the map, they stayed on it for almost a century – until finally an enterprising Frenchman called Louis-Gustave Binger went to have a look and found that they weren't there, an achievement for which he was awarded the highest domestic honours. But who could possibly perpetrate such an absurd act of cartographical chicanery? And how could they get away with it?

In the latter part of the eighteenth century James Rennell was something of a cartographical hero. His survey of Bengal was justly regarded as the most detailed and accurate yet undertaken, a feat achieved along newly scientific mapping principles. He was a pioneer too in the new science of oceanography, and he is remembered as one of the founders of the Royal Geographical Society. It was only to be expected, therefore, that any map he drew displaying a new discovery was not only believed but welcomed, particularly if it appeared in one of the most significant books of travel literature written in his lifetime.

And so it was: his most elaborate apparition made its debut in two maps published to accompany *Travels in the Interior Districts of Africa* by Mungo Park, the Scottish explorer's account of his quest to find the source and course of the Niger. (For more than a century, almost all African explorers' principal quests concerned rivers, and the ancient Greek conundrums of locating the source, the flow and the outfall – whether the White and Blue Nile, the Niger or the Congo.)

Park's challenge was set by the newly formed African Association, the London-based society established by Joseph Banks, William Wilberforce and others with the joint ambitions of intellectual and commercial conquest. Africa's gold reserves and the prospects for British trade were believed to be limitless, and although the coastline had been well-mapped by 1780, the interior remained largely a mystery. Park's journey through Senegal and Mali in 1795–97 was more circuitous and less penetrating than his second fateful mission a decade later (at the end of which he is thought to have drowned after a pursuit by spear-chucking natives), but his journals offer a vivid topography of a vanishing world on the eve of a colonising stampede.

The Mountains of Kong – 'a Chain of Great Mountains' – arrive on James Rennell's map in 1798.

James Rennell's accompanying maps are based on Park's written account, but also on additional information provided to the cartographer after Park's return to London. Rennell wrote an appendix to the book in which he explained how Park's discoveries had provided a 'new face' to the continent and had proved 'that a belt of mountains, which extends from west to east, occupies the parallels between ten and eleven degrees of north latitude, and between the second and tenth degrees of west longitude (from Greenwich). This belt, moreover, other authorities extend some degrees still farther to the west and south, in different branches ...' In his book, Park reported seeing only two or three peaks, but Rennell knitted them together. It wasn't by chance that the existence of these non-existent mountains reinforced Rennell's theory (vaguely suggested by Park) about the route of the Niger. He believed, erroneously, that it began in the mountains, travelled east-to-west along its range, but was prevented from travelling south and reaching the Gulf of Guinea ... by the mountains. He showed the Niger evaporating inland in Wangara.

He then explained how these 'other authorities', including the fifteenth-century Moorish geographer Leo Africanus, had previously referred to mountain ranges in the area but had failed to give them a name. But now they did have a name, inspired by Park hearing a native description of 'the Kingdom of Kong'. It was an intrepid and resolute act, the modern equivalent, perhaps, of drawing a thick contoured line through more than half of Western Europe and calling it the Mountains of Luxembourg.

And of course that wasn't the last of it. Mungo Park's account was a bestseller, and Rennell's maps had an immediate influence on others. The mountains were not just a dramatic obstacle; the legend grew that they glistened with gold. In 1804 the German map-maker Johann Reinecke produced what looked like a fluffy snow-covered range (titled Gebirge Kong) for a new atlas. A year later, the leading London engraver John Cary produced another map with the Mountains of Kong looming ever more menacingly over the plains (this time linked to the similarly fictitious Moon Mountains, the supposed source of the White Nile since the days of Ptolemy). Cary's work was titled, with some conviction, 'A New Map of Africa, from the Latest Authorities'.

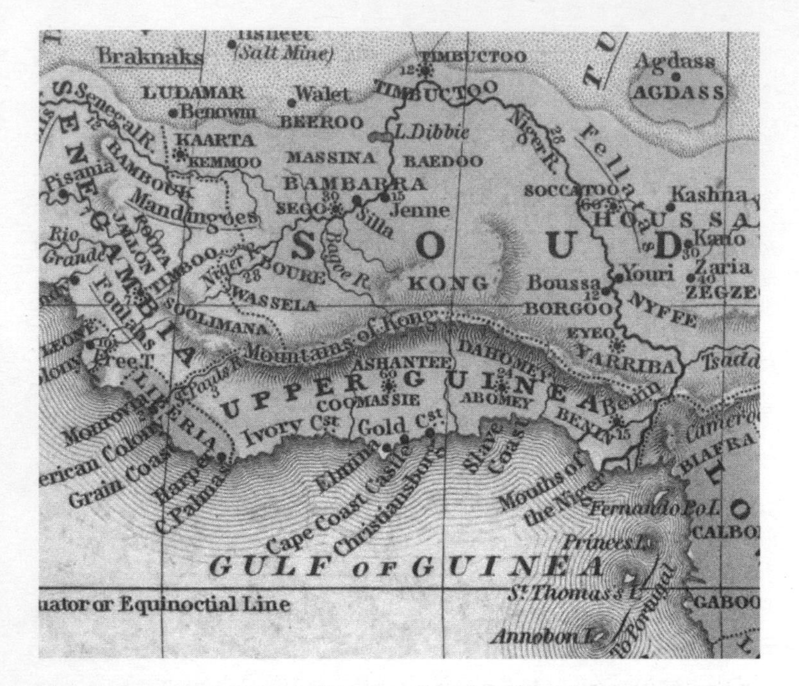

The phantom mountains stubbornly refusing to leave this American atlas of 1839.

How did the mountains remain standing for so long – at once both falsifiable and unverifiable? The American scholars Thomas Bassett and Philip Porter have identified forty maps which show the Mountains of Kong in various stages of development from 1798 to 1892, eventually forming a range the size of a small African state. Faced with a lack of evidence to the contrary, cartographers copy each other – we know that. But the fact that some of the most convincing representations of the Mountains of Kong appeared on maps many years after the Lander brothers confirmed that the Niger flowed into the Gulf of Guinea quite undermined the theory that we had entered a new scientific age. As Bassett and Porter found, cartographic knowledge in the nineteenth century was still 'partly based on non-logical factors such as aesthetics, habit, [and] the urge to fill in blank spaces …'

Rennell, one of England's most garlanded geographers (he was buried in Westminster Abbey), changed the cartography of Africa for ninety years. One needs look no further for a pristine example of the power of the printed word to confer status, or the power of the printed map to confirm authority. It was only in 1889, with the travels of the French officer Louis-Gustave Binger, that things began to change. In December 1889 Binger addressed a distinguished audience at the Paris Geographical Society, and recounted his previous year's journey along the Niger from Bamako (in present-day Mali) to the outskirts of Kong. What did he find? 'On the horizon, not even a ridge of hills!'

Binger's demolition job had an immediate effect: the Mountains of Kong disappeared from almost all maps as quickly as they had appeared. They last featured in Rand McNally's map of Africa of 1890, although as late as 1928 the esteemed Bartholomew's *Oxford Advanced Atlas* featured this in its index: 'Kong Mountains, French West Africa, 8°40 N 5°0 W.'

Charlie Marlow, the chief narrator in Joseph Conrad's *Heart of Darkness*, arrived in Africa just a few years too late to see the Mountains of Kong. But his regret lay elsewhere – in the fact that most of the white spaces he had seen on the map as a boy had been filled in. 'Now when I was a little chap I had a passion for maps,' he tells his fellow crew members as they sit anchored on the Thames estuary waiting for the tide to turn at the beginning of the novella. 'I would look for hours at South America, or Africa, or Australia, and lose myself in all the glories of exploration. At that time there were many blank spaces on the earth, and when I saw one that looked particularly inviting on a map (but they all look like that) I would put my finger on it and say, "When I grow up I will go there."'

Even though by Marlow's adulthood Africa 'had got filled ... with rivers and lakes and names ... It had become a place of darkness', he remains entranced by a map in a shop window of a snake-like river curving over a vast country, and he endeavours to join any enterprise that will get him there. Before his interview with an ivory company, he sits in a waiting room with another map, both shiny and colourful. 'There was a vast amount of red – good to see at any time, because one knows that some real work is done in there, a deuce of a lot of blue, a little green, smears of orange, and, on the East Coast, a purple patch...' He wasn't interested in any of this. 'I was going into the yellow. Dead in the centre.'

Charlie Marlow's apocalypse lay ahead of him, and maps wouldn't be of much use. But his concept of Africa becoming 'a place of darkness' is an illuminating one. Beyond the spiritual darkness of those he encounters, Marlow (and one presumes Conrad, who had himself travelled up the Congo in the 1870s) viewed the continent as dark when it was full: fully explored, fully colonised, fully mapped (and conceivably, one imagines, as was the fashion, full of the dark-skinned).

Most Victorian-era explorers and cartographers had an entirely different interpretation of dark. It was a term for the barbarian unknown, and the unmapped. When Henry Morton Stanley entitled his book *Through The Dark Continent* in 1878 (two decades before Conrad completed *Heart of Darkness*) Africa was still dark, despite the recent intentions of Mungo Park, Richard Burton, John Speke, David Livingstone and Stanley himself, to name the British alone. In fact, Africa was getting darker by the year: Stanley's follow-up, another bestseller, was *In Darkest Africa* (1890).

But there is an even stranger tale of light and dark, and it is unique to Africa: the story of how we consciously placed blank spaces on a map that was hitherto crammed with life and activity.

Among cartographers, the Irish satirist Jonathan Swift is known for these four lines from his long poem 'On Poetry: A Rhapsody':

> So geographers, in Afric maps,
> With savage pictures fill their gaps,
> And o'er unhabitable downs
> Place elephants for want of towns.

Certainly this was once the case. The Belgian map-maker Jodocus Hondius had a nice safari of elephants, lions and camels on his map of 1606, and in 1670 John Ogilby had an elephant, rhino and what may have been a dodo doing their worst in Ethiopia. But in 1733, when Swift wrote those lines, it was hardly true at all. Africa was emptying out. There were no animals, or such animals that did exist had been confined to the cartouche, alongside naked natives. This wasn't to make way for the latest geographical discoveries and a new topography, but the opposite: the interior was turning blank again. It wasn't just symbols and illustrations that were disappearing, but a multitude of rivers, lakes, towns and mountains, and it was a remarkable thing – one of those rare instances of a map becoming less instructive and less sure of itself as the decades and centuries passed.

Here are two examples. The first is Blaeu's popular *Africae Nova Descriptio* from the early 1600s. The outline of the continent is essentially correct, there are many recognizable kingdoms and lakes (alongside elephants, crocodiles and large frogs), and the map looks full. Partly this is a trick, with the text of coastal locations named by Portuguese explorers in the previous two centuries being turned inland, rather than the usual practice of displaying them towards the oceans. And partly it is wishful thinking, the interior topography a combination of Herodotus, Ptolemy, haphazard Portuguese expansion in the quest for gold, and hearsay. It is not wholly inaccurate, but there is a lot of supposition.

Compare this to the key map of the country made more than a century later in 1749. This is by the influential French cartographer Jean Baptiste Bourgignon d'Anville, who is notable chiefly for two things: the scientific accuracy of his maps elevating the art of cartography throughout Europe, and the fact that he hardly left Paris. His map of southern Africa is noteworthy for its extreme honesty; d'Anville rejected hearsay and plagiarism, and sought verification on every mark he placed; if there

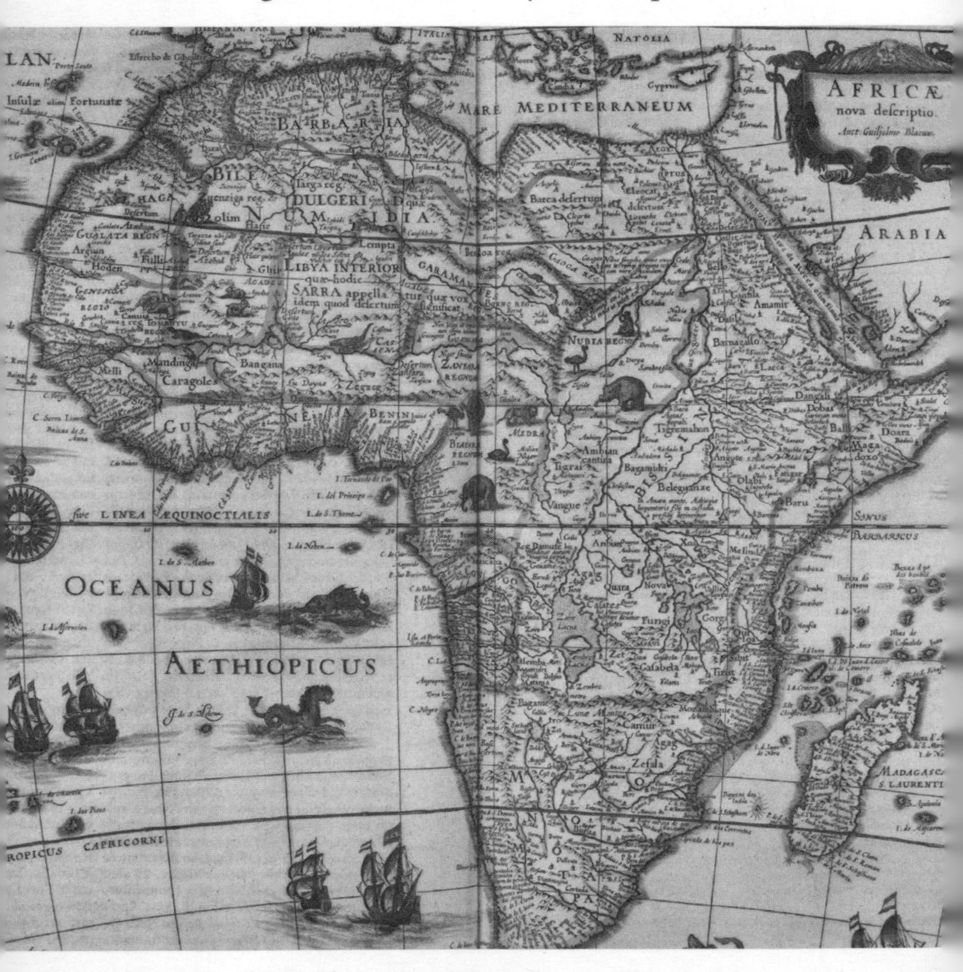

Africa full in Blaeu's *Africae Nova Descriptio.*

was no confirmation yet he believed a river or settlement to exist, he would duly note an uncertain provenance. D'Anville's map thus contains substantial details of three areas only. The kingdom of Congo on the west coast, the state of Manomotara and its immediate neighbours on the east coast, and the southerly tip by the Cape of Good Hope, 'Le Pays des Hotentots'. Madagascar is also well documented. But the rest of the country is a vast swathe of blank, a brave act for a map-maker.

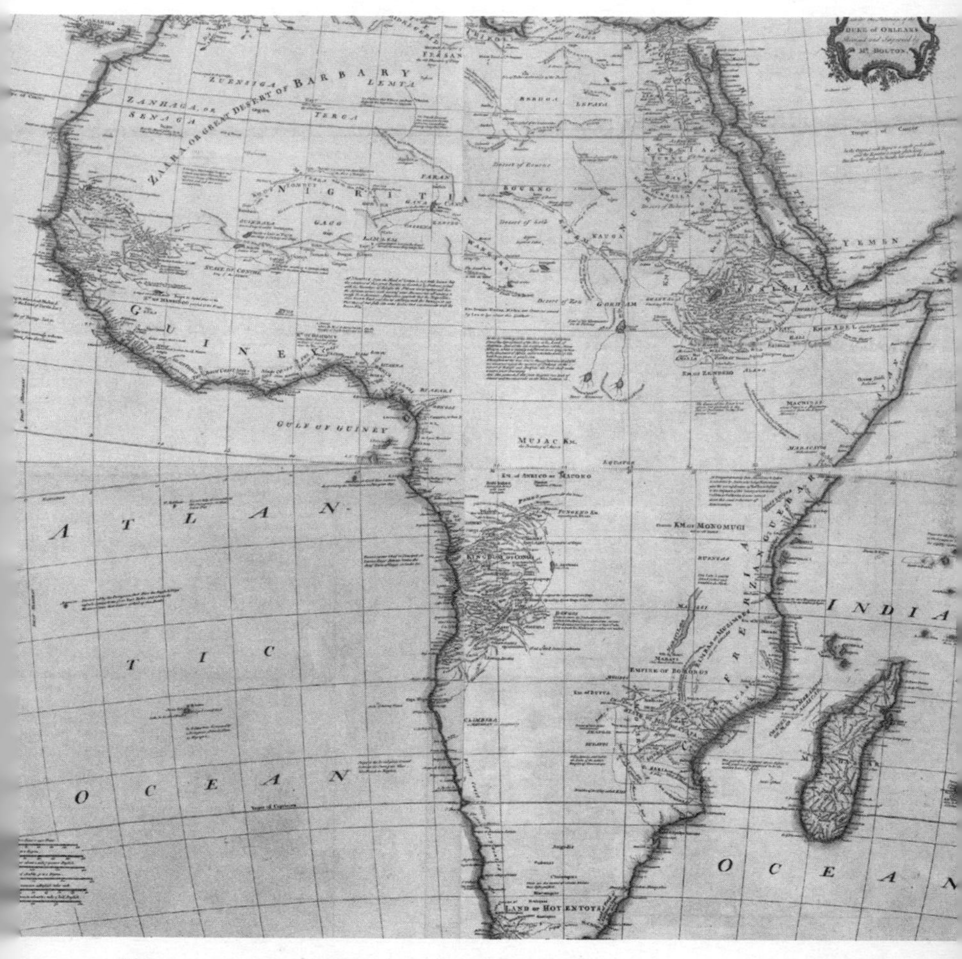

Africa empty in a 1766 map based on d'Anville.

The blanks fired intellectual curiosity; many regarded them as an insult to the enlightened age. But d'Anville's gaps also made huge political suggestions: the continent, universally known for its troves of slaves and gold, is wide open for conquest; the indigenous population, such as it exists, can have no claims on the unmapped territories, and will therefore present no resistance to subjugation. The blank spaces, swept clean of their inhabitants, were now all potentially white spaces too. Over the next fifty years, d'Anville's map became the dominant impression of Africa throughout Europe, and went through many editions unchallenged. And in this way science yielded to commerce and avarice. Did d'Anville have such intentions himself? Almost certainly not. But when members of the African Association gathered in London to gaze upon the map towards the end of the century (and potentates did likewise in Antwerp, Paris and Amsterdam), they must have been licking their lips.

The blank spaces didn't last long. In 1873, William Winwood Reade drew an engaging thematic map of the 'Literature of Africa', a textual display showing where the key explorers of the late-eighteenth and nineteenth centuries had travelled. 'David Livingstone', the first to traverse the central width of the continent, straps the map like a belt, while 'Mungo Park' and the French explorer René 'Caillie' both curve around the Niger. Stanley had located the fading Livingstone near the shore of Lake Tanganyika by the time the map was drawn, but he was yet to set off on his own big discoveries to Lake Victoria and beyond, and so barely features.

But Stanley does feature prominently in the one of the most brutal accounts of colonisation that we possess. In fact there are two such accounts. One is in Stanley's own hand,

Explorers make a name for themselves in 1873.

his bestselling journals of his violent sashays from the western mouth of the Congo to Zanzibar, alive with death as he hacks his way through the forests with his valiant and modern expeditionary force.

And the other is a map of equatorial Africa, the region once known as Congo Free State, which shows Stanley basically doing the same thing. Stanley's magnificent achievements as an explorer – not only the successful location of Livingstone, but the confirmation of Lake Victoria as the source of the White Nile – have been undermined by his participation in what may be the worst humanitarian disaster ever conceived by colonial hubris and greed.

Encouraged by Stanley's heroics along the River Congo between 1874 and 1877, Leopold II, King of the Belgians, coopted him to take part in a rather less 'scientific' venture. Leopold had seen the blank maps and wanted a piece for himself. In a period that saw Britain, France, Italy, Germany and Portugal carve up the continent in a wild imperial looting expedition, the conquest of land through a mixture of industrial ambition and religious divination might have seemed merely like the natural order of things. Leopold made his intentions clear at a geographical conference in Brussels in 1876, proposing the establishment of an international committee with the purpose of increasing the 'civilisation' of Congo natives 'by means of scientific exploration, legal trade and war against the "Arabic" slave traders.'

He claimed a higher goal: 'To open to civilisation the only part of our globe which it has not yet penetrated, to pierce the darkness which hangs over entire peoples, is, I dare say, a crusade worthy of this century of progress.' But his ideas of progress and scientific methods were cruelly unconventional, involving as they did brutal enslavement, a military dictatorship and the ruthless control over the ivory and rubber trade, an ambition only made possible initially with Stanley as his entirely respectable agent, buying up vast areas for Belgian

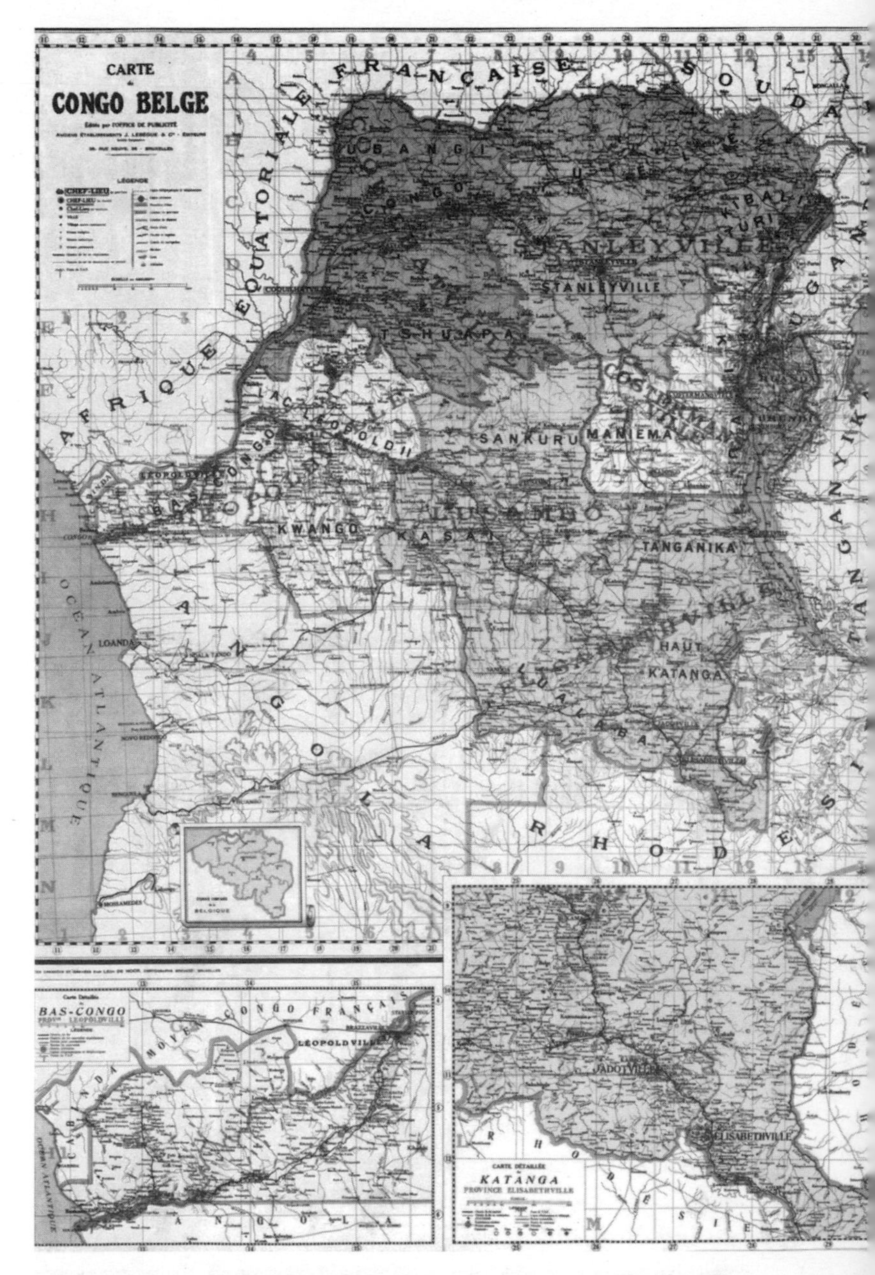

Belgian Congo – the darkest and bloodiest of colonial maps. Stanleyville sits at the top.

control with sweet-talk and trinkets. To what extent Stanley knew of Leopold's intended subterfuge has long been the subject of debate, but the king reportedly informed him, 'It is a question of creating a new state, as big as possible, and of running it. It is clearly understood that in this project there is no question of granting the slightest political power to the Negroes. That would be absurd.'

Leopold (and Stanley's) conquest of the Congo was one of the prime motivations behind Otto von Bismarck's Berlin Conference of 1884–5, an attempt to divide the rightful ownership of this recently blank continent. (In *Heart of Darkness,* Bismarck's Berlin Conference becomes a parody: the 'International Society for the Suppression of Savage Customs'.) The subsequent map looks colourful and ordered enough, and suddenly full again. But the new appearance of King Leopold's massive Congo Free State heralds one of the truly dark periods of colonial rule. And the bright new partitions on the rest of the map at the start of the twentieth century – French Algeria, Portuguese Angola, Italian Libya, German Cameroon and British South Africa – show only the ability of maps to conceal what's really there, and to mask the misery to come.

A quarter of a century after Conrad's *Heart of Darkness* first appeared, and in the year of the author's death, a private press published the author's own thoughts about the lightness and darkness of maps. Like Charlie Marlow, Conrad was a map fan. He had to be: he had led such a peripatetic life on land and sea that they were the only way he could find his bearings. In *Geography and Some Explorers* he wrote of how 'map-gazing, to which I became addicted so early, brings the problems of the great spaces of the earth into stimulating and direct contact with a sane curiosity and gives an honest precision to one's imaginative faculty.' He was aware he was living through a revolution in which 'the honest maps of the nineteenth century nourished in me a passionate interest in the truth of geographical facts and a desire for the precise knowledge which was extended to other subjects. For a change had come over the spirit of cartographers. From the middle of the eighteenth century on, the business of map-making had been growing into an honest occupation, registering the hard-won knowledge, but also in a scientific spirit, recording the geographical ignorance of its time. And it was Africa, the continent of which the Romans used to say "some new thing was always coming," that got cleared of the dull, imaginary wonders of the dark ages, which were replaced by exciting spaces of white paper.'

What really excited him about maps, he realised, was a simple thing: 'Regions unknown!' Not defined certainty, but the opposite – the mystery, and the life-enhancing possibility of discovery.

Pocket Map
The Lowdown Lying Case of Benjamin Morrell

The Mountains of Kong were far from the only fictional features of nineteenth-century maps. The Pacific was littered with more than a hundred imaginary islands, floating around happily for decades in every atlas. Then in 1875 a disgruntled British naval captain named Sir Frederick Evans began crossing them out. In all, he removed 123 islands from the British Admiralty Charts that he believed were the result of: a) mistaken coordinates, b) too much rum and nausea, and/or c) restless megalomaniac commanders longing for posterity. In his eagerness, Evans also removed three genuine islands, but it was a small price to pay for cleaning up an ocean.

One of the worst offenders, of the megalomania variety, was an American called Captain Benjamin Morrell. Between 1822 and 1831 Morrell had drifted around the southern hemisphere in search of treasure, seals, wealth and fame, and failing to find much of the first three, opted for posterity. The published accounts of his travels proved popular and convincing enough for his findings – including Morrell Island (near Hawaii) and

New South Greenland (near Antarctica) – to be entered on naval charts and world atlases, where they endured for a century. In fact, Morrell Island caused a westward diversion of the international dateline until 1910, and appeared in *The Times Atlas* as late as 1922.

The strange thing was, the true nature of Morrell's travels and deception had begun to unravel much earlier. In March 1870, the Royal Geographical Society in London gathered to discuss Morrell's claims. The debate was led by Captain R.V. Hamilton of the Royal Navy, who was a Morrell fan. He spoke of how the British had recently made great discoveries in the southern oceans, and claimed that Morrell, valiantly slicing through the ice in his schooner *Wasp*, had made the greatest

Benjamin Morrell: the world was his football.

headway. He explained that the story of his voyages were on the RGS shelves – not only in book form, but on new maps as well. Hamilton had recently placed the results of Morrell's discoveries on the new Admiralty charts, for they were 'curious and important'. His main regret was that Morrell's narrative was 'not as detailed as it might be.'

That was the understatement of the year. Even the least experienced captain will diligently log his progress through unfamiliar waters, noting his coordinates alongside sailing and weather conditions. But Morrell's Antarctic log contained blank weeks, and pages torn out. It observed no ice where others had seen nothing but; and birds of paradise suited only to the tropics. Other navigators at the meeting were rightly sceptical, chief among them J.E. Davis, who had followed Morrell's Antarctic 'course' sixteen years later, as a member of Sir James Ross's expedition. Davis concluded that Morrell's work not only lacked credence but resembled the fiction of Robinson Crusoe (Morrell had in fact sailed to the Juan Fernandez Islands (off the coast of Chile), where Alexander Selkirk washed up in 1704, inspiring the Defoe novel.)

However, it was half a century later before the matter of New South Greenland was finally laid to rest, when Ernest Shackleton, on his *Endurance* expedition of 1914–16, found that its supposed location was in fact deep sea, with soundings up to 1,900 fathoms. With Shackleton's reputation far stronger than Morrell's, off the map it came.

It wasn't the last of Morrell's fabrications to be removed: Morrell Island, in the Hawaiian archipelago, soon followed. However, the modern naval historian Rupert Gould has identified some useful and verifiable Morrell discoveries. Among them – and perhaps a fitting memorial – is Ichaboe Island off the coast of Namibia, which Morrell found to be rich in guano deposits from native seabirds.

Chapter 12
Cholera and the Map that Stopped It

On the morning of 7 April 1853, Dr John Snow was summoned to Buckingham Palace to attend the birth of Queen Victoria's eighth child Prince Leopold. Snow, a forty-year-old Yorkshireman, was one of the leading advocates of administering chloroform during labour (and during most other things too: in his lifetime he used the anaesthetic in 867 tooth extractions, 222 female breast tumour excisions, 7 male breast tumour excisions, 9 eyelid corrections, and 12 penis amputations).

The birth was a success, as was that of Princess Beatrice four years later, which Snow also attended. Snow had also used chloroform on the daughter of the Archbishop of Canterbury, and the combination of church/royal approval did much to popularise the use of anaesthesia in general. The procedures made Snow famous and rich, although these days he is largely remembered for something else: the use of a map to illustrate the infectious spread of cholera.

The map, which centred around London's Soho, was not recognised as anything extraordinary at the time. It was not the first map to display this fatal disease, and the rigour of

John Snow – the map doc

its science was found wanting. But it is now regarded as iconic, one of the most important maps of the Victorian age. And as a way of engaging young minds in the elementary detection of medical mysteries, it is a map yet to find its equal in either *Sherlock Holmes* or *House*.

Asian cholera first came to Britain in 1831, claiming more than 50,000 lives. A second epidemic killed a similar amount in 1848, a devastating figure for a country being told by its government that the new Public Health Act, passed in the same year, would transform the nation's sanitation. But cholera would prove a stubborn foe: by the time the third epidemic began to decimate Soho at the end of summer in 1854, there was still widespread disagreement about its cause. Most believed that cholera was miasmic (caused by airborne infection), a view supported by the two most prominent medical names of the day, Florence Nightingale and Sir John Simon, the Medical Officer of the City of London. But several leading epidemiologists had begun to suspect otherwise.

Snow's study of the disease, in his 1849 pamphlet *On the Mode of Communication of Cholera*, dismissed the idea that there was just something in the air. He suggested cholera was caused either by the human consumption of contaminated food or water, or by 'fomites', which usually meant infested clothes or bed linen. He claimed that the 1848 outbreak had been caused by the arrival in London of an infected sailor and his bedclothes from Hamburg, though he found this difficult to prove. Snow suspected a cellular structure to the cholera organism, but as he had not been able to show it beneath a microscope he proceeded largely on instinct.

In late-August 1854 Snow was examining how the water supply routes from the Thames may have affected a serious cholera outbreak in south London when he learnt that new cases had been reported just a few hundred yards from where he lived in Sackville Street, Piccadilly. He used to live even nearer, in Frith Street, where there had already been several deaths, and he believed that his knowledge of the area, and contact with local residents, might yield the clues he needed to support his theory. He did what doctors still did in those days: he made house calls. It was a brave endeavour: in his efforts to match human illness to human behaviour he appeared to put himself at grave risk, for if cholera was airborne, the inquisitive Snow would surely be one of its victims.

Sniffing out the causes of cholera: Robert Seymour's health inspectors on the trail in 1832.

In the first week of his investigations more than five hundred Soho residents would die. People began falling ill on 31st August, with a peak in fatalities occurring two days later. But by the third day, Snow believed he had found his cause: the public water pump where Broad Street met Cambridge Street. This was not only the main water supply for those living nearby, but also a common stop for passing traders and children. There were other fatally opportunist users too, including the local pubs who watered down their gin and whisky, and many coffee houses and restaurants. Snow noted subsequently that one keeper of a coffee house in the neighbourhood, popular with mechanics, 'informed me that she was already aware of nine of her customers who were dead'. The water was also sold in small local shops, 'with a teaspoonful of effervescing powder in it, under the name of sherbet.'

Snow tested the water from the Broad Street pump on 3rd September, but his results were inconclusive: he detected few impurities with the naked eye, although when he looked again the following day he saw an increase in 'small white, flocculent particles.' One resident also told him that the water had changed its taste. Seeing no other possible cause, and perhaps fearing that he was running out of time, he requested a list of the dead from the General Register Office. Eighty-nine people had died from cholera in the week ending 2nd September, and as Snow walked around with his list he immediately saw the pattern he had anticipated: 'Nearly all the deaths had taken place within a short distance of the pump.'

As Snow continued walking he found further confirmation of his theory. Only ten deaths had occurred within the vicinity of another water pump, and five relatives of the deceased told him that they always drew water from Broad Street as they 'preferred' it – presumably either its taste or the fact they thought it cleaner. Two out of five of the remaining cases were children who went to school near Broad Street. Snow argued that the outbreak couldn't be supported by

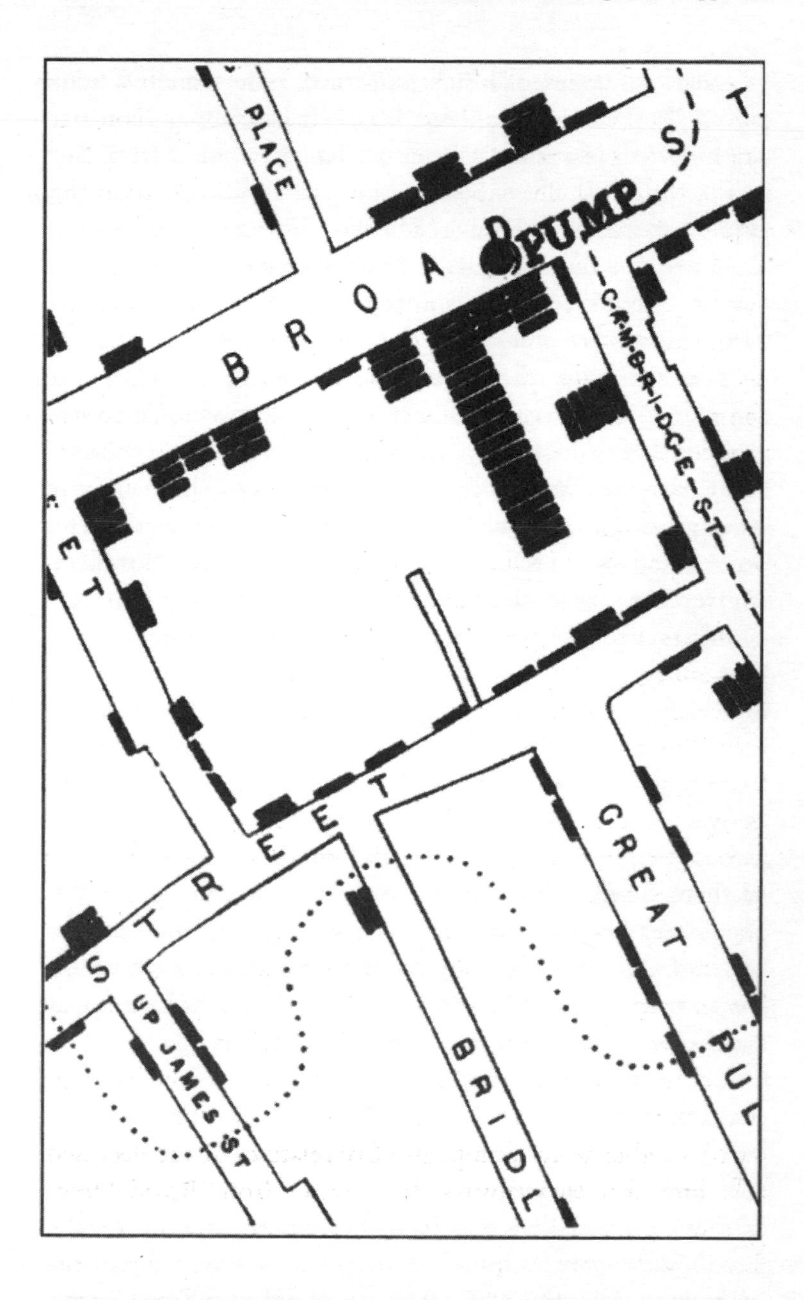

The ghost map: Snow's plan of Soho with the cholera-infected Broad Street (modern Broadwick Street) water pump at its heart

the miasmic theories (which associated disease directly with poverty) when he found that a nearby workhouse containing hundreds of people was not affected by cholera; it turned out they drew their water from their own well. The evidence now seemed overwhelming. On the evening of 7th September Snow met the local board of guardians and presented them with his findings. 'In consequence of what I said, the handle of the pump was removed on the following day.'

Like the water itself, the cases slowed to a trickle. The number of fatal attacks on 9th September was only eleven, compared with 143 eight days earlier. By 12th September there was one case, and by the 14th none. But this could not be attributed directly to the closure of the pump, as the cases were already decreasing in the days before. And as Snow himself reported, evacuation played a significant role – hundreds of inhabitants had already left the area in fear.

It was only now that Snow started putting together his famous map – an illustration of his findings rather than a cause of them. The base map itself was already available, printed by the nearby Holborn firm C.F. Cheffins, which also produced some of the earliest railway maps. The detailed scale was 30 inches to a mile, and the portion of the map Snow reproduced had Broad Street at its centre – the Jerusalem of Soho.

Snow added three key elements to the map. First were the locations of the water pumps – thirteen in all, the most northerly in Adam and Eve Court above Oxford Street, stretching down to Titchborne Street by Piccadilly Circus. Next, he placed a meandering dotted line over the area where the Broad Street pump would be closer for residents to visit than any other pump; this is now known as a Voronoi diagram, and Snow's version is the most famous early example. And finally he added small black dashes denoting deaths, like gravestones in a very crowded burial yard.

There were several areas of clustering: St Ann's Court off Dean Street had 24 bars; Bentinck Street off Berwick Street had 19; Pulteney Court by Peter Street had 10, with nine of them seemingly at one property. But the cluster around Broad Street is unmistakable: 82 deaths in the street alone, with many more marked close by.

Snow presented his findings – and his map – in a lecture to the London Epidemiological Society in December 1854. Soon afterwards, another report on the outbreak (conducted by local parishioners with Snow as their chief investigator) delivered an even more detailed breakdown of events, and another map. The findings were sickening, detailing not only extreme poverty and overcrowding but the sort of sanitation you wouldn't expect on a farm, with cellars and basement rooms layered with human faeces.

It was the miasmatics' claim that the fumes from these dungs would alone be enough to spread disease, and there was also a theory that the Soho area most affected by cholera in 1854 lay directly above a mass burial ground of thousands of casualties of the Great Plague of 1665. Despite the apparently glaring evidence displayed on Snow's map, his water-borne theory was still not universally accepted, and for a while a great riddle remained: how did the pump's water become contaminated in the first place?

The most convincing answer was provided by a local priest called Henry Whitehead, who, in the weeks that followed the Soho outbreak, had conducted a thorough investigation of his own. He began as a confirmed miasmatic, but began to have doubts as he did his rounds. At 40 Broad Street he met the police constable Thomas Lewis and his wife, and learnt that their daughter Sarah had died on 2nd September aged five months after a prolonged bout of diarrhoea. The mother spoke of the start of her illness on 28 August, and of emptying water containing the 'dejections' of the baby into the cesspool in the basement. When Whitehead examined this he found its faecal content leaking into the soil and evidently into the water supply. Snow was unaware of this first victim, and the baby is missing from his map, which shows the deaths at 40 Broad Street as four rather than five.

Snow's epidemiology wasn't the first of its kind. The concept existed long before the word was coined in 1802, and there was significant mapping of London's deaths following the Great Plague and Fire of 1665–66. And half a century before that, Bills of Mortality had identified the chief causes of death in London as dysentery and convulsion.

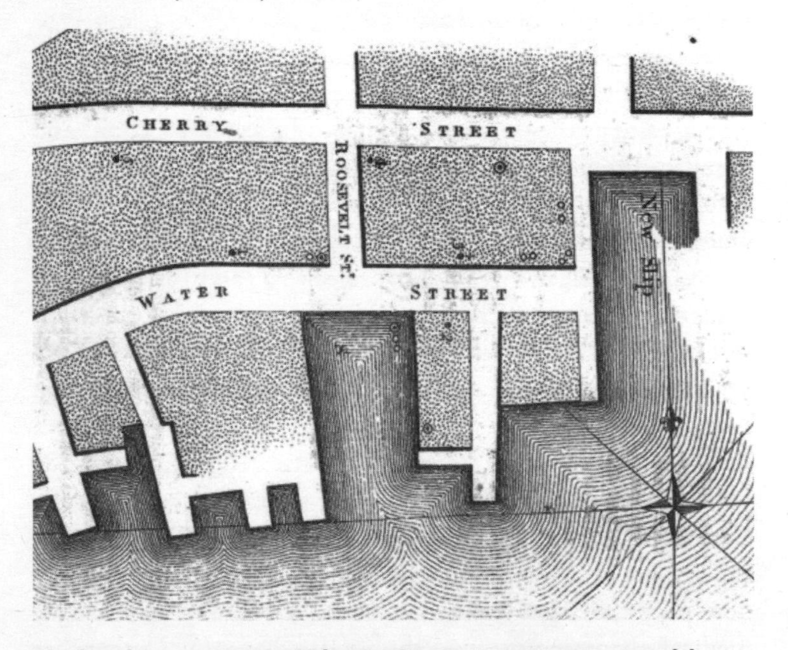

The first disease spot map: Valentine Seaman's 1798 impression of the New York waterfront, identifying incidences of yellow fever.

In America, the title of first medical cartographer should go to a man named Valentine Seaman, the New York public health official who drew the first proper disease 'spot' map in 1798. An outbreak of yellow fever in the Manhattan docks near Wall Street had caused hundreds of deaths when he began examining what he believed to be the principal cluster of cases and plotting them on two maps that were published in an influential new medical journal. Unlike Snow, however, Seaman was a confirmed believer in miasma theory, finding that many of the deaths had occurred close to what he called 'furry-fostering miasmata' and an open public toilet. In the absence of a pump handle, the New York authorities would in time learn to control tropical diseases with quarantine.

Seaman's maps remain both influential and timely. They amplified the emergence of a new trend, even a new science: medical geography. Its leading thinker was Leonhard Ludwig Finke, a German obstetrician who, in the 1780s, planned to construct an atlas based on disease. Heavily influenced by the writings of Hippocrates some 2,000 years earlier, he began to wonder about the overseas epidemics he had read about and wondered if they had a common thread. And then he found several, including the soil, vegetation, air, and methods of animal welfare. He believed that it was a particular district or country (rather than its people) that was diseased, and he constructed a scientifically rigorous three-volume explanation of his new geography – an early travellers' advisory of where not to go. His ambition for a disease atlas was quashed by high costs at the time of the Napoleonic Wars, but he did produce a world map of disease in 1792.

Finke's theories certainly endured. In 1847, Charlotte Brontë described Jane Eyre's search for a dwelling place in the Empire that was free of the terrible diseases that filled the newspapers, but her choices were extremely limited. The new medical geography seemed to justify what would otherwise pass for xenophobia. In *Jane Eyre* the Tropics were largely a malaria-ridden disaster zone, while West Africa was 'the plague-cursed Guinea coast swamp'. The West Indies, East Indies and much of the New World were similarly infested. But Bronte adopted a largely anti-imperialist stance, suggesting that the scarring of the physical landscape and the importation of pathogens were the definable and disastrous legacy of western colonialism.

In London, the fact that water was the prime cause of cholera was finally acknowledged in the 1870s, when Sir John Simon, who had left his post as London's chief medical officer to take on a similar role for the government, abandoned his miasma theory. It is impossible to estimate the number of lives saved by Snow's instinct and persistence, nor the impact on urban sanitation. He didn't do it all, of course: Henry Whitehead's work was equally significant, and health officials were well aware by this time that public sanitation had to be checked and improved, not least the water supply from the Thames (ultimately it was cholera itself, rather than any report or map, that spoke the loudest). Within four years of the Broad Street epidemic the engineer Joseph Bazalgette would be appointed to rid London of the Great Stink; his intricate network of underground sewers, completed in 1875, did more to cleanse London of cholera than anything else.

But Snow's work – and particularly his map – remains a thing of legend. The writer Steven Johnson has called it 'The Ghost Map', and it's an apt title: despite the anonymity of those slim black dashes, they are more than just statistics. Perhaps because Snow lived in the area and would have known some of these individuals as neighbours if not patients, and perhaps because the descriptions of their lives and dwellings in the texts by Snow, Whitehead and others are so vivid, we feel we have a stake in their lives, even after death. Charles Dickens may have something to do with this too – not least his pungent descriptions of London squalor from the late 1830s that Snow would have witnessed on his daily rounds as a young doctor.

And then there is Snow's map itself: its scale, the detail of the streets and alleys that many of us have walked down and enjoyed their noise and vibrancy. The map provides a degree of focus that was rare in public maps of the period, a focus not only on the streets but also the bustling and fetid activity they contained, something we are only now becoming familiar with again through the zoom of electronic mapping. But perhaps we remember Snow most of all because the story is so perfect: the map found the pump, and the removal of its handle stopped the epidemic. Neither is quite true, of course, but the map and its ghostly mortality has taken on an invincible life of its own.

Pocket Map
Across Australia with Burke and Wills

In 1860, Australia was a place of barren mystery. The shape of the continent had been mapped reasonably well, the coastal cities of Adelaide, Melbourne, Sydney and Brisbane were expanding rapidly, and a gold rush in the early 1850s had caused a great boom in the population and economy of New South Wales. But the interior was a different story. Was it arable and potentially profitable? Would an inland waterway make it navigable? Could a telegraph line be laid down to link it with the rest of the shrinking Victorian world?

The aborigines had no need for these questions, but white explorers from the newly established Royal Society of Victoria wanted to advance not only scientific and geographic knowledge, but also play catch-up with the rest of the world. Detailed scientific cartography had transformed the look of Europe; the belts of India and Africa were being traversed; the Northwest Passage was being opened up in the Arctic. But in Australia the map remained a carpet of blank.

A few explorers had made tentative inroads in the first half of the century. In 1813 Gregory Blaxland and William Wentworth ventured from Sydney to cross the Blue Mountains. In the mid-1840s Charles Sturt made a bold northward trek from Adelaide, but was forced to retreat after desert heat proved unbearable. A Prussian adventurer called Ludwig Leichardt travelled for thousands of miles along the north-east coast, but his journey into the interior in 1848 ended in one of those great Bermuda Triangle-style conundrums of exploration: he and his team were simply never seen nor heard of again.

But in 1860, a grander – one may say grandly ludicrous – expedition was proposed by the Royal Society of Victoria. This was a journey across the entire continent, from Melbourne in the south to the Gulf of Carpentaria* in the north; and then back again, replete with maps and journals. It was a journey that would nowadays be classified as Extreme Sports, except that most extreme sports are rather better organised and carry less air of folly. For this was not a civilised expedition in the manner of the Ordnance Survey; it was an adventure closer in spirit to polar exploration. And like polar exploration, the quest would inevitably be led by men who were either too ambitious or too crazed to know their limits, and the voyage would be remembered for the sort of tragic heroism that made young boys sit up straight in class and vow to become explorers themselves. And then there was a map summing it all up – a map notorious for its depiction of death rather than glory, a precursor to something we would soon see in the Antarctic.

Several contemporaneous maps relate the exacting journey of Robert O'Hara Burke and William John Wills, but the most dramatic was produced at the end of 1861, a few months after the two men and others in their party had met a sorry, starving end. It was published, on a scale of 1:3 million, by De Gruchy & Leigh, a flourishing lithographic company that had made its reputation with city maps of Melbourne. It was a long millipede of a map,

* The Gulf of Carpentaria is a large waterway to the east of Darwin. It was where Willem Janszoon – the Dutch explorer rather than the cartographer – became the first European to make landfall in Australia in 1606.

The most dramatic section of De Gruchy & Leigh's map of Burke and Wills' route to the Gulf of Carpentaria. The rings show the doomed explorers' final demise.

locationally unreliable, scientifically wanting, with little logic to its place names and observations, and a seemingly random start point, several months into the journey, halfway up the country by Cooper's Creek. As such, it was utterly compelling.

Burke and Wills' expedition – a party of 19 men, 26 camels and 23 horses – set off from the Royal Park in Melbourne on 20 August 1860, with some 15,000 inhabitants cheering them away. It was an inauspicious start: their equipment included six wagons, but one collapsed before it had left the park and two others didn't make it out of Melbourne. By 16 December, when the narrative on the maps starts at Cooper's Creek, all but four of the nineteen men had either left the mission or been deployed on back-up and food storage tasks.

The biographical details of those who remained read like the start of a bad joke: Robert O'Hara Burke was an Irishman

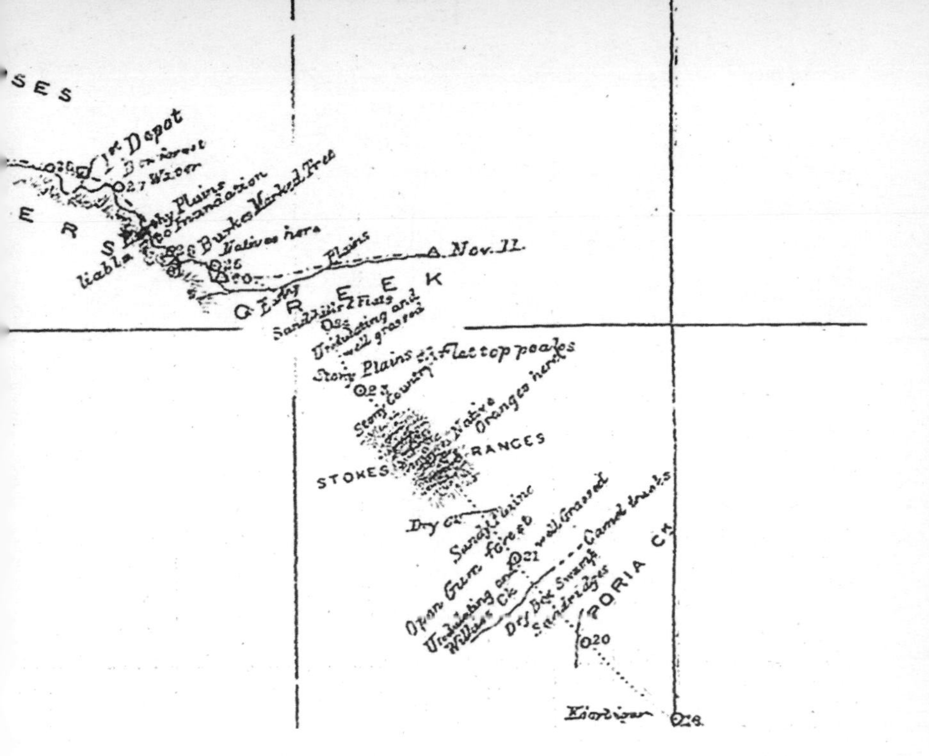

from County Galway who had once fought in the Austrian army; William John Wills was an Englishman from Devon, who had emigrated to South Australia and become a medical assistant and astronomer; John King, another Irishman, had gained some geographic experience in India; while Charles Gray's background was more of a mystery, beyond the fact that he hailed from Scotland.

The second half of the outward journey, from Cooper's Creek to the gulf, took a relatively uneventful two months. The scenic observations noted on the map are calm enough: *Suitable for Pastoral Purposes; Stony Desert; Brackish; Rivulet of Pure Water; Large Anthills 2½ to 4ft High; Tea Tree Spring – Mineral Taste Probably Iron*. Occasionally there is mention of the local inhabitants: by King's Creek they noted *Natives Inclined to be Troublesome*. Near the summit, which reached Flinders River just below the Gulf of Carpentaria, a line declares: *Burke & Wills proceed N on foot, leaving King & Gray in charge of camels*. And then, at the top of the map, the happy legend: *Burke & Wills, Feb 11 1861*.

But the return journey is a very different concern. Wills, the main surveyor, made his own piecemeal maps every few weeks, but he was in no fit state to draw the full route. The De Gruchy & Leigh map was compiled retrospectively, largely from field notes recovered from the earth months after they were buried by the travellers, eager to lighten their loads. Their increasingly desperate narratives are reflected on the map with markings both shorter and sadder: *Golah [a camel] Left Behind; April 16, Gray Died; Wills' Body Found; Burke's Body Found.*

How did they die? Another map entry provides the source for the answer: *King Found.* John King was the only survivor of the quartet, discovered camping with natives by a search party led by Alfred Howitt several weeks after the others had perished. His diaries provide a vivid account of the final days of the expedition as they were destroyed by a combination of extreme heat, cold, hunger and exhaustion. They staggered through parched land and swampland. They encountered snakes with girths like tree trunks, shot about forty rats each night near their camps, and found the native Yandruwandha tribe to be both generous and thieving.

'Is that Mount Hopeless up ahead?' Burke, Wills and King crossing the desert in an engraving by Nicholas Chevalier in 1868.

They sacrificed their camels to cure and eat their flesh, and went back and forth over the same land for weeks with the increasingly delirious dream of being rescued. One landmark mentioned frequently in the field notes was called Mount Hopeless.

King's account also revealed the tragi-comic circumstances of how the travellers just missed their support group (and the promise of rescue) on their return leg. Their back-up party had waited for the quartet to reappear at Cooper's Creek for four months, assumed they had all died, and on 21 April turned back to Melbourne. Just eight hours later, Burke, Wills, Gray and King finally made it to their depot. It was their last chance. 'From the time we halted, Mr Burke seemed to be getting worse,' King wrote after his lonely return. '... He said he felt convinced he could not last many hours, and gave me his watch ... He then said to me, "I hope you will remain with me here until I am quite dead – it is a comfort to know that some one is by".'

The funeral of the pioneers was held in Melbourne in January 1863, and attracted about 100,000 mourners. Within a decade a telegraph line had been laid from Adelaide to Darwin, and in 1873 Ayers Rock, in the middle of the continent, was mapped by the English explorer William Gosse. A decade after that, much of the Australian interior had been traversed from many angles.

But it is Burke and Wills' calamity that endures around the campfires. The voyage has become a movie – *Burke and Wills* – starring Jack Thompson as Burke, Nigel Havers as Wills and Greta Scacchi as the rather unlikely singing love interest; 'Runs short of plot even before the explorers run short of water,' said the *New York Times*. But more than that, the map and its narrative have etched themselves on the national consciousness and some distance beyond, so that they are now as much a part of the fantastical landscape as Ned Kelly, Dame Edna and Crocodile Dundee.

Chapter 13
'X' Marks the Spot: Treasure Island

'It's all about a map, and a treasure ...'
Robert Louis Stevenson, describing his new book, 1881

Why would anyone ever go to Trinidad? That's Trinidad out in the southern Atlantic, one of six islands in a barren archipelago, rather than the sandy white Caribbean Trinidad honeyeymoon isle, or the Trinidad in Colombia, Cuba or Paraguay, or even the Trinidads in California, Texas or Washington. All these places may have their charms.

But our Trinidad lies somewhere else. It is a place where even the most experienced sailors struggle among treacherous reefs to negotiate a landing, and, failing to do so, gratefully turn back. It is a place where sea birds attack from the air and armies of crabs attack on land. The island boasts little edible vegetation, but it does offer an angry volcano and sharks at its rim. In 1881, the British adventurer E.F. Knight called the place 'one of the most uncanny and dispiriting spots on earth', and vowed never to return.

But what if you came across a map of Trinidad with details of buried treasure – the genuine 'X marks the spot' variety, where the value of X makes the tomb of Tutankhamun look like a disappointing evening's metal detection on Clacton

Sands. Then wouldn't you be romantic and greedy enough to take up the challenge?

You wouldn't be alone. In 1889, the same Edward Frederick Knight who vowed never to return, returned. Lured by the prospect of shiny wonders from Lima (giant gold candlesticks, jewel-rimmed chalices, chests of coins, gold and silver plate), he was equally attracted by the dangerous narrative that piratical legend confers. He not only believed the treasure was there, but that divine intervention had selected him to locate it. God had wanted him to go there, he believed, because God had given him a map.

In *The Cruise of the Alerte*, the vivid and (one imagines) true account of his quest published in 1890, Knight explained that he had not known of the existence of treasure on Trinidad when he had first approached the island eight years earlier. On that trip, he had been yachting around the South Atlantic and South America with a small crew, and on a journey from Montevideo to Bahia faced unexpectedly strong headwinds. Steering eastward, he was about 700 miles from the east coast of Brazil when he first saw Trinidad, latitude 20° 30' south, longitude 29° 22' west. He decided to land, struggled desperately with the coral reefs, spent nine ghastly days there, and found the land crabs 'hideous'.

Four years after returning to England (he was a successful though presumably rather unsatisfied barrister) Knight read in a newspaper that a ship called the *Aurea* was leaving the Tyne bound for Trinidad with a skilled and determined crew aboard, and many digging tools. They were hunting for buried treasure, but were unsuccessful. Knight thought little more of this until three years later, when he heard that a few Tynesiders were planning to have another crack at the loot. He travelled to South Shields, found one of the men from the *Aurea*, and was told the following tale.

There was a retired sea captain living in Newcastle known, for the purposes of the story, as Captain P. He had been involved in the opium trade in the late-1840s, and on one of his voyages had enlisted a quartermaster known as The Pirate, so called because of a scar across his cheek. The Pirate was probably a Russian Finn, and had so impressed Captain P with his navigational skills that the two became friends. When The Pirate was struck with dysentery on a trip from China to Bombay, Captain P gave him special care, but there was little to be done – the man was dying. Nearing the end in a hospital in Bombay, The Pirate wanted to repay the captain for his care, and said he wished to reveal a secret. But first the captain would have to close the ward door. Having done so, The Pirate directed his visitor to a chest, and asked him to remove a parcel. It contained a piece of tarpaulin, and on the tarpaulin there was a treasure map of Trinidad.

The spot marked X lay in the shadow of Sugarloaf Mountain. Much of the bounty had come from Lima Cathedral when it was plundered during the war of independence. The Pirate had heard about it because he was indeed a pirate, the only survivor from a trip to Trinidad to bury the treasure in 1821. He had been unable to return since that date, and he believed all his fellow pirates had been captured and executed. He was certain that the loot was still many feet beneath the sand and rock.

As tall tales go, this was a whopper. In 1911, the American author Ralph D. Paine undertook a survey of 'the gold, jewels and plate of pirates, galleons etc, which are sought for to this day', culminating in 'The Book of Buried Treasure'. He found one strikingly common trait. There was always a lone survivor of a piratical crew, and he, 'having somehow escaped the hanging, shooting or drowning that he handsomely merited, preserved a chart showing where the treasure had been hid. Unable to return to the place, he gave the parchment to some friend or shipmate, this dramatic transfer usually happening as a death-bed ceremony.' The recipient would then dig in vain, 'heartily damning the departed pirate for his misleading landmarks and bearings,' before handing down the map, and the gried, to the next generation.

☠ ☠ ☠

Two years before Knight embarked on his own adventure, another account of buried treasure took hold on the popular imagination. Robert Louis Stevenson's *Treasure Island*, serialised in 1881 and then published as a book two years later, told the story of untold wealth on a far-flung isle, and a pirate's chest containing a mysterious oilskin. The oilskin contained two things, a logbook and a sealed paper, with the latter, according to the book's young narrator Jim Hawkins, showing 'the map of an island, with latitude and longitude, soundings, names of hills and bays and inlets, and every particular that would be needed to bring a ship to safe anchorage upon its shores.' The buried treasure was the property of the murderous Captain Flint, who, although dead when the book begins, had drawn instructions to help others locate it.

Treasure Island was about nine miles long and five across, and shaped 'like a fat dragon standing up'. It contained a central hill called 'The Spye-glass', and three red crosses, one of them accompanied by the words 'bulk of treasure here.' And so off they went, Jim Hawkins and his buccaneer friends, towards skulduggery and terrible betrayals, and a hoard worth £700,000. The map that accompanies the book, the most famous in all fiction alongside Tolkien's Middle Earth, was drawn by RSL himself and dated 1750. It is crowned by an illustration of two mermaids displaying the scale, while two galleons patrol the coasts.

Its origins are uncertain. Stevenson wrote the bulk of the book at the tail-end of a wet summer in the Scottish highlands, and among his fellow holidaymakers was one Lloyd Osbourne, his newly acquired twelve-year-old stepson. Osbourne later

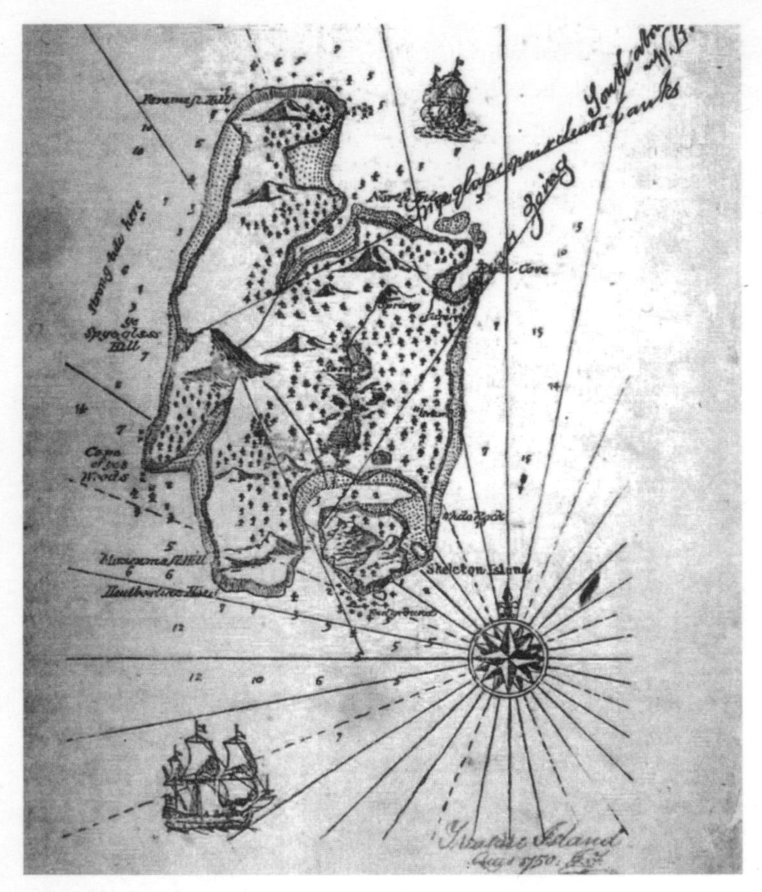

Irresistible – Robert Louis Stevenson's Treasure Island.

recalled a day when he 'happened to be tinting a map of an island I had drawn. Stevenson came in as I was finishing it, and with his affectionate interest in everything I was doing, leaned over my shoulder, and was soon elaborating the map and naming it. I shall never forget the thrill of Skeleton Island, Spyglass Hill, nor the heart-stirring climax of the three red crosses!'

Osbourne also remembers Stevenson writing the words 'Treasure Island' in the map's top right-hand corner. 'And he seemed to know so much about it too – the pirates, the buried treasure, the man who had been marooned on the island.' In what he recalls as 'a heaven of enchantment,' Osbourne then exclaimed 'Oh, for a story about it!'

Whether the story was already underway, or whether this map was the inspiration for the thirty-one-year-old Stevenson's first complete book (and who wouldn't claim to be the spark for one of the classics of world literature?), remains the subject of conjecture. Looking back at his creation for an article in *The Idler* magazine years later, Stevenson wrote of the enthusiasm of his stepson for the early drafts of his work as it appeared in early serialised form in *Young Folks* magazine, but he claimed the map as his own work. 'I made the map of an island. It was (I thought) beautifully coloured; the shape of it took my fancy beyond expression; it contained harbours that pleased me like sonnets ...' He also claimed it drove the plot: once he had constructed a place called Skeleton Hill, 'not knowing what I meant', and placed it at the south-east corner, he then directed the narrative around it.

Stevenson had created a moral fable of venality, vice and some virtue. *Treasure Island* is a coming-of-age story, and its obsession with disease and disability mirrors Stevenson's own battle with illness from childhood. Its swashbuckling narrative is also a rebellion against the high-Scottish moralising that almost suffocated its author in his youth. The book defined our communal mental vision of piracy, parrots, peg-legs, rum rations, and mutiny with Bristolian accents, and the treasure map within it not only drives the plot but stealthily does something else: it forms the basis of how we imagine treasure maps to this day – ragged, mischievous, curling and foxed, with not quite enough information to make those who hold it sure of their course, yet just enough to fire up a life-defining quest.

In 1894 Stevenson wrote that all authors needed a map: 'Better if the country be real, and he has walked every foot of it and knows every milestone. But even with imaginary places, he will do well in the beginning to provide a map; as he studies it, relations will appear that he had not thought upon; he will discover obvious, though unsuspected, short cuts and footprints for his messengers; and even when a map is not all the plot, as it was in *Treasure Island*, it will be found to be a mine of suggestion.'

☠ ☠ ☠

Maps in our minds are powerful things, and those we see as children may never leave us. We may already know the shape of E.F. Knight's Trinidad, as it was a model for Arthur Ransome's fictional Crab Island in *Peter Duck*, the third of his *Swallows and Amazons* adventures, published in 1932.

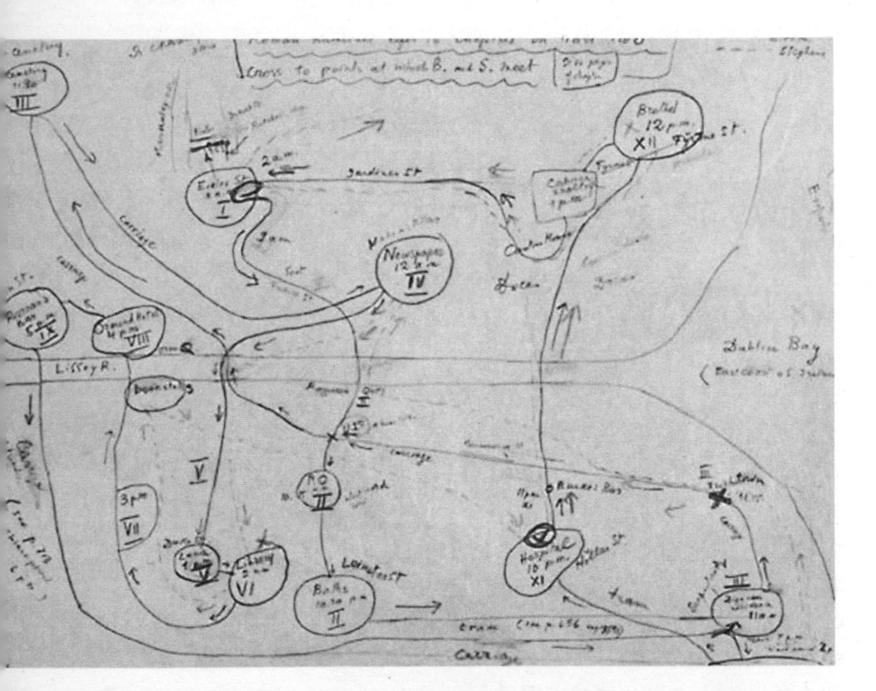

So that's left at Davy Byrne's Pub. Nabokov plots Bloom's Dublin ramblings in *Ulysses*.

Five years later, J.R.R. Tolkien's *The Hobbit*, another trail of treasure, was published with a map as its endpapers, and maps became a crucial feature of his *Lord of the Rings* trilogy. In fact, Tolkien's maps are probably the most influential in modern popular culture, having spawned, over the past couple of decades, a whole genre and generation of fantasy books, maps and games; they continue to wield a heavy influence over computer gamers.

Beyond children's fiction, a book's journey may inspire maps where none were intended – impossible ones such as Sir Thomas More's *Utopia*, or ones such as Nabokov's map of the journeys through Dublin of Stephen Dedalus and Leopold Bloom in *Ulysses* which he hoped would help engage his students. James Joyce once claimed that his novel was itself a practical map: if Dublin 'suddenly disappeared from the earth, it could be reconstructed from my book.'

A treasure map is the earliest form of map we have. They began appearing on cave walls in the Paleolithic age (take your spear to this place, a chalky arrow directs, and a woolly creature will be yours), and they surround us still, most often in disruptive digital form (click here, lucky competition winner, and you will be led astray). As a child we encounter them in literature, board games and easter egg hunts, and at school we learn how to fabricate them with coffee grounds to age the edges. As an adult, more of the same: intricately planned dives persist on Pacific coasts where galleons once spilled doubloons, albeit with sonar replacing the leathery chart. We like a puzzle and we like reward, and a treasure map, with its alluring powers to guide, reveal, perplex and make you instantly stinking rich, satisfies fundamental human needs.

If this appears fanciful, examine the archive of treasure maps in the Washington DC's Library of Congress. This contains a detailed list of a hundred-odd guides and nautical charts. It includes the official wreck chart from the New Zealand Marine Department showing twenty-five coastal wrecks between April 1885 and March 1886; and a list of 147 wrecks in the Great Lakes between 1886 and 1891, with descriptions of sunken vessels, approximate location of sinking, and value of unrecovered cargo.

Others are more romantic. There is a 'map of famous pirates, buccaneers & freebooters who roamed the seas during the seventeenth and eighteenth centuries ... from the coast of Central America east to Ceylon'. Or a 'charte of olde Choctawhatchee Bay and Camp Walton, showing those alleged locations of the buried & sunken pyrate booty of Captain Billy Bowlegs, a freebooter, as well as the lost treasures, sunken ships and riches of sundry other pyrates and sea rovers believed to have frequented these waters.'

This latter map, with details of treasure buried between 1700 and 1955, was published in 1956 and was available from 'Mr Titler, Wayside Miss, $1.00 postpaid.' But the key supplier is clearly one Ferris La Verne Coffman, who not only sold individual treasure maps of the Gulf of Mexico and the Caribbean, but also constructed an *Atlas of Treasure Maps*, containing 41 different spreads of the Western Hemisphere with 42,000 crosses for sunken and buried treasure. 'Of these I have been able to authenticate about 3,500,' the indefatigable Mrs Coffman claimed. Yours direct from the explorer, $10.

The catalogue of these maps comes with a slightly comic cautionary foreword from Walter W. Ristow, Chief of the Library of Congress Map Division: 'The Library assumes no responsibility for the accuracy or inaccuracy of the maps, and offers no guarantee that all who consult them will find tangible riches.' He explains that maps should be their own reward, a pleasure beyond greed.

☠ ☠ ☠

Robert Louis Stevenson loved not only treasure maps, but maps in general. He liked the idea of a map, and the feel of its folds in his hands. He liked the naming of maps, and the fact that a map could get you home but also get you lost. And he liked the fact that it would take him to places he had never been before, in reality or in his head. 'I am told there are people who do not care for maps, and find it hard to believe,' he wrote in *My First Book* in 1894, his account of the inspirations for *Treasure Island*. What else did he like about maps? 'The names, the shapes of the woodlands, the courses of the roads and rivers, the prehistoric footsteps of man still distinctly traceable up hill and down dale, the mills and the ruins, the ponds and the ferries, perhaps the Standing Stone or the Druidic Circle on the heath; here is an inexhaustible fund of interest for any man with eyes to see or twopenceworth of imagination to understand with!'

An odd thing about the map of Treasure Island is that it is not the one that he consulted when he wrote his book. The original sketch was forever lost in transit between a post office in Scotland and his London publishers Cassell. 'The proofs came,' Stevenson recalled, 'they were corrected, but I heard nothing of the map. I wrote and asked; was told it had never been received, and sat aghast.' So he redrew it, finding the experience both dispiriting and mechanical. 'It is one thing to draw a map at random … and write up a story to the measurements. It is quite another to have to examine a whole book, make an inventory of all the allusions contained in it, and, with a pair of compasses, painfully design a map to suit the data.' What he was describing here was, of course, the horror facing all students of real-world cartography. He did manage it, and produced the map we have today. His father helped by adding the sailing directions of Billy Bones and forging the signature of Captain Flint. 'But somehow it was never Treasure Island.'

The map he left us resembles the basic shape of Scotland, or at least the shape of Scotland as he knew it to exist in the eighteenth century. But its minuscule size and random details – coves, rocks, hills, forests and 'strong tide here' – suggest a more precise inspiration. One of the sparks may have come from the small island in a pond in Queen Street Gardens, Edinburgh, where Stevenson played as a child. Another may have been Unst in the Shetlands, Britain's most northerly inhabited island, close to the spot where his uncle David Stevenson built the lighthouse at Muckle Flugga in the 1850s to guide British naval convoys safely on their way to the Crimea. But it could have really been any treacherous region off the Scottish coast: his family formed the engineering dynasty known as the Lighthouse Stevensons, and had taken Robert on several stormy trips to inspect their creations. He himself had trained as a marine engineer before his poor health suggested a more sedentary career.

Moreover, Stevenson's engagement with maps was evidently an hereditary affair. Not long after his grandfather Robert Stevenson had constructed the monumental Bell Rock lighthouse off the coast of Arbroath in 1810, he published an 'Account' of his achievement, along with a map of the lighthouse and its surrounding area of rock (also known as Inchcape, and wholly visible only at low tide). Like a conquistador of old, Stevenson Snr was free to identify and name places on the map as he wished, plotting its narrative story much as his grandson would later plot his novels.

So there was Gray's Rock, Cuningham's Ledge, Rattray's Ledge and Hope Wharf – and nearby were Duff's Wharf, Port Boyle and The Abbotsford. There were almost seventy names in all, with every pool, promontory and outcrop christened, mostly with the surnames of those involved in the lighthouse's construction. However, Scoresby's Point was named after Stevenson's explorer friend Captain William Scoresby Jnr, who

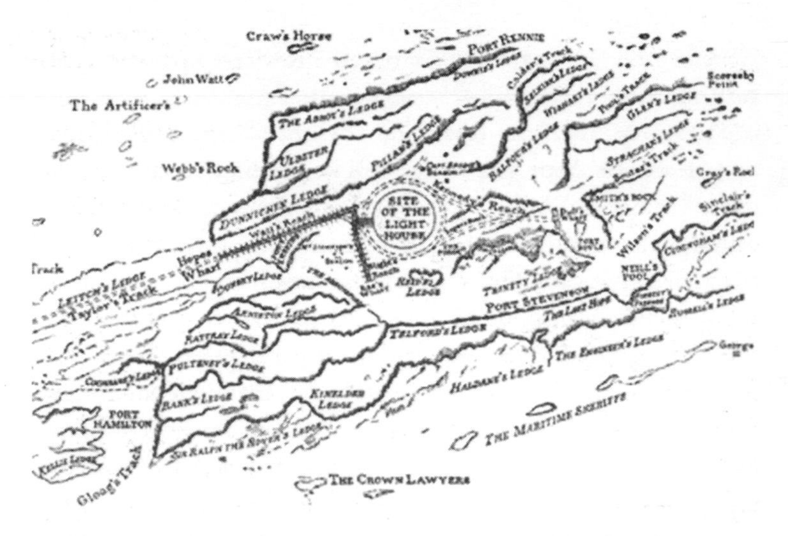

Mapping in the family: Robert Stevenson Snr litters the Bell Rock lighthouse with treacherous rocks named after lawyers.

opened up the Arctic, and on the south-western extremity, Sir Ralph The Rover's Ledge is named after the pirate who once may have carried away the rock's alarm bell. And scattered around the main rock lie other rocks, less visible but equally treacherous, which Stevenson was careful to name after meddlesome lawyers and civil servants.

☠ ☠ ☠

If any of this was known to Edward Frederick Knight in 1885 when he first heard of the Trinidadian treasure, he made no mention of it as he assembled a crew for his own voyage, took along the same map that the *Aurea* sailors had obtained (a copy of the map from Captain P, detailing the shape of the island, where best to land, and where the treasure lay), and set off in his 64-foot teak boat with 600 gallons of water, four sailing professionals and nine 'gentleman adventurers' each paying £100 for the privilege.

And why would we, as latterday readers, not wish him well? As Ralph Paine wrote two decades later, 'to be over critical of buried treasure stories … is to clip the spirit of adventure to a pedestrian gait … The base iconoclast may perhaps demolish Santa Claus, but industrious dreamers will be digging for the gold of Captain Kidd long after the last stocking shall have been pinned above the fireplace.'

The fantasy buccaneers of Stevenson's *Treasure Island* find their hopes dashed; someone had dug up the gold before them. E.F. Knight and his colleagues on the *Alerte* spent three long months on Trinidad in the winter of 1889, most of it spent digging. The map of the island that accompanied his published account the following year does indeed look like a challenging place, craggy from the coast and almost wholly mountainous on land. Along the entire northern and eastern shoreline are marked layers of lava rock ledges. There are no place names on land, but many on its shoreline: the Ninepin, West Point, The

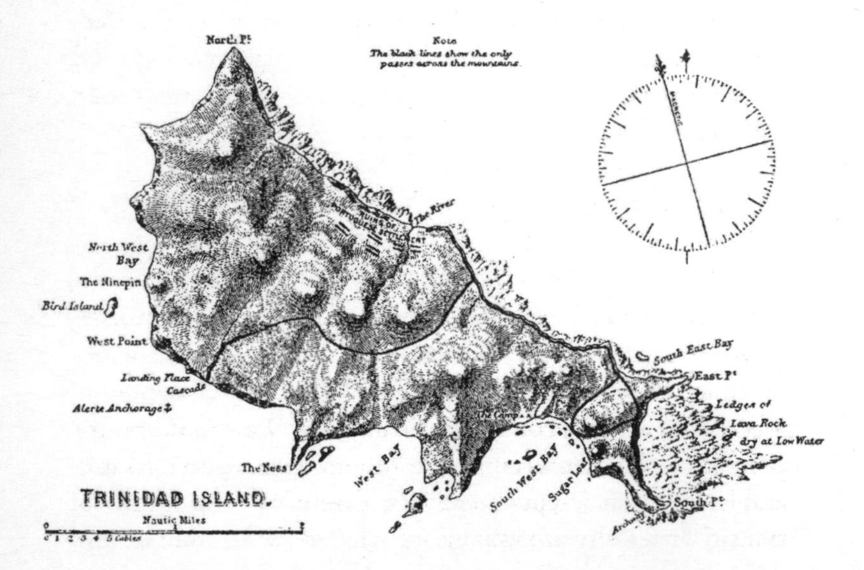

Knight's map of Trinidad from his book, *The Cruise of the Alerte*.

Ness, Sugar Loaf, East and North Point. Knight's anchorage for the *Alerte* is marked to the west, more than two nautical miles from the crew's camp on land, which was another half-mile from the digging site in the shadow of Sugar Loaf. The map is bisected by a red line marking the only passage through the mountains.

Knight had brought with him a drill, a hydraulic lift, many guns, several wheelbarrows, crow bars and shovels for every man, and with these tools they worked at the rock and earth, initially with optimism and finally with desperation. They had landed with difficulty, found the island quite as foreboding as they had anticipated, loathed the sun, erected fences to ward off the land crabs, and complained of 'a wreath of dense vapour' even on the clearest day.

But all would be worth it if they returned rich. The map led them not only to the ravine and remains of a craggy cairn where they were instructed to dig, but also to the rusting tools abandoned after previous attempts. As they dug, fragments of wreckage from stranded ships washed about them. They made trenches, cracked rocks, and began to resemble savages in their torn clothes and roasted skin. After three months they abandoned their quest. They had uncovered nothing and were close to starvation and collapse.

As they sailed for home, Knight took the valiant view, and grew ever prouder of his men, like Long John Silver in *Treasure Island*. Their real treasure lay in human experience; they had followed a map and a dream, and they had returned poorer but wiser. 'We seemed happy enough as we were,' he concluded. 'If possessed of this hoard our lives would of a certainty have become a burden to us. We should be too precious to be comfortable, We should degenerate into miserable, fearsome hypochondriacs, careful of our means of transit, dreadfully anxious about what we ate or drank, miserably cautious about everything.' It is possible he may actually have believed this.

A few years after his voyage, Knight became a war correspondent for *The Times* and lost an arm in the Second Boer War. An account of his death published in the *New York Times* in 1904 turned out to be greatly exaggerated; he lived until 1925. To the end he believed that the treasure of Trinidad was real, and there was only perhaps 'one link in the directions' that had led him astray.

Chapter 14
The Worst Journey in the World to the Last Place to be Mapped

On 10 September 1901, Ernest Henry Shackleton, Robert Falcon Scott and the crew of the *Discovery* decided to make an unscheduled stop on their journey to the Antarctic. They had been at sea for five weeks, had crossed the equator eleven days before, and now, after early problems with sails and leaks, decided to break their trip down the coast of South America with a day's fact-finding on an unknown island. Perhaps their food supplies would be increased with edible birds. At 10am three days later, Scott and his crew climbed into two boats and tried to land. In a subsequent letter to London he noted that the 'curious rocky promontory' chosen for his embarkation point had previously been observed by a man called E.F. Knight.

Shackleton gave no further details of his day ashore, but another voyage member, the second surgeon Edward A. Wilson, was keeping a personal diary intended for his wife Oriana. He was 'called some time before sunrise to see South Trinidad as I had asked. It was worth it … It was a most striking sight, this oceanic island, after so long seeing nought but clouds and sea and sky.'

Wilson's three-page entry on Trinidad – the legendary Treasure Island – is among the longest in his three-year diary, much of it concerned with the difficulty of disembarking and the ease of killing birds once they had done so. Wilson also noted dead and bleached tree trunks all over the slopes, 'white outside, red inside and rotten', which suggested either volcanic damage or some terribly ravenous creature. And then there was his most chilling description of all: 'The whole shore of the island was alive, literally alive, with a big vividly coloured red and green crab, a flat long pointed clawed beast.'

But of course Shackleton, Scott and Wilson were going further.

Antarctica has long been described as the last place on earth to be mapped, and romantically we still like to see it that way. We may never tire of its great and terrible stories, and if they grow more vast, heroic and mythical in the telling then so be it. In cartographic terms the stories are thrillingly recent, and it is odd to think that, a little over a century ago, the continent – all 5.4 million square miles of it – was still predominantly white and silent on the map.

The maps we remember from the age of Shackleton, Scott and Amundsen were not drawn by professionals, and they were not drawn by the polar superstars. Edward Wilson's record of the *Discovery* expedition is the most personal account from all the polar trips, and the most aesthetic. It is full of yearning for the woman he had married only weeks before setting sail, a state

The best known map of Amundsen's route to the pole was drawn by the Englishman Gordon Home from telegraph reports. It appeared in a book by the Norwegian a few months after the event.

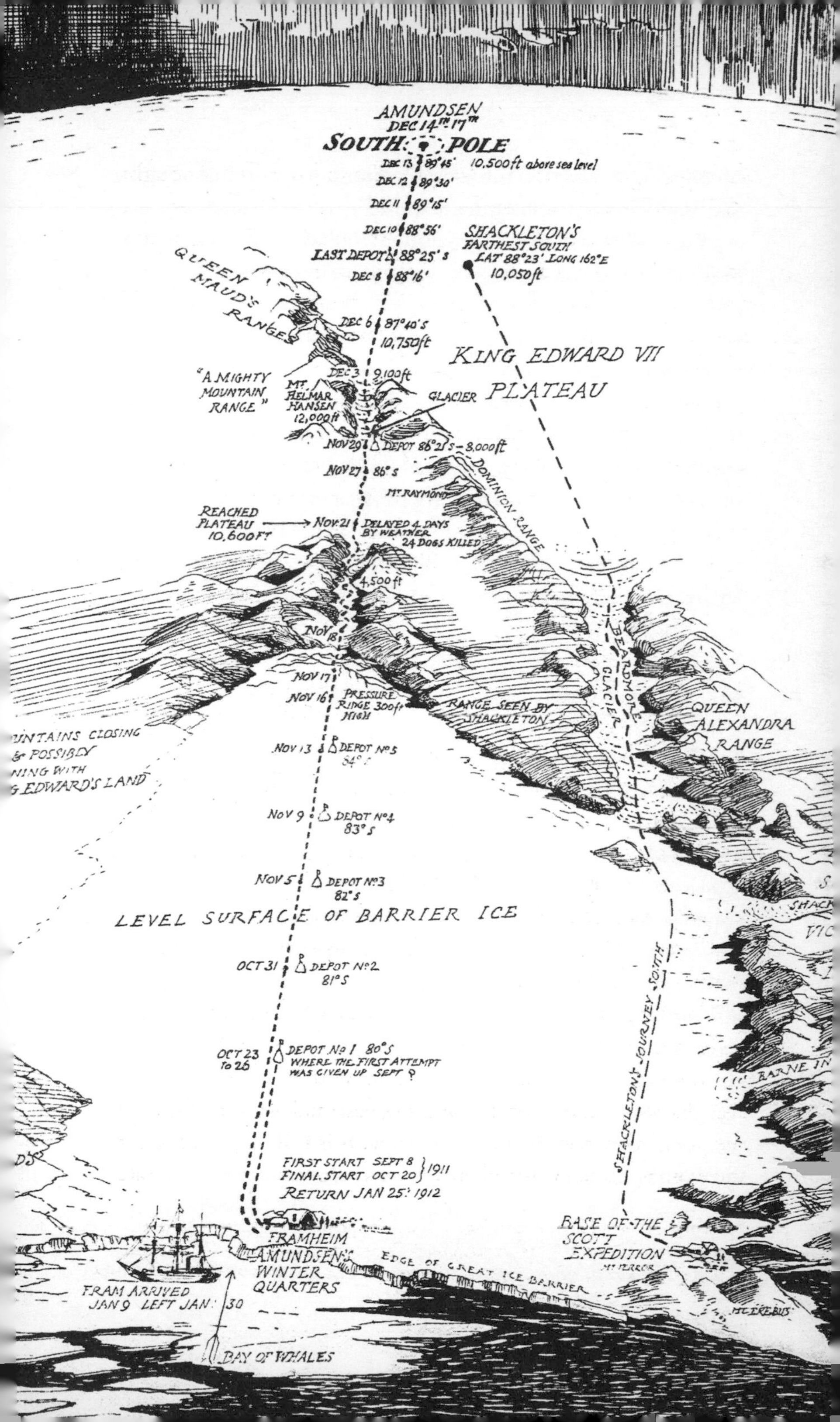
AMUNDSEN
DEC 14TH 17TH
SOUTH POLE
DEC 13 89°45' 10,500 ft above sea level
DEC 12 89°30'
DEC 11 89°15'
DEC 10 88°56'
SHACKLETON'S FARTHEST SOUTH LAT 88°23' LONG 162°E 10,050 ft
LAST DEPOT 88°25' S
DEC 8 88°16'
QUEEN MAUD'S RANGES
DEC 6 87°40' S 10,750 ft
KING EDWARD VII PLATEAU
"A MIGHTY MOUNTAIN RANGE"
DEC 3 9,100 ft
MT. HELMAR HANSEN 12,000 ft
GLACIER
NOV 29 DEPOT 86°21' S - 8,000 ft
NOV 27 86° S
DOMINION RANGE
MT. RAYMOND
REACHED PLATEAU 10,600 FT
NOV 21 DELAYED 4 DAYS BY WEATHER 24 DOGS KILLED
4,500 ft
NOV 18
BEARDMORE GLACIER
NOV 17
NOV 16 PRESSURE RIDGE 300 ft HIGH
RANGE SEEN BY SHACKLETON
QUEEN ALEXANDRA RANGE
UNTAINS CLOSING & POSSIBLY NING WITH G EDWARD'S LAND
NOV 13 DEPOT No 5 84°
NOV 9 DEPOT No 4 83° S
NOV 5 DEPOT No 3 82° S
LEVEL SURFACE OF BARRIER ICE
OCT 31 DEPOT No 2 81° S
SHACKLETON'S JOURNEY SOUTH
OCT 23 to 26 DEPOT No 1 80° S WHERE THE FIRST ATTEMPT WAS GIVEN UP SEPT 9
BARNE IN
FIRST START SEPT 8
FINAL START OCT 20 1911
RETURN JAN 25TH 1912
FRAMHEIM
AMUNDSEN'S WINTER QUARTERS
BASE OF THE SCOTT EXPEDITION
MT TERROR
EDGE OF GREAT ICE BARRIER
MT EREBUS
FRAM ARRIVED JAN 9 LEFT JAN 30
BAY OF WHALES

of mind that perhaps informs his unofficial notes of mishap and inglorious incidents not recorded in, say, Scott's lionised account. Wilson was a keen painter, and his medical and zoological training had made him not only an excellent skinner and preserver of birds, but also an accurate draughtsman.

His maps from the voyage are invaluable. Some are just little doodles within his notes, such as the map of the sleeping arrangements drawn at an ice-sawing camp at the end of 1903. The large tent slept thirty, split into two halves divided by a stove. In one half were six three-man sledging tents, in the other a huddle of single sleeping bags, marked by Wilson with the name of each occupant: Skelton, Royds, Hodgson and the rest. Supply boxes were scattered at heads and feet. The sketch suggests camaraderie, cramp and odour, something confirmed in Wilson's journal: 'There was never a healthier crowd of ruffians, than the 30 unwashed, unshaven, sleepless, swearing, grumbling, laughing, joking reprobates that lived in that smoky Saw Camp.'

Wilson's more conventional maps show the vast distances covered in this barren land. One tracks the route of the *Discovery* from Trinidad at the northwest of the map (as we see it), through Cape Town and then the Crozet Islands, onto Lyttleton in New Zealand before venturing deeper south. The to-and-fro of the winter quarters at the tip of the Great Ice Barrier are clearly marked, as is the southern sledge journey in the winter of 1902/03. The map is a scratchy specimen, hurriedly drawn with no care for posterity. One could mistake it for a spider's web, with tiny flies for landmarks. But as an historical document of a participant in a grand quest it can never be matched. And, of course, Wilson worked under the most trying conditions. 'Sketching in the Antarctic is not all joy,' he noted on 25 January 1903, 'for apart from the fact that your fingers are all thumbs, and are soon so cold that you don't know what or where they are … apart from this you get colder and colder all over, and you have to sketch when your eyes stop running, one eye at a time, through a narrow slit in snow goggles …'

This may explain why his most enchanting map was drawn from memory some seven years after he left the ice. It shows the area around Cape Crozier and Mt Terror, and because it is a map combining jotted notes, memory and imagination, it carries a lot of detail. Wilson was particularly keen to show the work he had carried out with the local inhabitants, marking with crosses an area of Emperor penguins nesting on sea ice, clearly distinguishable from the 'dots that indicate thousands of Adelie penguins in a rookery in an enclosed arena sheltered entirely from sky blizzards'.

The map is torn from a sketchbook and drawn in pencil, and uses classic Renaissance techniques to mark coastlines, cliffs, craters and inlets – contouring, chiaroscuro and cross-hatching that makes the land resemble the palm of a wizened hand. It really does look like a treasure map.

But it had a very clear intention. It was drawn for the next great assault on the continent in 1910, the *Terra Nova* expedition, in which Wilson would again attend in his capacity as doctor and naturalist, this time with Scott as captain. His writing on the map looks like a summons from Jack Hawkins: 'Old glacier of blue ice where we shall cut a cave for an ice house and from where we shall get all our drinking water,' he wrote just below the steep Cape Crozier rock cliffs. Close by was the 'Probable position of our hut on edge of a snowdrift.'

Such careful mapping for such a fateful voyage.

Of course, Wilson and Scott were just adding to what had gone before. We may remember how, in the second century BC, the ten-foot globe made by Crates of Mallus envisioned a quartered world of four gigantic islands split by a torrid ocean. Only one

– his own – was inhabited, but the others were believed to be perhaps equally hospitable. We have also seen that, around 114 AD, Marinus of Tyre, the great inspiration for Ptolemy, used the name Antarctic on his gazetteer of place names to signify the region lying at the polar opposite of the Arctic.

Geology now affords us a greater understanding – or at least advanced theorising – about how it got there. Antarctica may once have been a place of lush greenery and busy rivers inhabited by amphibians and large reptiles. Recent discoveries of a varied range of fossils on exposed rock suggest something closer to Amazonian conditions than icy wilderness, and the possibility of dinosaurs and six-foot-tall penguins. This ties in with the idea that Antarctica was once part of Gondwana, the southern 'supercontinent' that originally comprised South America, Africa, India and Australia. It is thought this once lay near the equator, and gradually began to drift south before being broken apart by the shift in the earth's tectonic plates. South America and Africa drifted off first, while India, Australia and Antarctica continued towards the South Pole, arriving about a hundred million years ago. The break-up continued. India and Australia moved north some thirty-five million years later, but Antarctica remained. It began to ice over between ten and twenty-five million years ago.

In our modern age, no region has been the subject of more conflicting hypotheses. The possibility of a fabulously abundant southern continent took hold in the West in the middle ages, where they were unaware of the (perhaps true) fable of the Polynesian sailor Ui-te-Rangiora, who is supposed to have sailed his canoe to the edge of Antarctica in about AD 650, to be met by a vast frozen ocean. But as *mappae mundi* gave way to true exploration things became a little less alluring, and Antarctica tended to disappear from maps altogether. In 1497, Vasco da Gama's journey around the Cape of Good Hope at the tip of southern Africa disproved once and for all the prospect of this southern continent still being attached to a temperate country.* Then in 1531, the French cartographer Oronce Finé published a famous woodcut of the world in two heart-shaped spheres, notable for showing Greenland as an island for the first time, and for a remarkably accurate estimate of the coast of Antarctica as it would appear to us now if free of ice. This was accompanied by a rather modest observation: 'Not Fully Examined.'

But for the next three centuries the mapping of Antarctica remained a confused mess of conjecture. It was perennially considered part of Terra Australis, a huge shape-shifting area in the southern hemisphere that, at various times, contained Tierra del Fuego, Australia, New Zealand and anything else floating in the Pacific Ocean that sailors encountered by accident. The word 'Australis' was simply the Latin word for 'Southern', and on maps of the seventeenth and eighteenth centuries it most commonly appeared within the phrase 'Terra Australis Incognita' (a shortened form of 'Terra Australis Nondum Cognita', which Ortelius spread over the whole base of his world map in 1570). The South Pole or 'Antarctic Pole' was often located on these maps, but in most of the great atlases by Blaeu, Mercator, Jansson and Hondius it is surrounded not by a white mass of land but a sepia or green one, made up entirely of ocean. The entire continent had seemingly disappeared, and so it remained until Captain Cook suggested otherwise.

* Vasco da Gama's journey round the Cape into the Indian Ocean was the most celebrated but not the first to follow this route. The Portuguese sea captain Bartolomeu Dias had made a similar journey round the tip a decade earlier, but then promptly turned back when his crew threatened mutiny. The news of his route to the Indies was not treated with glee by the Portuguese – indeed, quite the opposite: he had demonstrated just how far one had to travel to reap these potential new rewards. Cartographically, though, Dias inspired a breakthrough. The German map-maker Henricus Martellus took full advantage of the reports of Dias's expedition on a map from about 1490 and makes what may be considered a cartographic joke. It presents the world as it stood on the eve of Columbus's stellar voyage, and it is still based on Ptolomeic principles. But now it also includes an Africa with its southernmost tip unmistakably washed by a navy blue sea, and to emphasise the point he has extended it deep into the painted frame of his map. It's as if he's saying: *This is the News.*

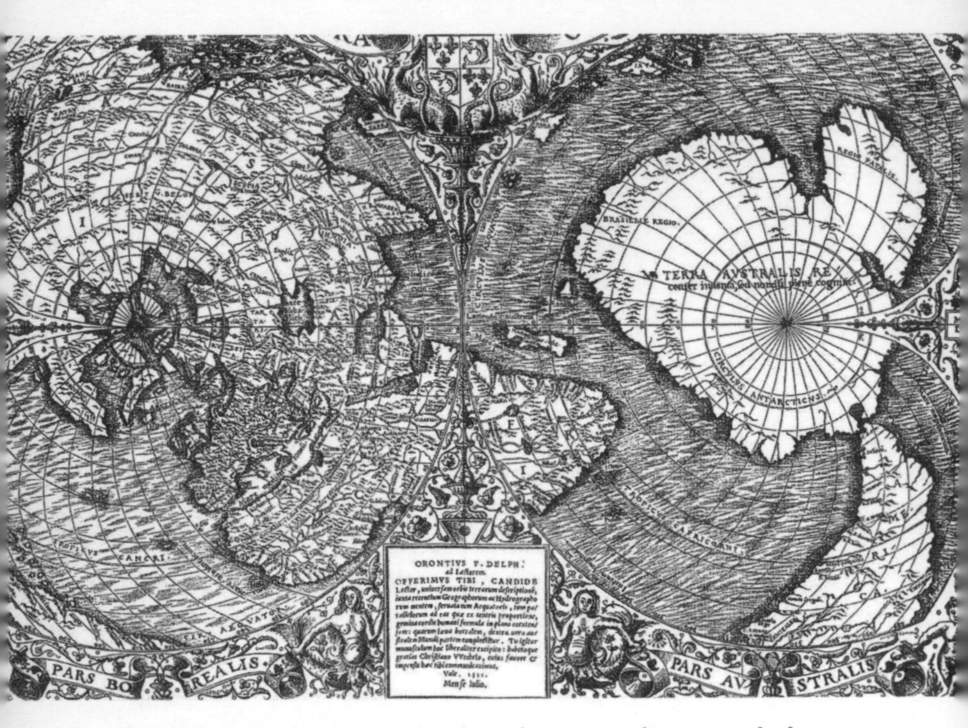

Oronce Finé's remarkably accurate map of Antarctica from 1531, which also shows Greenland for the first time as an island.

We have seen how much explorers and cartographers abhor a blank space, so perhaps we shouldn't be surprised that, faced with a lack of true knowledge, we began inventing things. Mythical islands appeared on the southern base of the map at regular intervals. Francis Drake got there first in 1578, when the *Golden Hind* was driven south by gales and he encountered what was in reality probably the Tierra del Fuego archipelago, which he instead named Elizabethides and claimed for his Queen. But Drake merely set a trend. Between the sixteenth and nineteenth centuries, Elizabethides was joined by Isla Grande, Royal Company Island, Swain's Island, The Chimneys, Macey's Island, Burdwood Island and Morrell's New South Greenland – all of which floated around Antarctica, all of which made their way onto popular maps, all of which were discovered by proud (mainly English) explorers, and none of which actually existed.

Between 1772 and 1775, on his second great voyage, James Cook undertook what may be considered one of the bravest and most brutally elemental sea voyages ever made. He entered the Antarctic Circle three times in dense fog, on each occasion being forced back by pack ice. His voyage was sponsored by the Royal Society, where the Scottish Hydographer Alexander Dalrymple had postulated that Terra Australis was not a hypothetical land, and lay somewhere not far south of Australia. (Cook had sailed to Australia in 1770, replacing the name on contemporary Dutch maps – New Holland – with the British 'New South Wales'.)

There was indeed land somewhere south, but it was far more southerly and far less welcoming than anyone at the Royal Society's London offices had dared imagine. 'I will not say it was impossible anywhere to get farther to the south,' Cook wrote after his third attempt, 'but the attempting it would have been a dangerous and rash enterprise...' He noted the sound of penguins but saw none, and envisioned land somewhere beyond the icebergs. The area spooked him, and he was pleased to be heading north – 'I, who had ambition not only to go farther than any one had been before, but as far as it was possible for man to go.' More than a century later, Captain Scott would sum it all up: 'Once and for all the idea of a populous fertile southern continent was proved to be a myth, and it was clearly shown that whatever land might exist to the South must be a region of desolation hidden beneath a mantle of ice and snow...The limits of the habitable globe were made known.'

Cook never claimed to have actually seen Antarctica, and he did not affirm its presence on the map. Precisely who saw the Antarctic Peninsula first is still open to conjecture, and it may well have been any number of uncelebrated and anonymous British mariners in search of seal fur. Or it may have been Sir Edward Bransfield, an Irish lieutenant in the Royal Navy, and his pilot William Smith. But the written records suggest that the first official sighting, forty-five years after Cook turned for home, involved a mythically grizzled Russian sea captain and a young fur trader from America.

In November 1820, Nathaniel Palmer, a twenty-one-year-old American sea captain, was considered sufficiently experienced by his New England sealing compatriots to pilot a small shallow-bottomed boat in search of new bounty in the Southern Ocean. The area around the newly-discovered South Shetland Islands was yielding such vast quantities of fur and blubber that a trawler with a large enough hull could make a fortune for its crew on just one voyage (if only it could avoid the stunning pale blue icebergs).* Palmer was out on watch duty one night on his boat, *Hero*, when he believed he heard voices in the fog. At first he thought it must have been penguins or albatrosses, but when the fog lifted the following morning it turned out to be the Russian frigate *Rostok*.

Palmer went aboard, and relayed what happened next by letter to his niece. He was ushered into the presence of the ship's commander, Fabian Gottlieb von Bellingshausen. He told him how far south he had been on his journey, and that he had sighted land. 'He rose much agitated, begging I would produce my logbook and chart.' When it arrived, the Russian, speaking through an interpreter, proclaimed:

> 'What do I see, and what do I hear from a boy in his teens – that he is commander of a tiny boat the size of a launch of my frigate and has pushed his way to the pole through storm and ice and sought

* In the 1970s one of these 'tabular' icebergs was reported to be the size of Luxembourg. Another, labelled Iceberg B-15, part of the Ross Ice Shelf, was reported to have been larger than Jamaica.

> the point I, in command of one of the best appointed fleets at the disposal of my august master, have for three long, weary, anxious years searched day and night for. What shall I say to my master? What will he think of me? But be that as it may, my grief is your joy; wear your laurels with my sincere prayers for your welfare. I name the land you have discovered in honor of yourself, noble boy, Palmer's Land."'

And there it sits on the map to this day, a long narrow slice of the lower half of the peninsula not far from the main body of the continent. The Bellingshausen Sea is there as well, but it buffets a westerly area around the largest Antarctic island that Bellingshausen could lay claim to, Alexander Island, which he named after the reigning Tsar.*

The prospect of unclaimed territory was an immensely appealing ambition for almost everyone who followed Palmer in the direction of the pole, and occasionally ambition undid them. Fans of the heroic age of Antarctic discovery will be familiar with the Weddell Sea, the treacherous area of drifting pack ice that crushed Shackleton's *Endurance* in 1915. The man after whom it was named, the British Royal Navy sealing captain James Weddell, is not remembered by colleagues as a universally upright character (he failed to repay his loans), and in cartographic respects he appears to hover between unreliable and fraudulent. His map of 1825 appears to have been copied straight from Nathaniel Palmer, merely changing names to include his sponsors, colleagues or friends. So Spencer's Strait became English Strait, Sartorius Island became Greenwich Island, and Gibbs Island became Narrow Island. These random transformations (there were more than twenty of them) remained on atlas maps for half a century.

* In 1949 von Bellingshausen's story changed, posthumously. Anxious about the possibility of losing out in the post-war land grab, Soviet claims on the Antarctic strengthened to compete with the Americans and British, and Bellingshausen had suddenly seen the peninsula first.

There is also the suggestion that in the reports of his voyages Weddell deliberately overstated just how far south he had travelled, claiming he had sailed six degrees of latitude nearer the pole than anyone before him, apparently making the journey in clear sea without pack ice. The apparent ease of his passage inspired many subsequent navigators to plot a similar course, wherein they encountered only thick, impassable ice and were forced to turn back. Admiralty maps thus changed their description in the 1820s from 'Sea of George the Fourth, navigable' to merely Weddell Sea.

In 1838, the mapping of the continent received a welcome new scientific impetus. At the eighth meeting of the British Association For The Advancement of Science, a discussion was devoted to the dilemma of Terrestrial Magnetism, the pull of the earth's magnetic force that had baffled sailors and their compasses for centuries. Those at the BAFTAS that year believed that Germany was leading the way in magnetic science, and, if unchecked, would lead to unfair dominance in the mapping, settling and trading of distant lands. So a committee was appointed, led by the indefatigable astronomer-photographer-botanist Sir John Herschel. Lord Melbourne, the Prime Minister, was to be kept continually informed, and it was recommended that a naval expedition depart for magnetic observations in the Antarctic seas. According to the meeting's resolutions, the expedition was to focus specifically on the 'horizontal direction, dip and intensity', measured hourly. The further south these observations could be made, the better. But who, among the scientific and naval personnel at the disposal of Her Majesty's Government, was up to this task?

Step forward Captain Sir James Clark Ross, a thirty-eight-year-old English naval officer, hooked on discovery since his teenage travels with his uncle Sir John Ross to find the North-west Passage. In 1831 James Ross was the first to find the North Magnetic Pole, the point where the earth's magnetic field points precisely 90 degrees south, a constantly shifting measurement due to changes in the earth's core. Surely the South Magnetic Pole was within reach too?

In the journal of his voyages between 1839 and 1843, Ross provides a dramatic commentary on his travels, and his magnetic observations are supplemented, almost incidentally, by notifications of great geographic discovery. But before the likes of St Helena and the Cape of Good Hope, an unscheduled stop on 17 December 1839: the island of Trinidad. No sooner had he clambered aboard than Ross observed the strangest phenomenon: his magnetic readings on the island oscillated wildly, with three separate compasses (placed far enough apart so not to affect each other) showing three degrees difference, and none of them accurately displaying what he believed to be the island's true geographical position.

Thirteen months later, in January 1841, Ross made one of his great discoveries – Victoria Land. The map illustrating his find, included in his published journal, provides one of the proudest, greediest, and most egocentric examples of colonialist cartography of the nineteenth century. Throughout the four-year voyage, the crews of the *Erebus* and *Terror* named every new southerly sighting after friends, family, heroes, statesmen and fellow crew members, as if they were cataloguing fossils on a beach. 'A remarkable conical mountain to the north of Mount Northampton was named in compliment to the Rev W. Vernon Harcourt,' Ross wrote in his journal on 19 January 1841. Harcourt was a co-founder of the British Association, and the mountain just to the south of it was named after Sir David Brewster, the other co-founder. And mountains nearby named Lubbock, Murchison and Phillips were named after the BA's treasurer, general secretary and assistant secretary.

There was nothing unusual in this practice, but it is rare to find an area of new geography so full of civil servants. One requires very keen eyesight to decipher them all on the original map published with his memoirs, the names written tightly on both sides of the coastline, capes on the left, mountains on the right. Even then we are looking at text resembling hairs on an arm: from Cape North we go vertically to Cape Hooker, Cape Moore, Cape Wood, Cape Adare, Cape Downshire, Cape McCormick, Cape Christie, Cape Hallett and Cape Cotter.

This is how maps were made: you see it and it's yours (or at your friend's, or the person who sent you on this mission). There are occasional loved ones too: 'This land having been thus discovered … on the birth-day of a lady to whom I was then attached,' Ross wrote on 17 January 1841, 'I gave her name to the extreme southern point – Cape Anne.' Romantic cartography, like romantic tattooing, can be a self-indulgent art, not to say a risky one: in this case, Anne did become his wife. In general, women got quite a good deal in Antarctica: Queen Mary Land, Princess Elizabeth Island, Queen Alexandra Range. Adelie Land and Adelie penguins were named for his wife by the French explorer Jules Dumont d'Urville, while Marie Byrd Land, a large chunk of West Antarctica, was named by the pioneering Antarctic pilot Admiral Byrd for his wife in 1929.

Names on the Antarctic map illuminate more than just devotion; they also show fear and disgust. Despair Rocks, Exasperation Inlet, Inexpressible Island, Destruction Bay, Delusion Point, Gale Ridge and Stench Point; at these southerly points, the explorers' scales had certainly fallen from their eyes. And we leave James Clark Ross in another bleak region – the Ross Sea and the Great Ice Barrier (later called the Ross Ice Shelf) – the area that rendered him immortal and froze the minds of all who followed him in the valiant and tragic subsequent British expeditions to the pole.

Apsley Cherry-Garrard set off on the ill-fated Terra Nova expedition with Scott and Co. Ltd in 1910, at the age of twenty-four, and he returned three years later almost blind and toothless, with post-traumatic stress disorder and depression to come. His stated role on the mission was principally to gather penguin eggs and supply the depots, but he had other uses too: he was indefatigably cheery, and he was, it turned out later, a brilliantly dramatic chronicler of events. *The Worst Journey in the World* (1922), his classic account of the expedition, still chills, and did much to realign our romantic vision of the age of ice and heroism.

Cherry-Garrard writes that one of the books passed around on the return voyage was a biography of Robert Louis Stevenson; and he was also aware of the quests of E.F. Knight. The first major adventure in his account, some forty days after setting off from Cardiff, concerns yet another landing at South Trinidad (Scott's second visit and his last), and he writes with displeasure at the land crabs and with anticipation of 'a very thorough search of this island of treasure'. Scott and his team were there partly for sport (they killed a lot of terns and petrels) and partly for research (they bottled a lot of spiders and labelled them for the British Museum). The island almost gets its revenge on the visitors when, on departure, many in the party are swept away onto rocks by huge waves; for a tempestuous while this looked like the end of the Antarctic party long before they tackled the pole, and they were only just rescued by ropes.

Cherry-Garrard's account is notable to us for another stark reason – three sketched maps. The first, showing McMurdo Sound, contains useful indications of points in the author's story, including a Fodder Depot, Safety Camp and Rescue Camp, but is otherwise unremarkable. The second shows the route of the winter journey from Cape Evans to Cape Crozier and back in search of an unhatched Emperor penguin egg. This was the 'Worst Journey' of the title, a dotted line with dates attached (June 28, July 15, August 1): five weeks of unyielding misery as three men hauled heavy supplies against raging blizzards and their chronicler experienced 'such extremity of suffering' that 'madness or death may give relief'.

But the last map is the one we remember – the terrible return trek from the South Pole to the 'safety' camp by McMurdo Sound, the camp that Scott's party didn't reach. A long dotted line passes over mountains and glaciers, and it seems to dominate them; for the first time, the route appears to be in charge of nature. But look closer and we see this is not the case – it is the line of a funeral procession. Every now and then there are familiar names marking landmarks that had not existed before: 'Evans Retd' is one, a small nick on the route between two supply depots. About 250 geographical miles further on (the scale is unreliable) we reach another nick by the Beardmore Glacier: 'Evans Died'. About 250 miles more there is another vertical nick, this one supplemented with a slightly longer horizontal line, the international sign of a cross for a grave or a church, and it says simply: 'Oates'. Eleven or so miles north along the line is another cross, marking the final resting place of Scott, Wilson and Bowers (Cherry-Garrard was in the search party that found the bodies six months after they died). On the map it just says 'Tent'.

Paul Theroux has made the point that great explorations demand fine writers to bring it all home – the fierce desperation, the unbound elation, the emotional

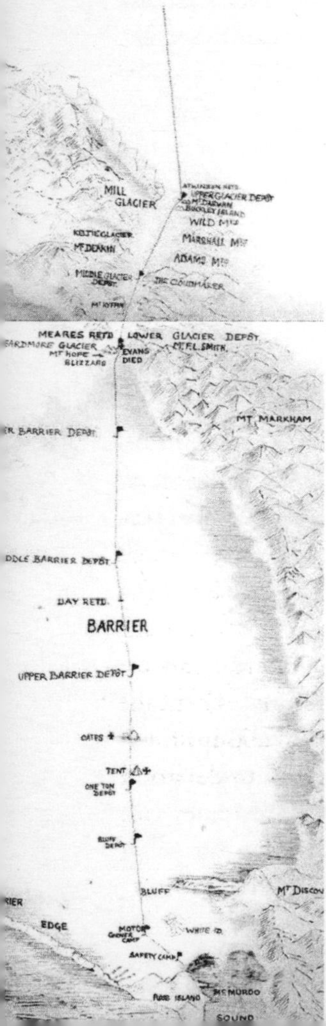

A long white trail of despair: Cherry-Garrard's map from 'The Worst Journey in the World'.

and humane mixed with the procedural. This explains why we know what cold feels like, but we don't really know what it's like to walk on the moon. A good sketched map delivers a similar bounty. We may detect the emotional state of the amateur cartographer through the graphite and nib of hand-drawn markings, and because we know we are witnessing history as it happens. In the introduction to his book, Cherry-Garrard expressed a sense of duty to pass on to the next set of adventurers as much systemised knowledge as possible to aid them on their travels, just as Cook had passed to Ross and Ross to Shackleton and Scott. Cherry-Garrard maintains that 'exploration is the physical expression of the Intellectual Passion', and the gradual filling-in of the maps is the most direct and literal way of reflecting progress in this field.

In December 1959, the map of Antarctica was made anew. Or rather it was settled anew, after twelve countries signed the Antarctic Treaty in Washington DC and agreed to use the continent for scientific and peaceful purposes. Weapons testing and nuclear waste disposal were banned, and the free sharing of information was encouraged, and when the treaty was renewed on its 50th anniversary, 36 other countries had agreed to its terms. Between 1908 and 1940, seven countries staked territorial claims to the land (Argentina, Australia, Chile, France, Great Britain, New Zealand and Norway), and the treaty officially refused to recognise or dispute them. When occasional spats – usually between Britain, Argentina and Chile – flare up over the possibility of exploiting the natural resources beneath the ice, the land-grab map is dusted off to reveal about fifteen per cent of the pie left unclaimed. It is widely understood that the United States will make a claim should the treaty fail, although it already seems to run the place.

In 2002, 165 years after Ross began to put the region on the map, and 93 years after Amundsen and Scott reached the Pole, the Americans built a permanent road there. Stretching some 1,400 km from McMurdo Sound to the Amundsen-Scott research station at the pole, it is a road through half the map of Antarctica. The South Pole Traverse is a strip of packed ice marked by flags, the coldest man-made roadway on earth. A trail of huge-wheeled vehicles pulls sleds of food, medical supplies, waste, communication cables and visitors, and since becoming operational in 2008 it has saved an estimated forty flights a year. This being an American enterprise, the traverse has another name too: the McMurdo-South Pole Highway.

It takes about forty days to make the trek. No dogs, horses or possessed explorers die en route, and those who passed this way less than a hundred years before would have perished from shock if they had been told of such a thing. A road from the tip of Victoria Land across the Ross Ice Shelf; a road connecting a runway and helicopter pad with a permanently manned research centre resembling a small town, where its scientists are engaged year-round in glacial geology and the monitoring of the drifting ice sheet, astrophysics and the monitoring of ozone, and a slightly less academic science known as the vigilant disapproval of tourists.

A little more than a century after Amundsen and Scott, Antarctica tops the wealthy traveller's bucket list. It's an expensive trip: the clothes alone will cost as much as a summer in the Med, and you will lose your dinner many times over as you chop your way south of Argentina through stormy Drake Passage. But more than 20,000 visitors a year now visit a continent that was once thought unattainable. Many who make the trip report not just on the cold, the penguins and the incredible light, but echo the views of one polar explorer – Robert Swan, who walked to the South Pole in 1985–86 – who observed that the experience wipes the slate clean, like a child's magic drawing pad.

Despite the influx, there is still romance to be had. When I met the writer Sara Wheeler for tea one day to talk about Antarctica (the book of her adventures, *Terra Incognita*, is one of the modern classics of polar literature) she brought along one of the paper maps she had in her backpack as she made her trip to the South Pole. Produced by the US Geological Survey, the map is a topographical collection of data around the Taylor Glacier near Victoria Land, an area first explored by the British expedition of 1901–04.

'It was one of the happiest days of my life,' Wheeler told me. 'I was following the map with my finger, trying to locate exactly where I was, and I came to this point here …' She unfurled the map, and half of it was a big white blank. The words marking the end of this cartographic endeavour simply read, 'Limit of Compilation'. 'I'd reached the end of the map,' Wheeler said with delight.

But she was travelling in the 1980s, and that map was from the 1960s. Thanks to satellites, Antarctica is all mapped now. The region is still predominantly unexplored, but the satellites have seen it, and the frozen wastes have digital coordinates. Perhaps we cling to a romantic notion of Antarctica because we are responsible for transforming it from the last great unmapped place on earth into something dotted with old huts and new research stations, and we face the sobering fact that much of the recent research conducted there points towards environmental disaster. The map is no longer white, and the challenge is no longer to reach the continent but to save it.

Pocket Map
Charles Booth Thinks You're Vicious

Are you vicious? Do you lurch? Have you ever thought of yourself as semi-criminal? Or are you just purple going on blue?

If you lived in London in the 1890s, Charles Booth, creator of the London Poverty Map, had a category for you – for what you were depended on where you lived. If you lived in a nice neighbourhood such as Kensington or Lewisham then you probably lived in a street coloured yellow and designated 'Upper-middle and Upper classes. Wealthy'. If you lived in Shoreditch or Holborn, you might well have a street address edged in black ('Lowest class, vicious. Semi-criminal.').

It was a slight, of course, and a slight generalisation. But that's morphological mapping for you, and it was just this mapping that changed the lives of millions.

Charles Booth was born in Liverpool in 1840, which meant he was perfectly placed to witness the effects of industrialisation on a city that didn't have the social infrastructure to cope with it. When he took the new steam train to London the picture was even more extreme: those who had been made wealthy by mass manufacture and foreign trade were erecting fearful mental barricades against those whose lives had seemingly gone backwards in the rush. The well-off had begun to segregate themselves in cities like never before, and swiftly became reliant on the new police force to maintain order. But just how big was the problem of the poor? And did domestic squalor necessarily lead to social disorder?

Influenced both by the Quaker philanthropic zeal of Joseph Rowntree and by his wife Mary's experience of deprivation in the East End, Booth decided to find out. And as the President of the Royal Statistical Society, he was clearly well placed.

Booth had studied the census from 1891, and had broken down the figures on earnings and dwellings into conclusions that at the time established a completely new understanding of how poverty influenced a geographical area. Then he went further, and suggested that where you called home may influence not only how well you lived, but also how well you behaved. A multi-volume report of Booth's work was full of notes, tables and jagged graphs, and encompassed not just poverty and housing,

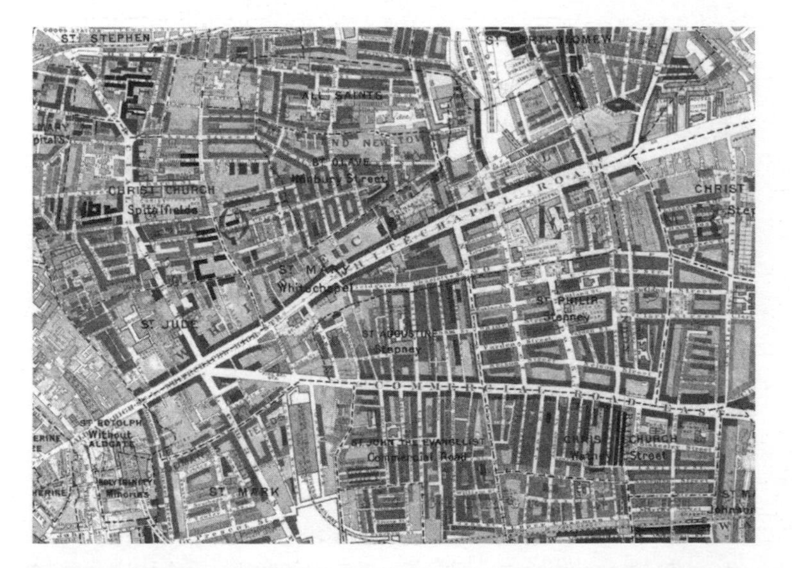

London's East End – well-to-do on the main roads but distinctly dodgy in the back streets.

but also industry and religious influences. Yet he knew from his earliest statistical work that the impact of his research rarely reached those directly affected by it. So he published his findings as maps.

Booth obtained the latest Ordnance Survey charts (on a scale of 25 inches to a mile), and instructed his assistants on hand-colouring. The streets of his first map of Tower Hamlets had six colour coded categories, but the large-scale map of London had seven:

Black: Lowest class. Vicious, semi-criminal.
Dark blue: Very poor, casual. Chronic want.
Light blue: Poor. 18 to 21 shillings per week for a moderate family.
Purple: Mixed. Some comfortable, others poor.
Pink: Fairly comfortable. Good ordinary earnings.
Red: Well to-do. Middle class.
Yellow: Upper-middle and Upper classes. Wealthy.

Some streets contained a blend of Booth's colours, but his findings were still stark. Just over 30 per cent of London's population was shown to be in poverty. His methodology set the tone for a new form of urban cartography that amplified a specific form of information in an aesthetically compelling way. But there was something else about Booth's maps: they made it look as if the city was moving, not unlike like a live traffic stream today. They weren't just about topography or navigations – they were about people.

The maps were first displayed at Toynbee Hall in the East End where Booth lectured, and they received instant acclaim. The *Pall Mall Gazette* called him 'a social Copernicus.' A closer look revealed far more than the dissection of London's rich and poor. They showed that the middle/merchant classes grouped around the large thoroughfares into the city – Finchley Road, for instance, as well as Essex Road and Kingsland Road. Extreme poverty settled by railway yards and canals, as well as cul-de-sacs and alleys; the common wisdom had it that criminal classes would find these labyrinths easier to hide in and ambush intruders. Nor did you want to live – or venture close to – the docks around Shadwell or Limehouse, areas we now would regard as warehouse-hip and 2012 Olympic.

Booth continued to expand and update his map coverage until 1903. He didn't work alone, and his many assistants gathered information from many sources, particularly school board inspectors, 'worthy' locals and the police. The descriptions that accompanied the maps were both startling and compelling. Chelsea, for example, was predominantly blue to black, its houses described as predominantly damp, crowded and peopled with 'lurchers' who never pay rent. Westminster was dark blue, a dirty, bad lot. Greenwich, red, was a little more des res, teeming with caretakers, police sergeants and works inspectors. The reports also detected what we may now call gentrification and the reverse, the formation of slums. As Booth put it colourfully, 'The red and yellow classes are leaving, and the streets which they occupied are becoming pink ... whilst the streets which were formerly pink turn to purple and purple to light blue.'

Booth's reports on the black and blue areas were less about poverty and more the degrees of crime. In the Woolwich 'Dust Hole' for example, blue and black on the map, the police refuse to attend incidents unaccompanied, and find that 'missiles are showered on them from every window when they interfere.' Elsewhere in the darkness, Borough High Street seemed to come straight out of *Nicholas Nickleby*: 'Youths and middle-aged men of the lowest casual class loafing. Undergrown men. Women slouching with bedraggled skirts. A deformed boy with naked half-formed leg turned in the wrong direction ...'

Booth's colour-coding had many limitations, as he himself acknowledged, not least condemning nameless inhabitants on account of their neighbourhood; it did nothing for the ghettoisation of the Jewish and Irish populations in the East

End. But the maps did lead to reform and the improvement of lives. In 1890, a year after his first map appeared, the Public Health Amendment Act prioritised the local provision of water and sanitation, while The Housing of the Working Classes Act of the same year enabled local authorities to purchase land for improvement and thus initiate slum clearance. Two of the causes Booth identified as the cause of poverty are so obvious to us now that they seem banal – low income and unemployment. But the third was more shocking in its generalisation: old age. Booth saw this last cause as the easiest to improve, and the statutory introduction of a non-contributory pension in 1908 owed much to his campaigning.

Booth's cartography had solidified a fairly novel theory in the way we live our lives, namely that where we live does indeed determine how we behave. The layout of the city – its morphology – was itself a prime cause of misdemeanour. Booth advocated the provision of more open green spaces and the eradication of the cul-de-sacs, courts and alleys – a prime force in the fairly new concept of urban planning rooted in social justice.

One looks at the Booth maps today (and they have a fine, searchable website) with a mixture of disbelief and awe. Has there ever been a finer depiction of a more vibrant city? Has there ever been a map where the population it portrayed gazed upon it with so much anxiety?

Chapter 15
Mrs P and the A-Z

In September 2006, a story appeared on the BBC News website celebrating the life and work of Phyllis Pearsall, the woman who made the London A-Z. Had she lived beyond her eighty-nine years, Pearsall would have been 100 that month, so it was a good opportunity to retell the story of her struggle to create what would become an iconic brand.

'Creating the first A-Z was a tough job,' the article began. 'Before satellite imaging or extensive aerial photography, Pearsall worked 18-hour days and walked 3,000 miles to map the 23,000 streets of 1930s London ... Her completed map was rejected by publishers, so she ran off 10,000 copies and sold them to WH Smith.'

It didn't happen quite like that, but never mind – empires are founded on less. And in Pearsall's semi-fictionalised biography, by Sarah Hartley, there is an even more romantic and far-fetched tale. One evening, Hartley writes, Pearsall was in her bedsit near Victoria getting ready to leave for a dinner party at the home of Lady Veronica Knott in Maida Vale. There was torrential rain outside and a power cut, forcing her to get dressed in the dark. Outside, her umbrella blew inside out and when she found a bus she alighted at the wrong end of Harrow Road, requiring a long walk. When she arrived, her fellow guests spoke of how tricky it was to negotiate one's way around London even in fine weather in daylight; only taxi drivers knew for sure where they'd end up. 'This conversation would nag at Phyllis all through the remaining duck and brandied-plum courses, and then through the night,' Hartley suggests. 'The very next day, she became determined to find a street map of London.'

At Foyles bookshop she found that the last Ordnance Survey map of London had been published sixteen years before, and called up her father in New York to say she was going to produce a new modern one for simple people like her. Her father said this was best left to the professionals. At the age of thirty, Phyllis Pearsall set out to prove him wrong. She began her quest – and the trudge along the city's 23,000 streets – at dawn the next morning.

In Pearsall's own account, *From Bedsitter to Household Name*, she describes a more sober beginning. There are no dinner parties and no rain storms, just a decision in early 1936 by a map-making colleague, the draughtsman James Duncan, to delay the publication of a new map of the United States in favour of an expanded map of London. This would cover the outer suburbs, and would base itself on the current Ordnance Survey maps (Pearsall recounts queuing up at Stanfords, the map and travel store, to buy them). As Duncan redrew and added to the maps, Pearsall visited thirty-one borough surveyors for their latest plans, as well as local estate agents. If she found discrepancies she would confirm a map's accuracy by 'checking on the ground' – ie a bit of walking. Her biggest task was indexing. She used file cards in shoeboxes, but disaster struck one summer's day when a pile of 'T's were pushed out of a window onto High Holborn.

Her map complete, Pearsall printed 10,000 and began the task of finding shops that would sell it. She found it impossible to push through the crowds at Hatchards in Piccadilly as Queen Mary had just entered. At Selfridge's she wouldn't be

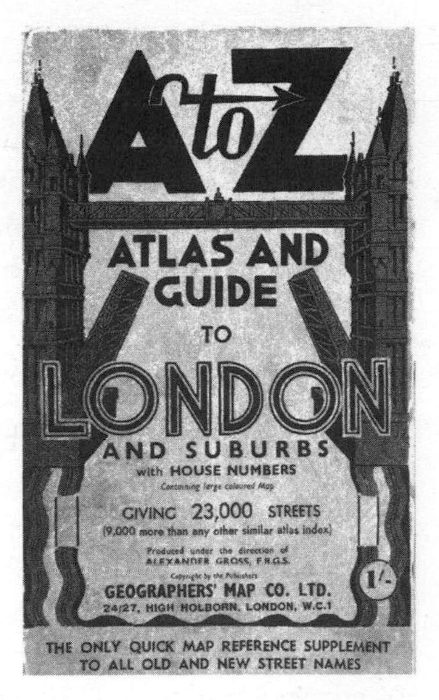

seen without an appointment. At Foyles, according to Pearsall, Mr Foyle looked at her map and said 'The map trade has been undisturbed for years … We're not going to let someone new upset it.' At Barker's she was asked what WC in her address stood for. Pearsall replied 'West City?' 'West Central,' her inquisitor replied. 'An inaccurate map publisher! Good day!'

Which just left W.H. Smith. Pearsall impressed a buyer at the head office in Holborn, who ordered 1,250 copies. She took them round in her wheelbarrow later that afternoon. She asked the buyer whether he thought they would sell them. 'If anybody thinks he knows what'll sell,' the buyer replied, 'he doesn't know the trade.'

But they did sell, and within weeks she was fulfilling orders from every railway station bookstall in the south of England. F.W. Woolworth took a few thousand as well. By 1938 the *London A–Z* was on the map.

For the Pearsall family it was pretty much business as usual. They had been involved in map-making since before the First World War. Phyllis's father Alexander Grosz, a refugee from Hungary, began his sales career around 1905, selling oil lamps and electric bulbs in Brixton. He did well, expanding to other premises, although a diversion into pornographic postcards was shut down by the police.

His brother, Frank Grosz, an amateur cycling champion, was also a salesman, distributing various companies' maps, atlases and globes to shops. When Alexander sold his lighting emporium to the expanding Southern Railway, he sunk the money into supplying his brother with new maps, becoming a publisher in the process. His first, which he varnished and mounted on rollers, was of the British Isles, 'produced under the direction of [the newly anglicised] Alexander Gross.'

In Phyllis Pearsall's privately published biography of her parents, she describes her father as domineering, boorish and radical. But his self-belief got him far. In 1908 he met Baron Burnham, owner of the *Daily Telegraph*, and the two began talking about the Balkans. 'Have you ever thought of reproducing maps to pinpoint news?' Gross asked him.

'No newspaper has ever tried it.'

'Then why not you? The Ottoman Empire is crumbling, The Balkan States are on the verge of revolt. I could supply you with the maps you need. At a moment's notice.'

Burnham agreed, and Gross supplied the *Telegraph* with a map of Bulgaria after Prince Ferdinand declared himself Tsar; Bosnia-Herzegovina when it was annexed by Franz-Josef; and a map of Crete when it gained union with Greece. With a five-year contract, Gross moved to Fleet Street, where he set up and expanded Geographia (named, alas, not in honour of Ptolemy, but after a photography shop in Berlin).

Beyond newspapers, Gross achieved a glowing reputation among pilots. With aviation still in its infancy, and the preserve primarily of keen amateurs, all modern maps were in demand, and Gross made frequent trips to Hendon aerodrome to discuss the pilots' needs (they were particularly keen on the large objects identifiable from the air, such as railway tracks). On one visit, he took along his young daughter, Phyllis, and a pilot lifted her into the cockpit for a spin. Gross refused to let her fly, lifted her out, and watched as the pilot took off, lost control of his plane after an explosion, and killed himself on impact.

In 1911, another disaster. Gross's maps were often produced in a great rush, and contained many errors. Their shortcomings were exposed when the *Daily Mail* offered a £10,000 prize in its Thousand Mile air race over England and Scotland. News of Gross's maps had spread to France, and one of the principal competitors, Jules Vedrines, had bought many of them for his challenge. After a great start, Vedrines lost the lead at Glasgow and told the press that he could not find his landing: 'It was wrongly marked on my map!' At Bristol it was the same story: 'Again my map was to blame. The landing place was marked on the right of the railway ... but it was on the left.'

But then there was a resounding success. The Balkans War of 1912 led to huge demand for a new *Telegraph* map of the region, and Geographia produced one to reflect the contested borders in record time (including a panicky last-minute re-etching of 'Meditterranean' to scratch off the extra t). Gross then began to make world maps for the *Daily Mirror* and a popular cartoon map showing an embattled Germany at the mercy of the British and Russians. With the income from his cartography, he moved his family to Hampstead Heath.

But when the First World War ended in 1918, Gross found the bulk of his map stock redundant. Only one publication fared well – his pocket *London Street Guide*, for sale initially at tuppence and then sixpence. And then, in a not uncharacteristic moment of hubris, Gross announced plans for a wildly over-ambitious new world atlas. In this, unlike with his newspapers, he faced serious competition. It led him to bankruptcy and flight to the United States.

Phyllis Pearsall considered herself a painter, but she was also determined to re-establish her father's cartographic reputation. Her London A-Z was more comprehensive and accurate than anything produced by the rival companies – Bacon, Bartholomew or Philips – though they all had the same base map: the Ordnance Survey.

The title was a stroke of genius. Her father had sent her dismissive telegrams from New York: 'Call the Street atlas The OK', Pearsall recalled. 'But involved day and night as I was in placing 23,000 cards of London streets into alphabetical order, A to Z seemed to me the only possible title.'

When his daughter handed him the finished book he began looking for mistakes. 'How could anyone in their senses leave Trafalgar Square out of the index?' he asked. Pearsall explained about the 'T' index cards falling out of a window; despite a dash to the street below, those that had fallen on car roofs were never found.

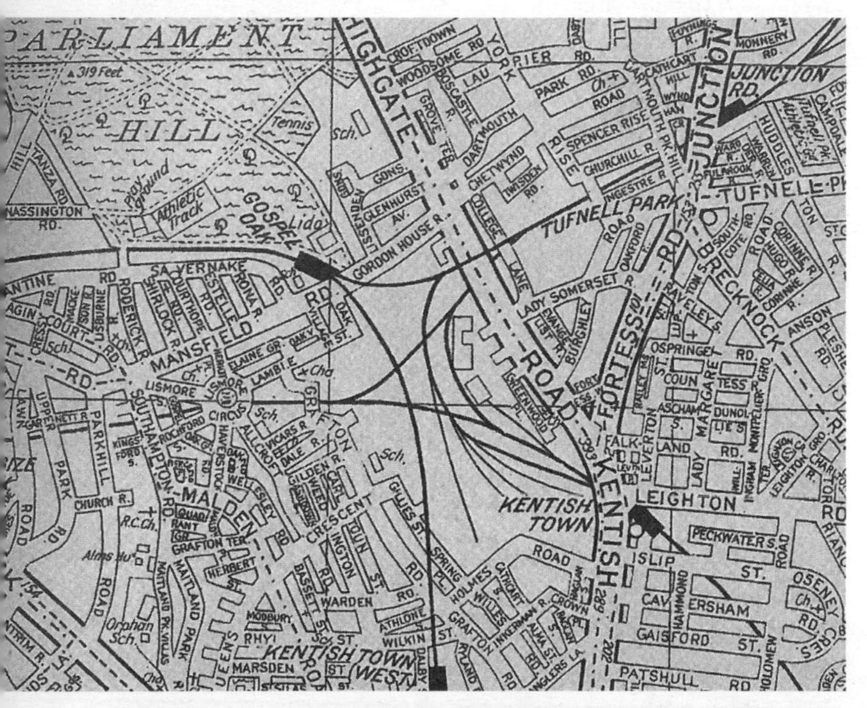

All of London, mapped and indexed in the first *A to Z* – but you'll be wanting a decent pair of specs to find your way.

The first A-Z was printed in black and white with a cumbersome title framed within a drawing of an open Tower Bridge: 'A to Z Atlas and Guide to London and Suburbs with House Numbers Containing Large Coloured Map Giving 23,000 Streets (9,000 more than any other similar atlas index).' The 'to' between the A and the Z was not yet a dash, but a neat linking arrow. It cost one shilling.

The guide contained no introduction, legend or glossary, and most of the text on the map strained your eyes. Still, the early editions did have one fascinating feature reflecting the rapidly expanding capital: Pearsall had obtained a list of some 2,000 name changes from the London County Council. You used to live in Albion Street, E3? That was now English Street. Broker's Alley WC2? That was now Shelton Street. Many of the updates simply replaced confusing duplicates: there were, for example, five Caroline Places, which became Sally Place, Carolina Place, Donne Place, Caroline Walk and Mecklenburgh Place. The ten Charles Streets became Aylward Street, Scurr Street, Greville Street and so on. Some new municipal projects had never featured on maps before, such as Lambeth Bridge, completed in 1932. And whole areas that had appeared as slums on Charles Booth's maps had been replaced by new streets that had not featured on a map before.

On the back of the revised map from 1938/39 was a list of all the other London publications already available from Pearsall's company: *The Premier Map of London*, *The Standard Street Guide to London*, *The Ever Ready Guide to London*, and others. Some of these were fold-out maps, some had photographs, and none covered the suburbs. But they do suggest that the A-Z was a natural development of these, as maps always are: building on what has gone before with a particular purpose in mind.

In 1939, Geographia's plans for an 'All In One War Map' were curtailed when the government made it illegal to sell maps with a scale of or under one inch to the mile. Pearsall got a job in the Ministry of Information, which enabled her father to claim that her daughter worked directly for Winston Churchill.

When the war was over, with paper restrictions still in place and London full of overseas troops, Pearsall placed an order for A-Zs to be printed in Holland. She was injured in a plane crash on her return from the printers to London, but her recovery was cheered by the unprecedented demand; the run of 250,000 copies sold within months.

The maps were updated every five years, regional versions appeared throughout the UK, colour was introduced in the 1980s (it was really the colour that made the maps iconic), and at its peak in the 1990s Pearsall's company was selling about a half a million maps a year. Both the A-Z logo and the distinctive look of the maps themselves achieved – by longevity and utility – the sort of brand recognition that results, inevitably, in teeshirts, mugs and an unspoken feeling of London pride. In a poll for London's Design Museum and BBC's *Culture Show*, the A-Z took its place on a list beside the Mini, Concorde and the Underground map. And of course in the biography of its founder that accompanied this accolade on the Design Museum website, Phyllis Pearsall is walking 23,000 streets for eighteen hours a day.

To get to the offices of Geographers' A-Z Map Company Ltd in Borough Green, Kent, one takes a train from London's Victoria Station and walks for about a minute. One doesn't need a map at all. The low building is neither a remarkable nor an attractive one, but the walls inside are brightened by some pretty watercolours of rural landscapes. They were painted by Phyllis Pearsall, and alongside the English idylls are examples of her other work – sketches of people hard at work on maps.

One of the sketches, from the early 1990s, drawn a few years before she died, shows the company's first computer in the corner of the drawing office, looking lonely. There are also two large armchairs entirely covered in material showing a London A-Z map, and a man called Norman Dennison who jokes that he is almost an exhibit himself, having worked at the company for forty-five years, most recently as joint managing director.

Like everyone at the company, Dennison called Phyllis Pearsall 'Mrs P'. 'You could near her coming right from the other end of the corridor. She could be difficult. The idea of the A-Z was to fund her love of painting really, and she wanted to re-establish her father's reputation. She was very poetic. She loved all the names, like Bleeding Heart Lane [in Holborn]. 'I think she did get up at 5 am and walk a lot,' Dennison maintains, 'but perhaps not 23,000 streets. Recently I had to go to a school in Dulwich, near to where she was born in Court Lane Gardens. They've named one of the school houses after her – Pearsall House. I always say: "In 1936 there weren't any easy maps to use, and she got lost on her way to a party one night, and came up with the idea of the A-Z and she walked the 23,000 streets." They do like to hear that.'

Dennison says that the company's best year for sales was 2004. When the new edition of the A-Z came out in that year the number of streets had grown from 23,000 to more than 70,000, and the company was also selling about 350 other publications – a huge range of city maps and atlases in many formats, a lot of A-Z branded merchandise.

But in September 2008 there was a new addition to the range: an *A-Z Knowledge Master* sat nav, the melancholy acceptance of a changing landscape. It included more than 360,000 streets, postcodes, and Point of Interest entries, and had all the usual madness: Nokia Smart2go software, 3-D vistas, speed camera alerts, an SD card covering the boulevards of Europe. The unit, which cost almost £300 (compared to the standard dead-wood edition at £5.95), was aimed primarily at London's cab drivers.

'You could hear her coming from the other end of the corridor...': Mrs P at the Geographia HQ.

'It's slowly gone down and down,' Dennison says three years later, referring to his company's paper map sales. 'A lot of other things have hit us too – the price of fuel, fewer petrol stations because of the supermarkets selling petrol, that's hit us because that was our main place to sell maps. Then of course Google Maps came along, and broadband, so you could download maps easily. Our reps in London tell us that they see people walking around with bits of paper they've just printed out, or looking at their phones for directions.'

The company is now selling about half the number of A-Zs it sold at its peak, but as Dennison points out, the bound paper map still has many advantages. It is cheaper and easier than a sat nav, more aesthetically satisfying and better indexed than a phone map. And there is a broader attribute. 'If we don't use maps we lose the idea of where we're really going. Where London lies in relation to Bristol or Newcastle is getting lost to the youngsters – you just enter the postcode now.'

Dennison takes me upstairs to the open-plan drawing room and introduces me to Ian Griffin, the graphic designer, and Mark McConnell, the chief draughtsman. The talk swiftly turns to Phyllis. 'She always said there weren't any other maps around, but there were,' Griffin says. 'We used to take the old Ordnance Survey county sheets and draw over the top of those. We enlarged the main roads so they stood out, and the B roads, and we added street numbers so you knew which end you were on a long road – with the full index, that was one of her major innovations.'

McConnell takes me through the old process with the old tools – tracing paper, pens with interchangeable nibs to vary the thickness of the line. After hand-lettering there was rub-down lettering: 'The great skill was in writing round corners, sometimes in italics, having to plan ahead like chess.'

He looks at an old sheet with trams on it, and compares it with a modern one. 'London has totally been redrawn,' he says. 'But what always amuses me most about this is that London's not really there at all. It's just the streets and the place names, but London as we know it, the houses and the shops, the people, the soul of the place isn't there, it's done away with.'

New editions of the A-Z now appear each year, and on one a few years ago there were more than 10,000 alterations and additions (predominantly minor things like building names and footpaths, but a new development by the docks would easily add 500 entries to the index). The handwriting and Letraset have long gone, of course. A man called Tim Goodfellow curves the road on a computer now, and new streets take seconds rather than days to appear. One small change will digitally reorientate the entire area, recognising which roads have priority in the colour scheme; on the A-Z sat nav, a new road block in the City will automatically recalculate the journey time from the Strand to St Paul's.

I asked Goodfellow if he was ever tempted to place some personal detail on the map – the name of a loved one, perhaps, easily excused as a way of protecting it from copyright infringement. 'We do have security roads within each publication,' he said.' We refer to them as phantoms. We would just add something and choose a name that's sympathetic to the area so that it doesn't look out of place. So if you have an estate where all the side-roads are flowers, we wouldn't put in a name of a stone.'

'But you can't invent whole roads?'

'You could. If it was a main thoroughfare then that would be wrong, because you'd be misleading, but you can add a small cul de sac. We have people calling up and saying "there isn't a road there," and then we tell them why. Often we'll then move it.'

The whole A to Z of London – complete with the odd phantom diversion – rests in a single file on Goodfellow's computer. It is driven, nowadays, by French software, supplied by Michelin. We gaze at his large computer screen. Even backlit, even digitised, even French, the map is still a beautiful thing.

BECK

Pocket Map
The Biggest Map of All: Beck's London Tube

In much of his correspondence, and in most photographs, Harry Beck doesn't seem like he's one for jokes. He seems more after respect, and perhaps a pay rise, and both would be his due. For it was Beck – an intermittently employed engineering draughtsman – who designed a new map for the London Underground that became one of the most useful items of the twentieth century. In its various forms the map has been printed more than any other map in history – perhaps half a billion times, and rising. Beck got paid a few pounds for it.

The London tube map is a prime example of how a designer can take a problem and simplify it – and inspire its users. It is a schematic and digrammatic map, which triumphs over its geography: in real life the stations are not, of course, all the same distance apart, the centre of London is not as large as it seems in relation to the suburbs, the trains do not travel on straight lines. But its concealments are its strengths, for it's only a map in the loosest sense. It is really a circuit board of connections and directions with no real-life obstacles in its way. The only part of the city to intrude is the Thames; the rest is a graphic interpretation.

The map has become one of the most durable symbols of London, partly due to its ubiquity and colour-coding, and partly due to its very neat trick of making a sprawling city appear ordered and manageable, in much the same way as the post-Fire engravings of London did in the 1670s. And although Beck never set out to be considered in the same breath as Ortelius and Ogilby (and indeed never considered himself a cartographer at all), he created a map that continues to prove influential worldwide.

What makes it so special? Its clarity, certainly, but also its aesthetic beauty. There were diagrammatic train maps before this – notably those for the London & North Eastern Railway by George Dow – but none that so convincingly combined so many separate lines. There was beauty before it too, not least the typographically luxuriant maps of MacDonald Gill from the 1920s and the curving spaghetti-style interpretation by Fred Stingemore that Londoners had in their pockets the year Beck came on the scene. But Maxwell Roberts, a psychology lecturer and a tube map obsessive (he spends many spare hours designing improved subway maps of the world for his own amusement), has defined some key elements that tend to define all the best maps: simplicity, coherence, balance, harmony and topography, and Beck's map scores on all but the last. He argues that his map was special not because it used straight lines but because it had so few corners.

But could the genius take a joke? He was, apparently, keen on lampoon and satire, but how far would this stretch? Would he have appreciated the plethora of faux-London tube maps that have appeared in recent years? The most famous of these is Simon Patterson's *The Great Bear* from 1992, which now hangs proudly at Tate Britain and in many reproductions at the London Transport Museum shop. Patterson treated the map as pop art, replacing its stations with a witty selection of footballers (the Jubilee line), philosophers (District and Circle Lines) and Hollywood film stars (Northern Line).

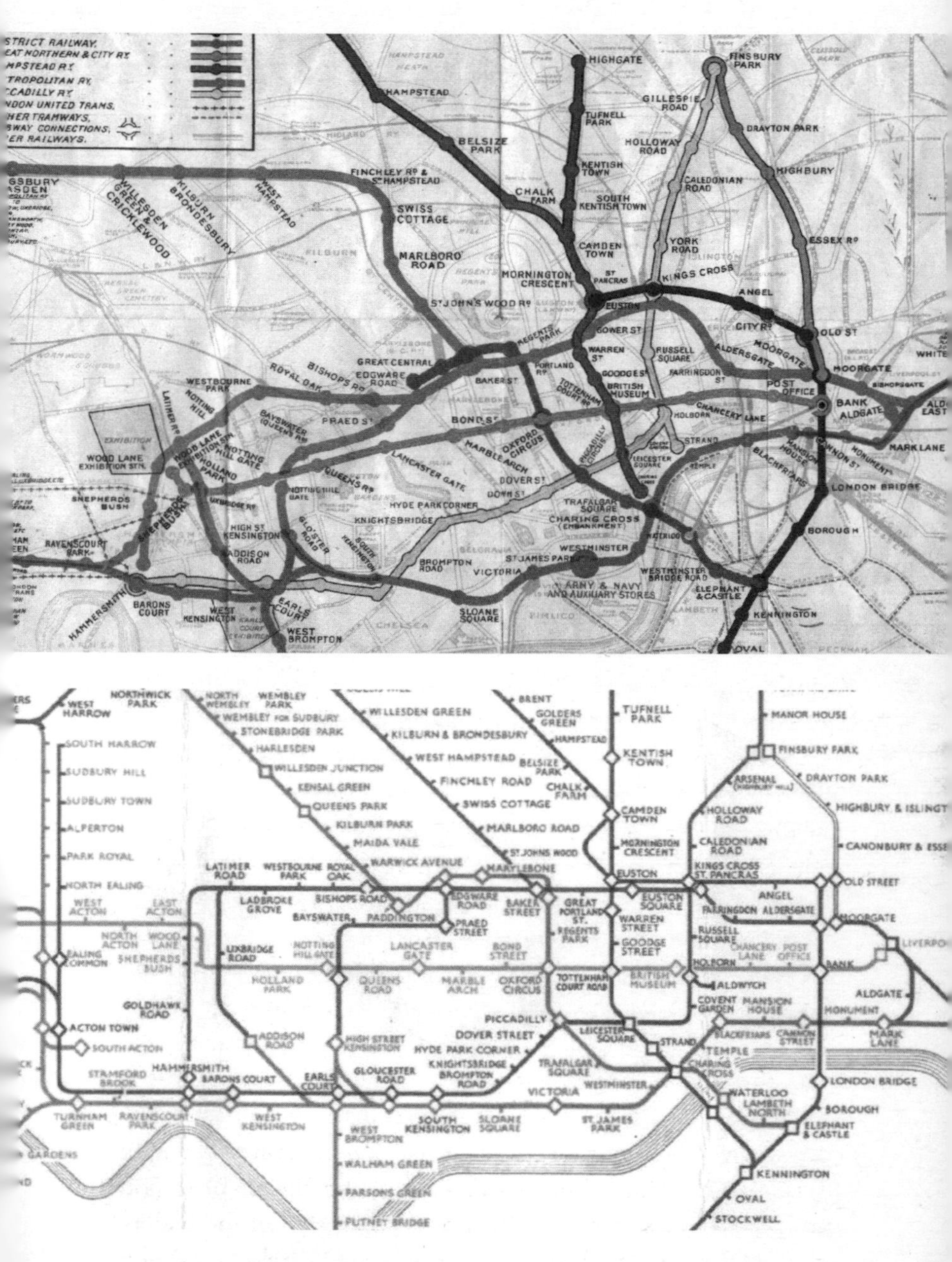

The London Underground – before and after Beck's 'circuit diagram' overhaul of its geography.

A more recent take on the same theme is the *Daily Mail Moral Underground* (not for sale in the London Transport Museum shop but viewable in full on The Poke comedy website), whose lines and stops reflect the apparent fears and obsessions of middle England. The District Line becomes a line of *Nuisances*, including Twitter, Sat Nav, Student Scum and 24hr Drinking, while the Bakerloo is *Medical Scares*, including Obesity, Cataracts and Deep Vein Thrombosis. The Northern Line (*Arch-Enemies*) is worth riding too, for you can alight anywhere between Guardian Readers, Single Mothers, Scroungers, Immigrants and The French.

We will never know where Beck would have told these parodists to get off, for he died in 1974, two decades before the personal computer made mash-ups of his map a possibility and a joy. But there was one that he would have been aware of, a satire that appeared in 1966 on an official London Underground poster designed by Hans Unger. This was a precursor of Patterson's work, depicting a section of the map as art movements, with stations labelled Op Art and Abstract Expressionism.

Of the maps that came after, some have been genuinely useful, overlaying information rather than disrupting it, including the map that shows which parts of the Underground are actually underground (only about 45 per cent), and the map that explained the distances between stations so that passengers would know if it would be faster to walk (the few hundred yards between Leicester Square and Covent Garden, for example, are almost always quicker on foot).

Then there are the elegant and artistic ones, such as Eiichi Kono's type map, on which hundreds of popular typefaces are placed upon a line according to their family (Futura and Bell Centennial on the *Sans Serif Northern line*; Georgia and Walbaum on the *Modern District line*; Arial and Comic Sans on the *Ornamental Overland line*). There is Barbara Kruger's emotional map, with stations named Betrayal, Compassion and Arrogance. Or the translation work of H. Prillinger, who dared to think how

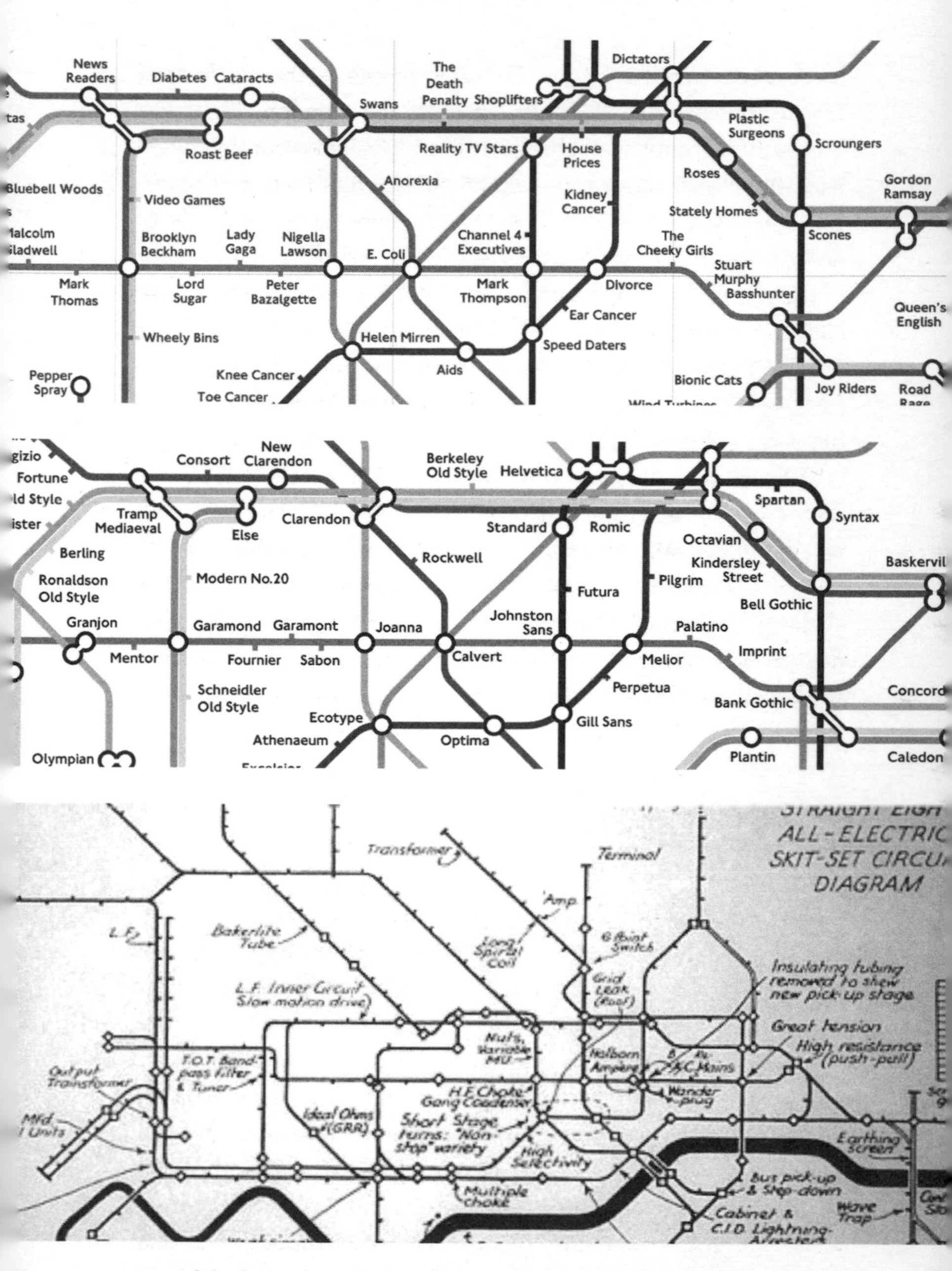

Two of the finest faux-tube maps – the *Daily Mail Moral Underground* and Kono's type map – and Beck's own parody.

Londoners would get about if the Germans had won the war (we'd go from Wasserklo to Konigskreuz and from London-brucke to Morgentonnencroissant). And then there is the global tube map, Mark Ovenden's construction of a Beck map that has taken over the whole world, with each station a different metro system, which is proudly displayed as this book's endpapers.

How would our man in the Brylcreem and heavy glasses have liked all of this? Probably rather well. It turns out that the first parody of Beck's map was by Beck himself. In March 1933, just two months after his map was first released, Beck (or someone impersonating his style and copying his signature) made fun of it in the pages of the London Transport staff magazine. People had remarked how much his map resembled an electronic circuit diagram, and so he mocked one up, an almost workable plan to make a radio in the form of the map. The *Bakerlite Tube* replaced the Bakerloo line, while stations were replaced both by Earth and Aerial. There was resistance, terminals and amps, and beneath it all the Thames flowed on.

Chapter 16
Maps In All Our Hands: a Brief History of the Guidebook

For many of us, our first significant experience of using a map has been abroad – in the pages of a travel guide. And so it has been for quite some time. Guidebooks are almost as old as maps. The Romans had *periplus* documents, noting ports and coastal landmarks, and *itinerarium*, which listed road stops, and in the second century, Pausanias created an impressively complete guide to the most interesting sights of the Ancient Greek world.

But for the first tourist guide worthy of the name we need to look to the year 330, when an anonymous traveller embarked on a pilgrimage and wrote an account called *Itinerary from Bordeaux to Jerusalem*. This was also the first boring postcard, a long listing of where he'd stayed and how long it had taken him to get there. The writer noted the number of times he had to change his transport, and found he could do Europe on two or three donkeys a day.

On the first leg – Bordeaux to Constantinople – he made 112 stops (or 'halts'), 230 changes of donkey, and had travelled

2221 miles. The closer the traveller got to his destination, the more excited he became. His remarks became more fulsome, the sights he listed more beautiful, and his tales taller: just beyond Judea he observed Mt Syna, 'where there is a fountain, in which, if a woman bathes, she becomes pregnant.' The manuscript contained no maps, but the 'A to B to C' descriptions served in their place. The Roman roads had mile markers, and pilgrims who followed in our traveller's footsteps would have little trouble finding their way.

Maps have guided the tourist long before the notion of tourism began (the word itself stems from the Greek word *tour*, meaning movement around a circle; middle-English took it further, to a trip or journey that ends up in the same place it began.) The Hereford Mappa Mundi served as both a geographical and spiritual tour for pilgrims, and John Ogilby's strip maps escorted travellers through Britain in the seventeenth century, promising a wayside inn or famous church every few miles. But tourist maps as we understand them today had another beginning, a birth allied to the cheap portable guidebook and the beginnings of popular travel in nineteenth-century Europe.

Before the 1830s, the Grand Tour of Europe required an educated local guide (a cicerone) and considerable amounts of money. But from 1836 onwards, the landscape was different, for you could travel through Holland, Belgium, Prussia and Northern Germany independently, armed only with a copy of what soon came to be known as a *Murray*. This was the first truly modern guidebook and it enabled travellers to venture where they wished, absorb as much information about a dusty monument as the heat would allow, and be assured of getting a decent dinner and bed at the end of it. Apart from the tiny point-size of the type, the *Murray Handbooks for Travellers* were largely the same things we might buy at a railway station or airport today: a nice bit of history, scenic and literary descriptions, detailed walking routes, passport and currency requirements, a check-list of packing essentials, recommended hotels, a few maps, and some elegant panoramic pull-out strips with a plan of a museum or the layout of Tuscan hills. The other main difference was, they were better.

Within a few years, the handbooks had become as essential a companion for the inquisitive and educated English traveller as an umbrella and emergency rations from Fortnum's. With their success allied to growing Victorian prosperity and the rapid spread of the railways, they also achieved something else: they made travel possible for independent women, as tourists and even as guidebook writers. And with this, women discovered maps like never before. Before now, maps had primarily been a male affair, indispensable for exploration, crucial for the military, essential for planning and power. But now women began to experience the value and joy of maps not just for travel but for perusal and possibility. It was the era of maps for the masses.

John Murray III, the latest member of a flourishing London publishing dynasty (his father had secured the firm's reputation by publishing Lord Byron and Jane Austen), had been travelling in Europe in the late 1820s when he noticed a paucity of anything that might help him make the most of his days. In Italy he found a book by Mariana Starke particularly useful (Starke was also published by his father), but elsewhere Murray found that he would arrive by brand new steam train or stagecoach, and had no idea what to do next. So he resolved to write such a thing himself, recommending the good experiences and castigating the bad. He laid out his terms for a successful template, noting that the guides had to be factual, devoid of fancy writing, and selective. 'Arriving at a city like Berlin, I had to find out what was really worth seeing there,' he explained.

So he would provide a guide rather than an encyclopaedia, as he was keen on 'not bewildering my readers by describing all that MIGHT be seen.' It was the early Victorian equivalent of the *What's Hot* list, and it was received by a hugely enthusiastic readership keen to discover a new Europe after the French Revolution and Napoleonic Wars had curtailed travelling for twenty-five years.

Murray wrote the first few guides himself, covering Holland, Belgium, Germany, Switzerland and France, but was prevented from covering Italy when the death of his father demanded that he concentrate on publishing duties in London. And so he commissioned others, most of whom were already experts in their assigned regions.* The series ran to some sixty titles in as many years – stretching as far as imperialist India, New Zealand and Japan, and returning to concentrate on British counties. The guides soon received the ultimate sign of fame, a parodic paean in *Punch*:

> So well thou'st played the hand-book's part
> For inn a hint, for routes a chart,
> That every line I've got by heart,
> My Murray.

('My Murray' was the phrase Lord Byron used to refer to his publisher. Byron was also responsible for popularising the word 'guidebook', practically unheard of before it appeared in his *Don Juan* in 1823.)

The majority of the early handbooks consisted of routes to be accomplished in a day, but there were few maps to

* Many of Murray's writers were already established in other fields, such as academia or literature. The writer of the 1855 guide to Portugal, for instance, one John Mason Neale, had composed the lyrics of the carol 'Good King Wenceslas'. The most famous of his writers was Richard Ford, who wrote a magnificently eccentric handbook to Spain (1845), the information gathered largely on horseback, its idiosyncracy best defined by his useful language section which included such phrases as vengo sofocado ('I am suffocated with rage').

accompany them. Those that were included were usually secured from civic sources and then updated to show new railways and other developments. Indeed, the spread of the railway network through Europe from the mid-1830s is elegantly tracked on the Murray maps like nowhere else. But the main visual treats were unexpected, such as a fold-out engraving in the 1843 guide to Switzerland of the chain of Mont Blanc, or the pull-out of the pyramids in late-1880s Egypt, and the occasional Greek phrase-book in an envelope glued to the cover.

But great ideas tend not to profit in isolation. In Koblenz, Germany, Karl Baedeker, the son of a printer, knew a good idea when he saw one. He was a fan of the earliest Murrays, recognising just the things he had wanted to publish himself. He had issued his first guidebook to accompany the popular Rhine cruises a year before the first Murray, but he had

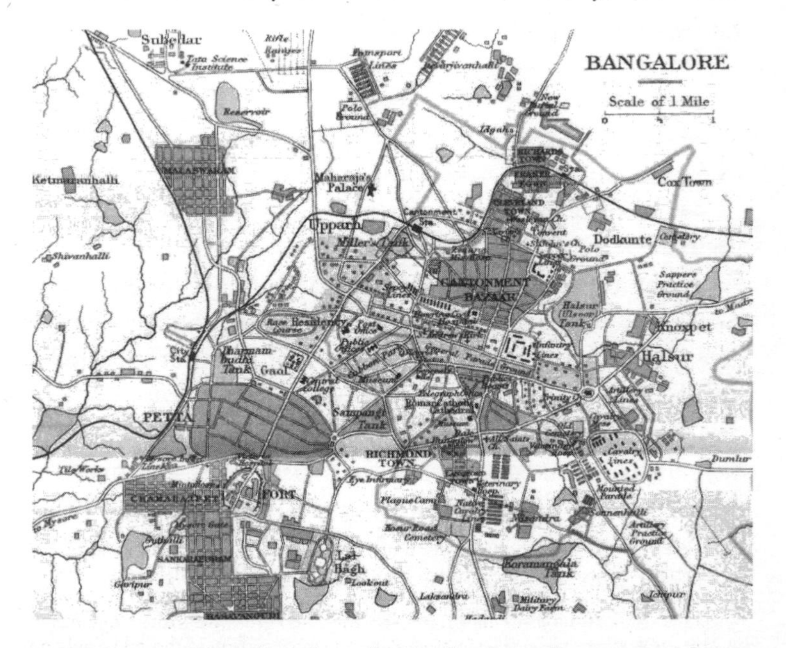

Murray's map of Bangalore, from the *Handbook to India, Burma and Ceylon*, 1924.

acquired it by default when he bought a bankrupt publishing house. For his own guidebooks he adopted not only Murray's red covers with gold lettering, but also large swathes of Murray's text, sometimes garbling the translations. Murray's description in an early Swiss guide of a place where 'the rocks ... are full of red garnets' became, in the Baedeker, 'overgrown with red pomegranates'.

Despite the instances of plagiarism, Baedeker and Murray became friends, each agreeing not to publish in each other's language – a promise that held good until the early 1860s, when one of Baedeker's sons, also called Karl, couldn't resist the opportunity to expand the market. But by 1860, one could argue, the Baedeker had perfected the guidebook form, outgunning even Murray to become the accepted byword for the failsafe travel companion. By the end of the century, one could easily tour the world without leaving home: in 1883 the series extended into Russia, and a decade later it had penetrated the United States. The writing was strict, unequivocal and trustworthy, the information current, the routes exhausting but fulfilling, the tastes tailored to a demanding yet non-academic readership. The whole experience was intellectually and spiritually uplifting.

Baedekering became a verb, while a Baedeker came to mean any reliable and comprehensive guide to anything (*The Joy of Sex* was once reviewed as *A Baedeker of Bedroom Techniques*.) The Baedeker style (with many parentheses denoting subsidiary yet important information such as cab fares) became influential, and Baedeker also developed the star rating system – ubiquitous now as shorthand before any arts or leisure review. Places he thought unmissable – the Tribuna in the Uffizi, for example – would get two stars for its Raphaels, but other places that met his disapproval, including Mont Blanc, were awarded no stars ('the view from the summit is unsatisfactory').

The guides inevitably attracted flak. In *A Room with a View*, E.M. Forster observed how they closed one's mind rather than opened it, directing the traveller up or down

Maps and panoramas folding out luxuriantly from Baedeker's *Rhineland.*

a pew in regimental fashion and acting as a protective veil against authentic emotion. Later and more damagingly, the guides were adopted by the Nazis, who noted areas cleared of Jews, and Baedeker's Britain served as a template for Hitler's deliberate cultural destruction in the so-called 'Baedeker raids', when German bombers were sent to wipe out star-sites to demoralise the enemy.

But for the fan of cartography, the classic Baedekers of the nineteenth and early twentieth century remain dazzling. Far more generous in number than Murray, the Baedeker maps covered both city and rural walking routes, and were particularly strong on ancient sites and mountain passes. Each new edition usually brought new maps, and just as well, for their propensity to tear, crumble and become detached from their bindings – they came at you from all angles – was as much of a feature as their topography. From one single map accompanying a reprinting of the Rhine guide in 1846, there were suddenly seventeen in the edition of 1866 and seventy in 1912. There was one lonely Swiss map in 1852, but eighty-two in 1930.

The maps began as simple engravings, but appeared in two or three colours from 1870. It was fitting, given its pedigree as the centre of so many medieval maps, that the first colour map was of Jerusalem (in the volume on Palestine and Syria). The colours chosen became lodged in the mind as no cartographic branding before it. The inland and dense city areas appeared in an ochre reminiscent of the clay dust one is unable to wash away from holiday sandals, while the green looked arsenical and the washed-out pastel blue made the coastal regions and lagoons look dry and oddly unappealing.*

The Baedekers – in their classic editions – disappeared after the Second World War. Murray's Handbooks had been sold some years before, in 1910, to Stanfords, before being acquired in 1915 by two Scottish brothers, James and Findlay Muirhead. The Muirheads had actually worked at Baedeker's, editing the English language editions, until the outbreak of the First World War found them out of a job. They continued the learned, encyclopedic traditions of the two grand guides as the *Blue Guides*, which for a decade or two operated an Anglo-French alliance, before *Les Guides Bleus* went their own Gallic way in the 1930s.

* In the US, the closest homegrown guides to Murray and Baedeker were those produced by D. Appleton & Co. in New York. Appleton was a successful general trade publisher of encyclopedias and fiction (its biggest hit was probably *The Red Badge of Courage*), but it wasn't slow to realise the impact of railway and steamboat travel on its readers' vacationing habits. Its *Southern and Western Travelers Guide* from 1851, for example, took in the tourist attractions Virginia Springs and the Mammoth cave in Kentucky, and included maps of the Ohio and Mississippi rivers, plans of Cincinnati, Charleston and New Orleans, and three folding engraved maps of the Western, North Western and South Western States. The latter were beautifully hand-coloured, and as artifacts of the opening up of the West are now valuable items in themselves.

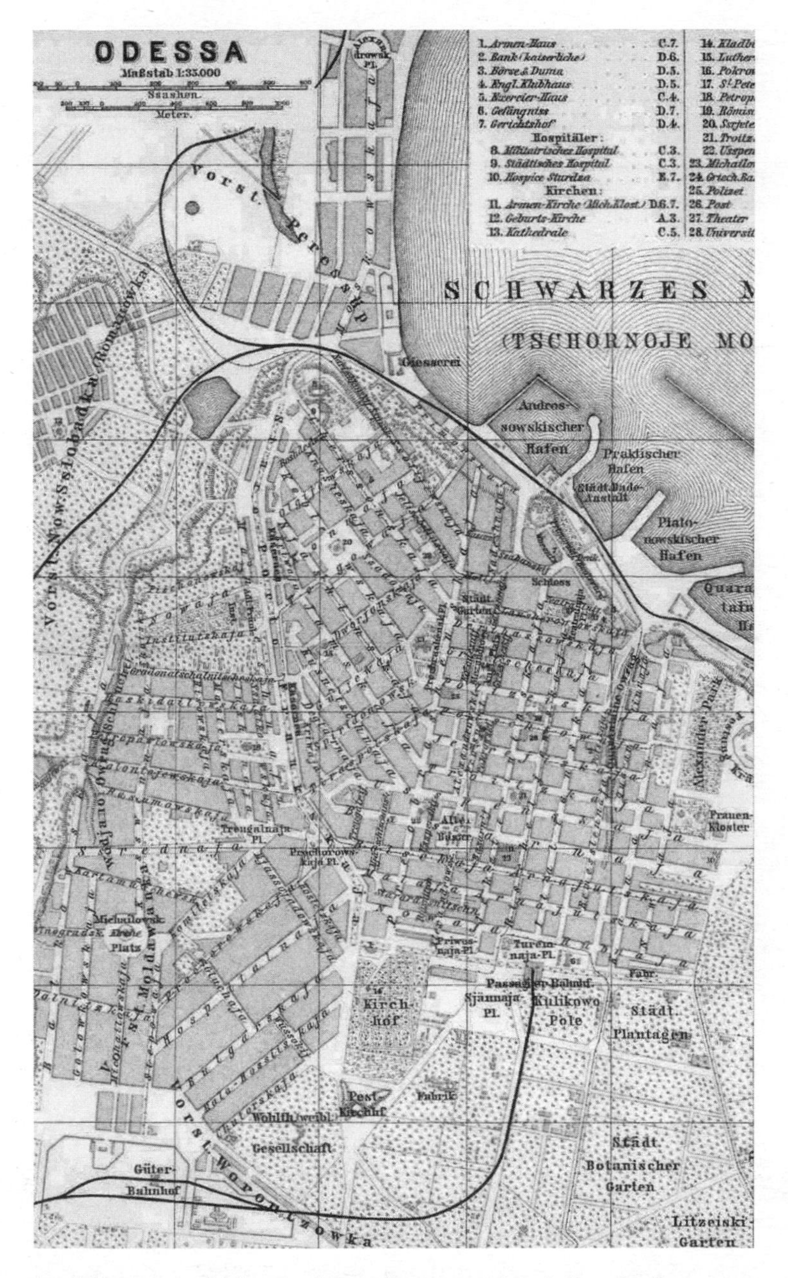

Baedeker's map of Odessa – delicately engraved in shades of yellow, ochre and black – from the 1892 edition of *Russland*.

The *Blue Guides* – both English and French – maintained the mapping traditions of their predecessors, albeit with less extravagance: the fold-outs were scaled back and their focus became more scholarly, with maps devoted principally to archeological sites and church plans. But the decades after the Second World War were a thin time for tourism and guidebooks, as an impoverished Europe holidayed largely at home. And for Baedeker, you might argue, the glory days had been those before the *First* World War. The novelist Jonathan Keates observed that a 1912 Baedeker from a town in south-eastern Europe boasted the Grand, the Europa and the Radetzky hotels, some old mosques, and some fine shops in the Appelkai selling carpets and inlaid metalwork. And then, two years later, a car carrying an Archduke passes along the same road in Sarajevo and is met by a man with a revolver, and 'in the echo of the shots he fires we hear the portable paradise of Baedeker and Murray vanishing into air.'

Things were rather brighter for guidebooks and maps in post-war France, where the cartographic future was entrusted to the tyre company Michelin, whose maps and guides began in 1900 and flourished like no other. That's because there was no other guidebook like them, nor any with maps with such a particular purpose. They began as a promotional wheeze to sell pneumatic tyres, sold as much to cyclists as motorists (in fact for the first few years the maps and books were given away free). Maps became central to the operation in 1910, guiding pleasure seekers to repair garages and petrol fill-ups, and increasingly to approved board and lodging (the three-star grading system, initially used for hotels with a restaurant, was introduced in 1931). Many pictograms in the guide and on the maps were esoteric, including a tilted shaded square indicating a hotel with a darkroom for photo developing, a scale of justice to indicate the availability of a solicitor after an accident, and a U-shaped mark to show where a driver may descend into a pit and get under the car.

In just over a decade there were Michelin guides and *cartes* not just for France but for large areas of Europe and beyond, cheering the motorist in *Grande Bretagne*, *L'Espagne* or *Maroc* towards engine oil and olive oil. The firm swiftly expanded into specialist maps, including, from 1917, unique Battle-ground Guides for pilgrimages to Verdun and elsewhere (marketed as *un guide, un panorama, une histoire*). And in the next war the maps served as an Allied tool, when the *1939 Michelin France* was reprinted in Washington DC in 1944 and handed to the troops sweeping through Cherbourg and Bayeux as they liberated the country after D-Day.

With many Europeans too impoverished to venture abroad in the post-war period, the Americans began to re-invent tourism, armed with their own new guides. Chief among

Michelin's numbered series – mapping France and beyond, and guiding the troops on D-Day.

these were the series invented by Eugene Fodor and Arthur Frommer.

Frommer served as a GI in mid-1950s Europe, where he compiled a budget guide for his fellow soldiers, itself a bargain at 50 cents. When he was discharged he beefed up the handbook for civilians, and *Europe on $5 a Day* was born. His no-nonsense, no-rip-off mentality appealed particularly to fellow Americans setting foot in Europe for the first time, and the series has run to more than fifty editions (though the prices obviously kept rising: by 1994 Paris would cost you $45 a day, while by 1997 New York was $70).

Eugene Fodor, born in Hungary, served in the war as an American soldier, although he had already written what he called his 'Entertaining Travel Annual' for Europe in 1936. His aim was twofold: to appeal to the American middle-classes with an eye on their purse, while also breezily expanding their cultural and historical horizons in a way he believed other guides were not – a less haughty, more flippant Murray for the 'modern' generation. But these new American guides, although much more informal than their Victorian-era, British predecessors, were in many ways far more conservative. They sent their readers on a narrowly-defined circuit of Europe, which comprised mainly the chief cities and sights. And their maps reflected this. The tinted, survey-style engravings of Murray or Baedeker, which you could have relied upon to ride across the continent, were replaced by crude sketches of city centres, noting the main attractions and hotels. All the art and detail of mapping had disappeared. It was as if guidebooks were ushering in a new dark age of cartography.

And by the 1970s, Fodor and Frommer had in their turn become very mainstream institutions, out of touch with a new wave of popular travel that followed in the wake of the hippies. Suddenly there was mass tourism again in Europe, much of it on a shoestring – hitchhiking or on 'InterRail' passes – as well as an opening up of places like India and Thailand,

Mexico and Peru. The new wave soon spawned its own travel guides: the Australian-based *Lonely Planet*, which arrived with an overland guide to South East Asia in 1974, and the *Rough Guides*, which began covering Europe in 1982.

These two series presented a new attitude to tourism, although their readers would inevitably prefer to be thought of as travellers, and travellers with a conscience to boot. The guides were a little too efficient to be regarded as hippy, but they did have an authentic, back-to-the-earth mentality. Above all they would take you to parts of the world not yet spoilt by the other guides, and, if they were, would tell you where to go to meet like-minded travellers who wanted to do something about it (the first *Rough Guide to Greece* was dedicated to a non-nuclear future; eco-tourism is more the goal now.) They were written in a colloquial, chatty style, respectful of local customs but wary of officialdom, and their maps were reassuringly primitive. They were often hand-drawn out of necessity – both companies were shoestring operations in their early years and often the only map, say to a village in Nepal, was the one the guide's researcher drew up on a napkin.

The original *Rough Guide to Greece*, complete with a dark age map on the front.

As Lonely Planets and Rough Guides developed, however, alongside other rival series (including a plethora in Germany), a respect for mapping re-emerged. It was like watching Murrays and Baedekers all over again. Lonely Planet would feature 100 maps in its India guide, Rough Guides would introduce a further fifty to towns not previously mapped since the Raj. In fact, as Rough Guides originator Mark Ellingham recalled, 'we often used the old Murrays and Baedekers as our source material – nothing superior had been published since. It was just a question of adding new areas of the city and changing the street names. And, of course, introducing a rather different array of interests: local music clubs and bars and bike rental places, instead of the old Thomas Cook bureaux and poste restantes.'

The 1990s proved to be a golden age. Lonely Planet realised its aim to cover every country in the world, and did so in ever more detail and with growing sophisitication as digital mapping replaced hand-drawn, schematic plans. But the same digital mapping was also to hasten thir decline.

Then as the new millennium began, the bubble burst. Suddenly, all the world's information was available on the Internet and, emboldened by cheap flights, and travelling often just for a few days, travellers did their research themselves. You might still buy a Rough Guide for a trip to Peru or Morocco, but for a few days in Italy or a weekend in Hungary you found your hotel through TripAdvisor and printed out a Google map. Or perhaps you had that Google map on your phone, allowing you to see not just the location of your hotel or intended sights, but yourself, a dot moving slowly towards them. In such a new world, why would anyone buy a book that was out of date the moment it was published?

But do we miss the graphite enchantment of those crinkly concertina engravings of Swiss mountains and Egyptian pyramids? I think we do.

Pocket Map
JM Barrie Fails to Fold a Pocket Map

A map is not like a well-pleated skirt; it does not readily return to the folds intended for it when you bought it. This realisation came to J.M. Barrie, the author of *Peter Pan,* when he was twenty-nine, and not yet famous. And why value Barrie's opinions on maps? Because he wrote the simplest and most enchanting map direction in the world: 'Second to the right and straight on till morning.'

That, at least, is how it appears in Barrie's original play, first performed in 1904. When the Disney movie appeared in 1953, Peter's directions had changed slightly ('Second star to the right and straight on till morning'), and the studio also produced a map of Neverland (in associaton with Colgate) to chart the territory: Crocodile Creek, Pirate Cove, Skull Rock, and the rest. The thin paper map folded out to about 3ft by 2ft, and you needed three packs of soap and 15 cents to get one. It also had a heart-breaking inscription: 'This map is a collector's item of limited use.'

The map is unlikely to have pleased the dramatist. In September 1889, long before Peter Pan took his inaugural flight, Barrie took against maps in general. He was living in Edinburgh

The Colgate-Disney map of Neverland – tragically, 'of limited use'.

when he noticed a trend in bookshops along Princes Street. An assistant would frequently offer him a new map of the city while tying up his purchases.

'Anything special about it?' he would ask. 'Well yes,' the bookseller would reply. 'It is very convenient for the pocket.'

'At the words "convenient for the pocket" you ought to up with your books and run, for they are a danger signal,' Barrie advised readers of the *Edinburgh Evening Dispatch*. 'But you hesitate and are lost.' Almost every house in Edinburgh contained a map that the whole family, working together, cannot shut, he went on. 'What makes you buy it? In your heart you know you are only taking home a pocket of unhappiness.' At the end of his diatribe, Barrie offers a list of negative advice born from terrible experience. This includes, 'Don't speak to the map', 'Don't put your fist through it', 'Don't kick it around the room' and 'Don't blame your wife'. And if, by sheer fluke, you do succeed in folding the map, 'don't wave your arms in the air or go shouting all over the house "I've done it, I've done it!" If you behave in this way your elation will undo you, and no one will believe that you can do it again. Control yourself until you are alone.'

Chapter 17
Casablanca, Harry Potter and Where Jennifer Aniston Lives

Here is a moment of cartographic joy from *The Muppets*, the 2011 nostalgia fest in which Kermit and friends get together for one last show to save their old theatre. As the frog drives across the United States, picking up old Muppets flung far and wide, a problem looms. Miss Piggy is working for *Vogue* in France, and they just don't have the time or money to fly there from the US. But then Fozzie has an idea, something he's seen in other films: 'We should travel by map!'

In the movies, travelling by map is the best way to travel. In *The Muppets*, Fozzie pushes a button on the car's dashboard marked 'Travel By Map' and a map of the world appears before our eyes. We watch as a thick line moves across it to the required destination, and we are transported along with the line, from New York to Cannes, as smooth as mercury in a thermometer. There are no delays, no queues, no passport checks, no customs laws. There are no detours and no misdirections. The journey across the Atlantic takes only a few seconds, but it would have taken precisely the same time had it been state-to-state or city-to-next-door-city. Sometimes, in place of a map, a little aeroplane symbol tracks the route across the globe. Either way, we have changed scenes and locations in one of the oldest clichés known to the movies, the hoary rival to the wavy-line dream sequence. It is how some of us learn our geography.

Trying to name the first cinematic journey by map is a fool's errand, for there will inevitably be something obscure in the vaults, probably Russian.* But we can all name the most famous. In 1942 Michael Curtiz made *Casablanca*, a film about love, loyalty and escape that starred Humphrey Bogart, Ingrid Bergman and cartography. Never before had maps played such a pervasive role in such a major film. From the opening titles to the end credits, and through several of the film's key interior scenes, maps fill the screen with their lure and their possibility. But this being wartime, the maps also set harsh limits: the borders are closed, the distances greater, the exit visas hard to come by.

The film opens with credits placed over a thick-lined map of Africa and the plodding blows of the Marseillaise. The map dissolves and is replaced by a globe spinning in clouds. 'With the coming of the Second World War,' the sonorous narration begins, 'many eyes in imprisoned Europe turn hopefully, or desperately, to the freedom of the Americas.' The earth still revolving, we zoom in over Europe, and at this point, close-up, the globe becomes a contour map – apparently Plasticine applied over a rubber ball. We hear that 'Lisbon became the great embarkation point. But not everybody could get to Lisbon directly, and so a tortuous roundabout refugee trail sprung up.'

* Early British and American contenders include the animated lines drawn over maps in two documentaries. The first appears over a map of the Pacific Ocean in *Among the Cannibal Isles of the South Pacific*, the film tracking Martin and Osa Johnson's 1918 adventures to the Solomon Isles; and *The Great White Silence* (1924), Herbert Ponting's film of the British heroic/tragic quest for the South Pole, in which moving black lines show the different paths across the Great Ice Barrier pursued by Amundsen and Scott.

Casablanca – how some of us learnt our geography.

The globe dissolves and we begin travelling by map. A heavy line marks our journey over land, a dotted one over sea. 'Paris to Marseilles, across the Mediterranean to Oran. Then by train or auto or foot across the rim of Africa to Casablanca in French Morocco …' Soon afterwards we are in the Moorish section of the city and Rick's Café but maps continue to cast a symbolic shadow over many scenes, not least in Renault's office as Rick and the officials moralise over the demands of love and the call of duty. And one of the greatest lines in romantic cinema ('Of all the gin joints, in all the towns, in all the world, she walks into mine') suggests both the vastness of the spinning globe and our helplessness within it.

★ ★ ★

Most who watch *Casablanca* fall under its spell, and young film directors are no different. Steven Spielberg's *Indiana Jones* series was his loving tribute to the Saturday afternoon heroes of his childhood, but it was inevitably influenced too by James Bond and cinematic Nazis. *Indiana Jones and the Last Crusade* opens with an adventure from Indy's youth, but for the first adult scene he's in his Ivy League classroom, with maps of archaeological digs on the wall and a nice thought for his doting students. 'Forget any ideas you got about lost cities, exotic travel and digging up the world,' Professor Jones tells them. 'We do not follow maps to buried treasure, and X never, ever marks the spot.'

And then the viewer follows a map to buried treasure. We are off to find Sean Connery in plundered Venice, and we travel by map. We track a red line from New York, stop at St John's for refuelling, cross the Atlantic, and hover over Spain to Italy. The map is superimposed over pictures of an airborne plane and Indiana turning pages of the Grail Diary. The Grail Diary has many maps of ancient sites, and we can vaguely make out the southern region of Judah by the Dead Sea, but before we can read it properly we are off again by cinematic map, this time a short red curl from Venice to Salzburg, where we have an overhead shot of a Nazi lair in a castle and another great movie cliché – people pushing counters on a vast table plan of Europe.

The map and the globe have never gone out of style in the cinema. Unless you are making *Strangers on a Train*, *Titanic* or *Snakes on a Plane*, the process of travelling is usually a drag to watch, and is seldom shown in real time. These days the only choice for directors is whether to employ the device straight (*Indiana Jones*) or ironically (*The Muppets*)*. The issue has even attracted academic discourse, and in 2009 the *Cartographic Journal* devoted an entire issue to the subject. Some of this was heavy going ('Applying the Theatre Metaphor to Integrated Media for Depicting Geography') but one essay, by Sébastien

* There is another nice parody in the ever-aware *Family Guy*. Peter and Brian float over the Middle East in a balloon and the countries appear below them as coloured slabs. "Huh, so that's what it looks like from up here," Brian observes.

Caquard of the University of Montreal, was startling. It suggested that a large proportion of the advances in digital cartography that we now take for granted – the zooming facility and shifts in perspective on digital maps, the layering of traditional maps with photographs and satellite views – all happened first in the movies, where the technology prefigured and inspired real-life cartographic possibility.

There are many examples. In 1931, Fritz Lang's *M* introduced a map that combined several features we would regard as digital and modern. A girl has been murdered by a serial killer in early 1930s Berlin, and an empty sweet bag is found at the crime scene. The police decide to investigate nearby confectionery shops, and their widening search is shown in a map sequence that changes in perspective from an oblique angle to an overhead 'God-shot', much as we may tilt the angle on a

'You can't fight in here, this is the war room!' The cartographic control centre in Dr Strangelove.

computer map or virtual globe. It may also be that *M* contains the first example anywhere of a map overlaid with sound – the talkies meeting cartography for the first time, another digital precursor, this time to the guidance/sound effects we have when we view maps on sat nav.

And for the first appearance of sat nav itself we may look – where else? – to James Bond. In 1964's *Goldfinger* Bond has placed a transmitter in Goldfinger's car and tracks him from a round green screen in his Aston Martin. The image and sound are as much submarine sonar as TomTom or Garmin, but the idea is an enduring one almost fifty years on: you get in your car and you're guided where to go. Another Cold War classic appeared in the same year. The operations room in *Dr Strangelove* has a backdrop of menacing moving dots showing the path of American B52s towards their Russian targets. The dots stop just in time, a darkly comic indicator of real-time remote military mapping we would later see in genuine conflicts.

* * *

Sébastien Caquard's theory makes good sense, and why wouldn't it? Why wouldn't the world of modern cartography be influenced by movies the way the rest of us are? But how does the theory stack up against *Harry Potter and The Prisoner of Azkaban*?

In 2004, Hogwarts welcomed a magical new plaything, *The Marauder's Map*. Presented to Harry by the Weasley twins, the map is not initially impressive. 'What's this rubbish?' Harry asks as he unfolds a large piece of rectangular parchment. It is completely blank. 'That there's the secret to our success,' the twins explain. George Weasley taps the map with a wand as he proclaims, 'I solemnly swear that I am up to no good.' And with that, the blank parchment is gradually overwhelmed with writing and illustrations.

An authentic replica Marauder's Map – good on the folds, even if it lacks the magical insults.

Why is the map – filmed fairly faithfully from the book – so useful? Harry takes a moment to realise. It is a real-time map of Hogwarts, and those are Dumbledore's footprints pacing in his study. Harry is still astonished. 'So you mean this map shows …' The twins interject: 'Everyone, where they are, what they're doing, every minute of every day.'

In its own way, we are looking at another Mappa Mundi, a mischievous world on calfskin. The map is expansive, and folds many times, coming to rest at approximately 2ft by 7ft. It shows practically the whole of Hogwarts, the classrooms, the ramparts, the corridors, the staircases, the cupboards. Harry will use it to locate the entrance to One-Eyed Witch Passage in Honeydukes sweetshop in Hogsmeade, and to find that Peter Pettigrew, widely considered dead, may not be. At the end of each session the phrase 'mischief managed' returns the map to blank parchment; if it falls into the hands of strangers, it reveals only insults in brown ink.

Reassuringly, this too has a modern real-life equivalent. 'The Marauder's Map clearly embodies the surveillance potential of digital cartography,' Sébastien Caquard posits. The ability to know where everyone is at any moment 'resonates strongly with the military concept of dominant battlespace awareness (DBA).' The question of whether J.K. Rowling infected the world's military, or whether it was the other way around, is still open to discussion.

★ ★ ★

Not long after Humphrey Bogart and Harrison Ford saw off the Nazis, two new popular tourist trails sprung up that show no sign of waning. The first was set-jetting: a trip to a film's location. Madison County to see the bridges perhaps, or to Paris for the Templar-Masonic locales of *The Da Vinci Code*. This can be fun – and many of us received our first mental maps of London, Paris and New York from the cinema – though we are wise enough to know that these cities only rarely look the way they do to Richard Curtis, Claude Chabrol or Woody Allen. And we are also aware that most Hollywood movies are not actually made in the places they purport. Better to travel the world the simple way, with a studio tour of the Universal or Warner Brothers lot.

Or of course we can cut out the movies altogether and go stalking. The second post-war post-movie crush has been our desire to see filmstars' homes – and maps have helped us on this quest since Johnny Weismuller lived at 423 N Rockingham, Brentwood and Gregory Peck put out his trash at 1700 San Remo, Santa Monica.

In the 1960s Mitock & Sons of 13561½ Ventura Blvd, Sherman Oaks, California, sold *The Movieland Guide to the Fabulous Homes of Movie, Television and Radio Stars*, a map which featured pictures on its cover of Lucille Ball and Desi Arnaz, Liberace, Bob Hope and Marilyn Monroe. And the map delivered. Not only could you find out where these stars lived, but you could drive to their houses, park by their gates, and presumably, in those innocent days, not be escorted away if you hung around. There were locations for Clark Gable (4545 N Pettit, Encino), Henry Fonda (600 Tigertail, Brentwood),

Errol Flynn (7740 Mulholland Drive), Rudolf Valentino (2, Bella Drive) and the last home of W.C. Fields (2015 De Mille Drive, Hollywood).

This was clearly a time when the stars were stars, and most of the maps (for San Fernando Valley, Santa Monica, Brentwood, Bel Air and Hollywood) are hand-drawn and authentically primitive, the sort of lines one might make to direct someone to a petrol station. But they are also clear and precise, the street names in black capitals, the houses of the famous marked in red, and the thick red stream of Sunset Boulevard running through the centre of most of them. The maps work well not just as location devices, but equally as social documents. Hollywood never seemed so inviting or self-contained, nor so stellar. Hardly a street seems to be without its share of glamour, and a diligent postman delivering scripts could cast a double feature on his morning round.

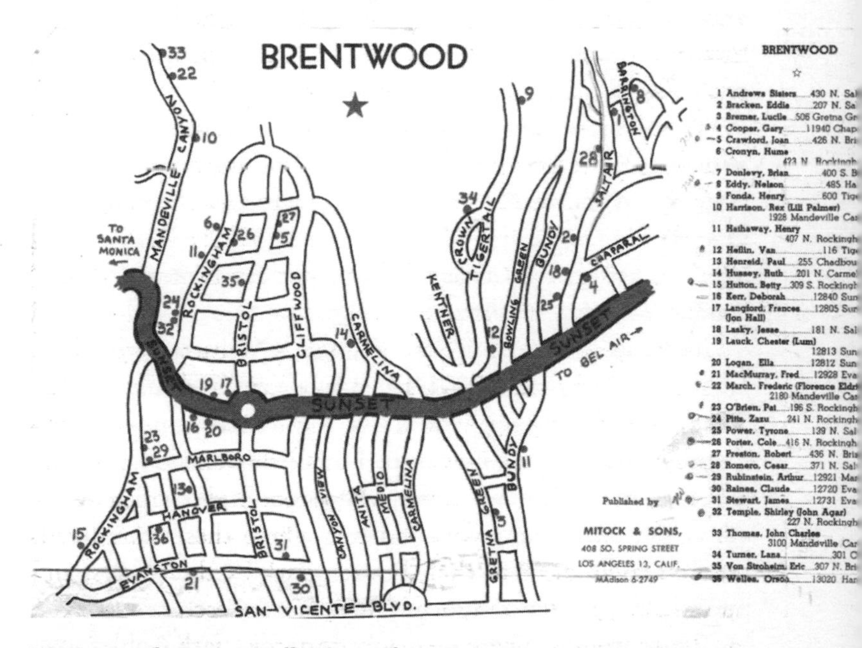

Movie stars' homes – a stalker's map from more innocent times. James Stewart is at no 31 opposite Claude Raines at no 30.

That was the sixties. In 2012, a hut on Santa Monica pier sells another map, *Movie Star Homes and Notorious Crime Scenes*, and the only thing it shares with the older map is a fascination with Marilyn Monroe. On the sixties map she was alive, and now she is dead, and the locations are increasingly prurient – her orphanage (815 North El Centro), the place she stayed after her bust-up with Joe DiMaggio (8336 De Longpre), and where, at the Mauretania hotel on North Rossmore, she had a thing with JFK. Can two maps, some fifty years apart, better sum up a downward society?

'Visit shocking crime scenes straight from the headlines,' the cover implores. 'You get details and prices of these incredible homes!! This is the map the Stars don't want you to have!!!' If there were any more exclamation marks there wouldn't be room for anything else!!!! But the map, which folds out to cover a small dining room table, is an extraordinarily efficient and compelling work. We learn, for example, where Hugh Grant got caught with a prostitute, and where Phil Hartman, the voice of Clinton on *Saturday Night Live* and Troy McClure on *The Simpsons*, was shot dead by his wife in 1998 before she turned the gun on herself. The map is a nightmare of electric colours, but it has very clear signage and markings, and an enviably simple legend: a red star signifies a crime scene, a pink one shows an actor's home, and a bullseye with flames around it marks a celebrity nightclub, boutique or deli.

The biggest change from the 1960s is Malibu. Once home to the relatively obscure (Dennis O'Keefe, Turhan 'Turkish Delight' Bey, Gregory Ratoff), it is now a living version of *People* magazine, and an attempt at traditional mapping would be out of date as soon as it was printed. So a new star map has presented itself, the gossip-bursting ninety-minute StarLine tour along Pacific Coast Highway conducted in a small open-top bus by a middle-aged woman named Renee.

There are twelve of us in the car park by Santa Monica pier, including a family of four from Dortmund, and we have each paid $39 for access to the map in Renee's head. We start at the Casa Del Mar hotel, where she says Al Pacino used to stay. Renee once saw Al Pacino in real life, and he was driving a convertible red Ferrari. We roll down towards the coast. 'This is where Will Rodgers lived, the actor and cowboy, he was killed in an aircrash in 1935. You know who was in court today? Lindsay Lohan, about that bracelet. On the left is Moonshadows, where Mel Gibson had a few drinks with the ladies. An hour later he got pulled over for drink driving.'

After fifteen minutes there is a photo opportunity at Jeff and Beau Bridges' house, and it is apparent that we are not seeing the beautiful beach-front homes in their best light. The most attractive vista is from the beach, from where you can see the decks, and maybe the superstars prone and oiled upon them. Unfortunately what we see is predominantly garbage bins and garages, and the occasional jogger carrying Evian.

'And the brown garage is Ryan O'Neal's house,' Renee says, 'and Ryan lived there with Farrah Fawcett Majors before she died. I saw their child last week. And there is the Osbournes' House – that was put up for sale this week, you'll see the Sotheby's sign ... That's Leonardo DiCaprio's house – the blue and white modern one. This is David Geffen's, with three garages. Number 22148, that's Jennifer Aniston's house, beautiful, both the house and the actress.'

The bus rolled on: it was still cartography of sorts.

Pocket Map
A Hareraising Masquerade

In 1979, a book appeared featuring fifteen colour paintings whose hidden clues led to a location in Britain where you could dig for treasure. The book itself was a mental map, but its novelty was that it demanded the construction of a real, practical treasure map from its readers.

The book was called *Masquerade*, a title which belied the anguish created for its devotional followers. If your map was accurate you'd uncover a ceramic box, within which you'd find an intricately turned golden hare with bells and jewels attached to its feet, and the sun and the moon dangling from its body. That was the prize, but it wasn't the fun. The fun was the quest – a race against others, a hunt for treasure that would make you feel like a child again. The back cover of the book stated that 'the treasure is as likely to be found by a bright child of ten with an understanding of language, simple mathematics and astronomy as it is to be found by an Oxford don.' It became an international sensation, with hopeful groups setting off from all points of the compass.

The concept, paintings and bejewelled hare were all created by Kit Williams, a folk artist with a proper beard and a vaguely Luddite mistrust of the modern world (the hare was in a brown clay pouch topped up with wax to thwart metal detectors). The book, a marketing department's dream, jumped effortlessly from the children's book pages to national news bulletins, and sold some 1.5 million copies worldwide.

The story wasn't really the thing – a tale about the moon falling in love with the sun and giving the messenger hare a love-token which somehow got lost in the stars. However, very slowly, if you were the sort of person who liked cryptic crosswords, you'd look long enough at the pictures and a location would reveal itself. Then you'd reach for your Ordnance Survey stash to start narrowing things down, and after that you'd write to Kit Williams, hopeful for affirmation.

But for a long time, nobody found the treasure. Instead, for almost three years, Kit Williams received more maps in the post than anyone had ever received before – hundreds every week. Everyone was an amateur detective, and although some readers would start digging before corresponding, most would await the nod from Williams to tell them they were somewhere close. The maps and diagrams were hand-drawn with various degrees of dexterity. Many would show a patch of the English countryside with fields, trees, nearby roads and other landmarks, although deciphering them was an arduous task.

Then one day in August 1982, Williams received a letter, inside which was the map he had been awaiting. 'It was almost

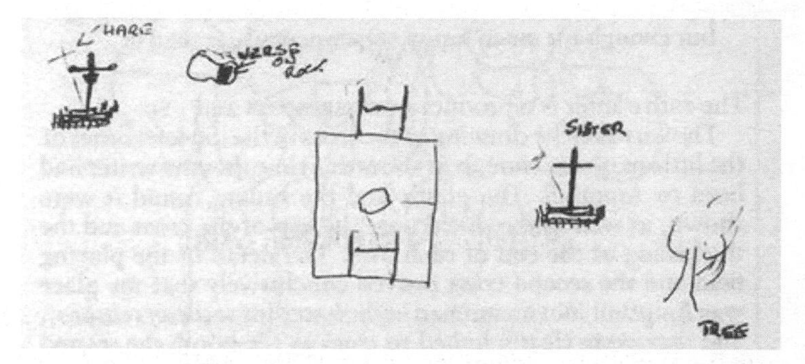

Ken Thomas's map, 'solving' Kit Williams's *Masquerade*.

a childlike map,' he recalled, 'but it was describing exactly where the jewel was.' The sender was a man named Ken Thomas, who duly set off to dig up the hare – which was buried in a park in Ampthill, Bedfordshire, off Junction 13 on the M1.

Ken Thomas told Williams that he had been searching for the hare for a little over a year but had uncovered the true location largely by accident. He had solved only a few of the clues, which led him to the region, but by chance his dog peed on the base of one of the two white stone crosses in Ampthill Park and he noticed its inscription – which turned out to be the key to everything. It was a disappointing conclusion, and an odd one. Solving *Masquerade* was huge news, yet Thomas was publicity shy.

The solution was then revealed in the paperback, and it was simpler than anyone had imagined. You followed the eye-line of every creature in every picture through their longest finger or biggest toe, picked out the letters they pointed to on every border, and eventually you'd get the words 'Catherine's longfinger over shadows earth buried yellow amulet midday points the hour in light of equinox look you.' It was a vertical acrostic, the first letter from each word or phrase arranged in a list producing 'Close by Ampthill'. You went to Ampthill Park on midsummer solstice, past a clock in the town that looked very like one in the book, waited for the sun to hit the top of a tall memorial cross erected for Catherine of Aragon (who is referenced more than once in the book), and then started digging where the tip of the shadow hit the grass. Williams had intended it to be 'like a cross on a pirate's map.'

How wonderful. Unfortunately, the charm of the story was subsequently tarnished by the revelation that Ken Thomas was really called Dugald Thompson and – unknown to Kit Williams – was the business partner of a man who lived with his ex-girlfriend. He had uncovered the hare not by a process of riddle-solving and map-making, but through her recollection of a picnic in the park. A rather modern spot of piracy.

Chapter 18
How to Make a Very Big Globe

About five miles into the journey, Peter Bellerby fixed the sat nav to his windscreen and tapped in the postcode. It was the usual procedure: you set off in the car from a familiar place, and only when you get a little nervous do the satellites take your hand. Bellerby, a forty-five-year-old maker of globes, was on his way to Chartwell in Kent, once the country home of Winston Churchill. The journey would take a little over an hour from his house in Stoke Newington, out of London on the M11 and then the M25, completing the journey on the A21 – not that one really had to know any of this anymore.

It was a clear, cold November day. Because he no longer had to concentrate on his route, Bellerby could concentrate on his story. He was going to Chartwell to see an exceptionally large globe. He had seen it for the first time a few weeks earlier, but on that occasion his trip had been cloaked in mystery. 'I had read about the globe,' he explained. 'And so I called Chartwell, and asked, "do you have the globe there?" And they said, "No, there's no globe here." '

Bellerby explained that the globe was given to both Churchill and Roosevelt during the war, and he understood that although it had been delivered to Downing Street, Churchill had taken it to Chartwell. 'No, there's no globe here whatsoever.' Bellerby described the dimensions: fifty inches in diameter. 'No, it wouldn't fit. We couldn't get anything like that in Chartwell.'

Bellerby then spoke to a woman at the Cabinet War Rooms who also denied having the globe. Then he tried the National Maritime Museum at Greenwich and drew a blank there too. He wrote to Downing Street just after David Cameron's election. 'Two days later I got a letter back. A man in the Prime Minister's office said he had also spoken to Chartwell, and they told him "absolutely not".' He also made enquiries at Chequers and other places, but nothing came of it. He then wrote to the Royal Collection and they also replied with a shrug.

'I was thinking of going to Washington to see Roosevelt's copy of the globe. Then I spoke to a woman from IMCoS, [the International Map Collectors' Society] and she said "It's definitely at Chartwell. I know it's there, because I saw it last

Roosevelt contemplates his globe – identical to Churchill's.

month. So unless they've moved it ..." But it's not the sort of thing you'd move, it would just fall apart.'

So Bellerby went down to Chartwell during normal tourist opening hours, and there it was. 'I spoke to a man at the grand entrance who said, "Oh, the Churchill globe...". He went on about it for ten minutes, and when I introduced myself he got a little bit sheepish. But he did give me free entry, and then said, "I think we're a bit coy about it because we're worried someone else might claim it. We certainly don't want any of the other museums to pinch it off us." But then on my way out he said, "Oh, we've established we own it so it's not a problem."'

Bellerby was keen to see the globe again because he wanted to copy it. Or at least he wanted to copy the idea of it – its size and impact, its in-your-face geometry – while updating its surface with a more modern map, rather than the one made in the 1930s, before Churchill, Roosevelt, Stalin and Hitler turned the globe on its axis.

Bellerby has the air of a down-at-heel toff. He looks very like the actor Steven Mackintosh. Before he turned to globemaking, he worked in globe rolling, running a London bowling alley called Bloomsbury Bowl. Bellerby arrived initially only to install the wooden lanes, but he then agreed to manage the place. It was full every night, a big party hit, but after three years the novelty had long worn off.

At about the time that he was getting bored with it all, the need to buy his father, a retired naval architect, an eightieth birthday present threw up a new opportunity. 'I just thought it would be nice to get him a globe. But I went to some shops and looked online, and the range was either expensive antiques that would go for tens of thousands, or new ones that were made in a factory and looked like it; many of them had lights inside them. There didn't seem to be anyone making really good hand-made globes in the whole country.'

So in 2008 Bellerby thought he would try his hand at a new profession, hoping that at last he had found his true calling. When he was growing up, he learnt that he was possibly related to the great missionary explorer David Livingstone. His great-great-grandmother was called Marion Carswell Livingstone, and she presumed that the man who had opened up the African interior was a cousin. Bellerby never felt any desire to verify the claim, but only now did the true value of not doing so become professionally useful.

When Bellerby embarked on his mission for his father, the market for globes seemed largely untapped. It was a tiny fraction of the market that existed in Livingstone's day, when there was a globe in every classroom and Great Britain governed half of it. But Bellerby believed – incorrectly as it turned out – that every self-respecting company director would want one in the boardroom. Globes would also make ideal retirement presents, or impressive pieces of furniture for the country home.

The closer the day got to his father's birthday, the more the possibility of making a globe in time for it receded. But his father's globe was now just one of hundreds in his head, the perceived demand growing each day. Why not globes in airline departure lounges? Or corporate globes with branding tastefully placed amidst the oceans? But how to make such a thing? That would be harder than he imagined, and would cost him his rather nice Aston Martin DB6.

We had a mile to go, and Bellerby's eyes had switched from the sat nav to the road signs. 'When sat navs first came out I thought, "Why would anyone want one?", he observed. 'But

'Making anything round is just a nightmare': Peter Bellerby, modern-day globemaker.

then my girlfriend started map reading for me and I thought, 'no, that's not the way forward ..." So I bought one when we went to Greece, and short of taking us on a 150 mile detour through France it was absolutely fantastic.'

At Chartwell we looked for the tradesman's entrance. The house had recently closed to the public for the winter, and although the gardens remained open there was hardly anyone else around. We were met by Nicole Day, one of Chartwell's stewards. She escorted us past the laden apple trees and magnificent views, and led us into a small painting studio that Churchill had converted from a garden summer house.

It was pretty much as he had left it. There was an easel and pots of paint, and the walls were decorated with his handiwork in oils. There was the obligatory half-smoked, half-chewed cigar on an ashtray on the table, as if its owner had just popped out for a bathroom break (you'll find these half-cigars at practically every Churchill shrine). And there, in one corner, roped off, was the fifty-inch globe.

We moved the ropes and a red leather easy chair by the globe's side, and a sign that stood on top of it (on top of the North Pole in fact) that read *Please Do Not Touch*. We could touch it very tentatively, Nicole said, if we really had to, and we could feel the areas that had been worn away, including northern France and New York, and the area around the equator that appeared to have suffered some sort of damage from the glue that joined the two hemispheres. But on no account were we allowed to try to turn it. If we tried, the whole sphere might disintegrate.

It was extraordinary to think that we were breathing over something that had played a small part in the outcome of the war. There are photographs of Churchill with his hand on the globe, and there are worn areas over strategic theatres of war. Here, perhaps, was the first tactile impression of a new offensive on Guadalcanal, the British victory in the Barents Sea above Norway and Roosevelt's plans to block Rommel's supply lines in North Africa.

Some of the globe's history may be divined from a framed letter hanging on the studio wall. It was sent from Washington on 12 December 1942 by General George C. Marshall, Chief of Staff of the US Army.

> My dear Mr. Churchill:
>
> We approach Christmas with much to be thankful for. The skies have cleared considerably since those dark weeks when you and your Chiefs of Staff first met with us a year ago. Today the enemy faces our powerful companionship which dooms his hopes and guarantees our victory.
>
> In order that the great leaders of this crusade may better follow the road to victory, the War Department has had two 50-inch globes specially made for presentation on Christmas Day to the Prime Minister and the President of the United States. I hope that

> you will find a place at 10 Downing Street for this globe, so that you may accurately chart the progress of the global struggle of 1943 to free the world of terror and bondage.
>
> With great respect,
>
> Faithfully yours.

'It clearly hasn't been restored, and quite right too,' Bellerby observed as he examined it. 'It's a bit like I'd just blown up a much smaller map and put that on a ball. You'll see how few cities there are on it, and I get the impression that it was a bit of a rushed job.' Nicole Day had her own observation: 'Why put that much detail into a map when you know you're soon going to be playing with the boundaries?'

In fact, the globe had been restored in 1989, after years of neglect. It was driven to the British Museum, where it came under the care of Dr David Baynes-Cope, the man who had fixed the mould on the Mappa Mundi with pyjama cord. But there was only so much he could do: the colours had faded, the varnish had worn away and there were dents, maybe from the globe being moved around Churchill's various residences.

Its scale is 1:10,000,000, which resulted in a thirteen-foot equator. It was based on a standard map readily available before the war, and contained no special consideration of disputed wartime borders. There were approximately 17,000 names, and if a few of the towns in the United States were small and unfamiliar, that is because everyone who worked on the map at the OSS made sure to include their home town.

How to transform the map into a globe? By the time-honoured tradition going back to the sixteenth century. The printed map was divided into gores, the acutely tapered triangular sections that had become a staple of globe production for centuries. These would be 3ft-long, with a width tapering from 4.5 inches to zero; had they been steel, these spears would have been lethal. Each hemisphere consisted of thirty-six 10°-wide gores, and mounting them required great precision. This task fell to the Chicago-based company Weber Costello, rivals to the other great Chicago map-makers Rand McNally. The first choice for the spheres was aluminium, but this was practically unobtainable during wartime. So they settled on hoops of laminated cherrywood. The halves were dowelled every six inches to limit expansion and contraction from changing temperatures. They were bolted together within themselves to maintain rigidity, and then, when the southern hemisphere was mounted on the northern, screwed together with rods from pole to pole.

The globe weighed about 750 lbs, and had it been made a decade earlier would likely have rotated on a pool of mercury. But mercury had become regarded as a health hazard, so a platform of three hard rubber balls was chosen instead, with this contraption concealed in a steel base that held the globe like an eggcup. At Chartwell, the base of Churchill's globe was painted black, and was in good shape. But it was believed that the rubber balls within had started to rot; certainly the 'easy action' had long gone.

Shipping a globe of this size is an arduous task in the best of circumstances. But with a war on, the most direct channels were restricted. And then there was the winter. The initial idea was to transfer the globe on a special flight from Maine, and then to Greenland and England in time for Christmas. But the weather was so bad at Maine that an alternative route was planned through South America, St Helena Island, Accra and then Gibraltar, which would mean that the globe would have seen almost half the area it covered. It had an escort throughout, a US army captain named B. Warwick Davenport. When the globe finally arrived at 10 Downing Street on 23 December, it was no longer a surprise. 'Where the hell have you been, Davenport?' Churchill snorted as it came through the door (according to a published account by Davenport, who seemed thrilled to be insulted by such a famous man). Churchill posed for a photo with the globe on Christmas day, with a cigar in one hand and the other on the northern hemisphere somewhere near Japan, and on the following day he sent General Marshall a telegram: 'We have marched resolutely through this past difficult year, and it will be of deep interest to me to follow on the Globe the great operation all over the world which will bring us final victory.'

The globe's attractions today – historical, ornamental, educational, monumental – may not be that removed from its value to Churchill at the end of 1942, nor even the earliest German examples from the end of the fifteenth century, when a globe was seen as God's molecule, a repository of knowledge and discovery, and status. And, of course, it was once a significant navigational tool, and before that a scientific one, designed to explain the rotation of the earth around its axis.

To Peter Bellerby, the globe at Chartwell was merely a stopping off point. He was hoping to make something with the same arresting visual impact, for he knew that even in a savvy and jaundiced world the wonder one feels when encountering a huge globe for the first time remains. His own version, which he naturally intended to call *The Churchill*, would have the latest gores, and include far more political and demographic information than the original. And it would be made from fibreglass, with a base cast from aluminium that would resemble an aerodynamically shaped housing nacelle of the type found on a Rolls Royce aircraft engine.

Some of these updates were Bellerby's ideas, and some came from his client, a man he would only initially refer to as David The Wealthy Texan, who was willing to pay him £25,000 to have one in his home. The two were developing a firm friendship, though it was not without its trials. The first globe that Bellerby had sold the Texan – the regular 50cm *Perano* – had not fully satisfied. 'He opened up the packaging,' Bellerby explained, 'and he said "it's the most beautiful globe I've ever seem, but unfortunately there's some damage".' One of the internal strengthening struts had snapped, and some internal cladding used to dampen the sound inside the globe (should any plaster flake off) had gone missing. The globe also had some holes drilled into it, and what may have been knife marks.

'It could have happened in two ways,' Bellerby believed. 'It could have exploded in the air due to the pressure, which I think is unlikely. Or they could have hacked into it with an instrument.' By 'they' he means US customs officials.

As we drove back from Chartwell to London, Bellerby observed that he now had to stop talking about Churchill's globe and start building it. In the three years since he had first imagined himself a globemaker, Bellerby had learnt that it was a tricky task. 'You know what?' he asked. 'Making anything round is just a nightmare.'

Making things round – and maps in particular – has been a problem since at least 1492, the year that Martin Behaim of Nuremberg made, or at least commissioned, the oldest terrestrial globe still in existence. Behaim, an ambitious merchant, had learnt of the new trade routes being opened up by the Portuguese, and his globe was an attempt to demonstrate to his German sponsors the potential value of establishing a new route westwards towards China. It shows the world at the precise time that Christopher Columbus got out his compass to look for Japan, making it an invaluable historical and scientific bridge between medieval map-making and the golden age of exploration. The biggest gap in the land surface on the globe was the very one to be 'filled' by Columbus.

The Behaim is twenty inches in diameter and spins on its correct axis within an elegant metal frame. It is slightly blistered in places but otherwise beautifully preserved, possibly because it has hardly been outside Nuremberg since it was made. When it was constructed, the real globe had not yet been circumnavigated (that was still thirty years away), but the amount of cartographical learning displayed by Behaim was staggering: the Americas are absent, but the Arctic and Antarctic circles are there, and between them lie the discoveries of Marco Polo, Henry the Navigator and other Italian and Portuguese explorers making their way around Asia and Africa.

And the detail is compelling. The globe contains some 1,100 place names, eleven ships being rocked by mermen, sea-serpents and seahorses, more than fifty flags and coats of arms, and almost as many intricate representations of kings upon thrones. Four saints are honoured with full-length portraits, while among them parade leopards, elephants, ostriches, bears and our old friend the sun-shielding sciapod. Behaim called his globe the *Erdapfel*, the Earth Apple.

Inevitably, Behaim and his chief draughtsman Georg Glockendon made mistakes, the errors as intriguing to us now as the many things

Behaim's *Erdapfel* globe.

they got right. Western Africa is mis-shaped; Cape Verde is in the wrong place; many place names appear twice. There are also curious omissions: no mention of Antwerp, Frankfurt or Hamburg for instance, crucial centres of trade and shipping. This is all the stranger given the globe's fascination with contemporary narrative discoveries. 'In Iceland are handsome white people,' one piece of text begins, 'and they are Christians. It is the custom there to sell dogs at a high price but to give away the children to merchants, for the sake of God, so that those remaining may have bread.' This information likely had a political slant, attempting to justify the piratical actions of kidnapping Icelandic children to be used as slaves.

The Icelandic text also features an early dietary pointer towards longevity: there were men 'eighty years of age who have never eaten bread, for corn does not grow there, and instead of bread they eat dried fish.'

Globes became something of a craze in the sixteenth century, an easy symbol of power. Miniature versions were particularly popular, with a round earth option often encased in a celestial shell. The fashion for engraved copper and hand-painted manuscript editions continued well into the eighteenth century, even though by then the method of stretching printed gores over a sphere was already established as a far cheaper technique (and would eventually ensure there was a globe in almost every European classroom).

The maps and style of the globes varied according to country of manufacture – the late-seventeenth century Italian globes of Vincenzo Coronelli, for example, were particularly decorative, while the German globes produced a few decades later by the likes of Homann and Doppelmayr tended increasingly towards the exact and scientific. But the biggest difference was in the choice of meridians. A globe's most functional use beyond the classroom was navigational, with measurements of longitude calculated from a ship's home port or capital city. So Cassini's French globes had their meridian through Paris, while the first American globes chose Washington. London's choice of Greenwich only became the global standard at the end of the nineteenth century.

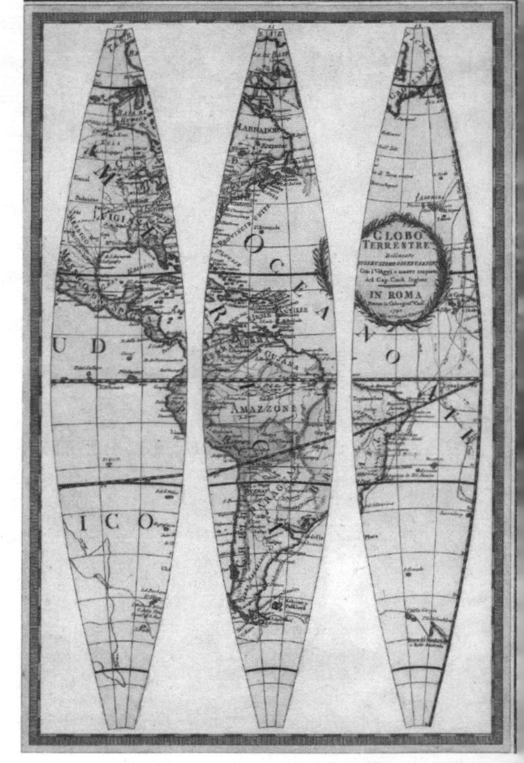

Three gores from Cassini's classic *Globo Terrestre*, 1790.

In 1850, Charles Dickens' *Household Words* contained an article in its 'Illustrated Cheapness' column about the popularity and construction of globes. The article explained the straightforward method upon which all globes were now made – as systematic, it claimed, as the process of making a Lucifer match. It estimated that about one thousand pairs of globes were sold each year (terrestrial and celestial), with sizes ranging from 2-inch pocket spheres to 36-inch giants, with prices from six shillings to fifty pounds. 'The number of globes annually sold represents to a certain extent the advance of Education,' the article reasoned, although unlike maps, globes – both more durable and costly – tended to be replaced infrequently, thus rendering them a far less accurate teaching tool at a time when the extent of the British Empire seemed to be broadening monthly.

Dickens' detailed description of globe manufacture – the many layers of paper required to be glued and dried, how to locate the correct axis – was evidently not lost on Ellen Eliza Fitz, the prominent American globemaker from New Brunswick who had an unlikely bestseller in 1876 with her *Hand-book of the Terrestrial Globe*. Much of this followed the Dickens template. 'A globe is made of pasted paper,' she explained, 'eight or ten layers of this being applied successfully to a mould prepared for this purpose. A turned stick of right length, with a short wire in each end for poles, is now introduced, one end in each hemisphere ...'

Globes attracted other successful women to the craft, not least Elizabeth Mount from Long Island, whose 'All States in the Union' sphere from about 1820 is now regarded as a key cartographical landmark. But the first successful commercial American globemaker was James Wilson, who built up a hugely popular business in Vermont and Albany in the early 1800s. Prior to Wilson the majority of globes in America had been imported from England (including models favoured by Thomas Jefferson during his presidency). Wilson was a self-taught, self-made man. In his youth he had admired an English globe made by Samuel Lane, and believed that a process of trial and error would enable him to make his own. His story has at least one modern parallel.

In mid-winter, Peter Bellerby's workshop in Stoke Newington, north London, feels like Iceland, or perhaps Greenland. The workshop doubles as a shop to catch the passing trade, with the front almost all glass, so that pedestrians can observe the archaic practices within. But people do not generally buy boutique globes on a whim, so Bellerby and his small team are seldom distracted from their task.

The three main work areas (front room, store room and courtyard lean-to) suggest that the modern globemaking process is very far from the state-of-the-art industry it was in Victorian times. As well as globes in various stages of completion, there are half-finished support legs, packing materials, sacks of plaster powder, metal rods, old globes from other manufacturers, chisels and other tools. Maps and sketches are pinned to the walls, and wet painted gores swing on pegs. Almost everything is covered in a layer of white dust.

Being neither geographer, historian nor cartographer, Bellerby learnt about making globes through trial and error. In 2008, two years before he embarked on the *Churchill*, he had more modest ambitions: the *Britannia*. This was his first globe, a 50-cm diameter model, which would cost £2,390. He began by buying the copyright to a multi-coloured political map with light blue seas, and he stripped this down on a computer to its coast lines, leaving only its most important rivers and place names. He then paid someone to write a computer program that would transform the rectangle into gores. 'That was a nightmare,' he remembers, but it wasn't just the goring that sounded like a trial. 'At the beginning I was having real trouble making the balls. Ours were just not round. We had this huge bulge all around the equator. And I had to learn how to manipulate paper to a much higher degree than you can imagine. The whole map was trying to change direction. That was about £60,000 or £70,000 pounds in, and I was, "Oh my God, we can't even do this basic thing."'

He says he tried 'about two hundred' methods of goring before he found one that stuck. 'I'll tell you one of the secrets,' he says. 'Not all paper stretches. The paper that does stretch will only stretch in one plane. I have a sheet of paper where the gores are printed horizontally across the paper, and if they're printed the other way round, they'll rip.' He uses modern inks, believing they will last for two centuries beneath UV varnish and acid-free glue.

The *Britannia* was named after the font designed by the typographer James Mosley, who happened to pass Bellerby's workshop one day and suggested – after much discussion and a visit to the National Maritime Museum – that his lettering would suit the look Bellerby was aiming for. The first edition (globes are 'published' in the manner of books) looked rather too modern, as if a schoolroom map had been removed from a wall and made spherical. It did not look like the sort of globe that would appeal to the market Bellerby was targeting – the boardroom, the retirement market. This globe required the patina of the antique, the appearance of an heirloom. The map would still be contemporary – with Belarus and Uzbekistan and a united Germany – but it would be painted in such a way as to look as if there was still all to play for in the Crimea.

The bigger the globe, the bigger the blank space in the Pacific. Painter Mary Owen touches up the Churchill at Peter Bellerby's workshop.

Bellerby then turned his attention back to the *Churchill*, and the amount of oceanic space he wanted to fill with information. The larger the globe, the vaster the Pacific, so the plan was to fill the sea with all sorts of information, including data on the most popular religions (Christianity, Islam, Hinduism), the most practised languages (Mandarin, English, Hindu, Spanish, Arabic) and a table of cities by size of population, starting with Mumbai and Shanghai. He also wanted to include an extensive list of world leaders and heads of state, and possibly members of the European Community and UN Security Council. At the beginning of 2011 there were many changes to consider in Egypt, Tunisia and Syria. Beyond this, there was the particular problem of printing all this information on a curved surface, with every second or third letter requiring bespoke spacing.

'I'm doing loads of editing every day,' Bellerby explained. 'I need to change Sudan, and I'll check any border changes. When we bought our map in 2008 there were so many absurd errors, and now I just don't trust companies selling maps. Dar es Salaam was down as the capital of Tanzania; Tel Aviv as the capital of Israel. We're talking a hundred and fifty quite major errors. They had Tasmania as a country!'

I first met Peter Bellerby at Stanfords, the traveller's shop in Covent Garden. He had brought along a couple of his globes to display, but they looked out of place next to the cheaper, mass-produced models. It was a busy Saturday afternoon, and people were more concerned with buying maps and travel guides. But he did have a conversation with a man called James Bissell-Thomas, who was over in London from the Isle of Wight. Bissell-Thomas had the air of a man in authority, and he was not overly pleased to see Bellerby's globes. This was because he was a globemaker himself (of the company Greaves & Thomas). He began berating Bellerby for several things, not least the size of the paper caps or 'calottes' that served to hide and secure the tips of the twelve gores as they met at the North and South Poles (and inevitably obscured the poles themselves).

Bellerby was clearly taken aback with the force of the attack, and in his defence he said that he believed his own globes were of superior quality and represented better value for money. This spat, I discovered later, had a bit of previous. Two globemakers in the same spot on English soil was a rarity, and they were both battling for a small and specialised market. It wasn't like the nineteenth century, when several British manufacturers led the world in the production of globes for offices and schools. Now the schools and offices weren't interested because they had Google Maps instead, and there were only two proper bespoke British globemakers battling for the shrinking market during a recession, and they were within fist-throwing distance.

A few weeks after they met at Stanfords, I emailed James Bissell-Thomas suggesting a chat and a visit to his own workshop in the Isle of Wight, and he replied that he was 'a little wary' of my association with Bellerby. He claimed that Bellerby had broken open one of his globes and copied his method of construction, a method that previously hadn't been employed for four hundred years. 'Despite the above,' Bissell-Thomas reasoned, 'I welcome him to the world of globemaking ...'

On one visit to Bellerby's workshop, he told me that business was good and he would soon be moving to larger premises. He looked around him, ticking off his inventory. 'This one's going to Dorset, this one is potentially going to Taiwan. For the fourth *Churchill* I'm thinking of putting in ocean currents, because the person who ordered it, who is German, has just done a big sailing trip. He wants to have it in his house and talk to his grandchildren as he spins it round and tells them, "This is where we were and this is the route, and these were our tradewinds."'

Bellerby then considered the role he enjoyed least – selling. He believed he needed just one piece of good editorial to gain the momentum he was looking for, which would then lead to the elixir of all successful bespoke businesses, the waiting list. The *Financial Times*' 'How To Spend It' magazine was interested in running a piece on him and had again made Bellerby wonder how big the market for globes really was.

Bellerby thought he had the answer. It depended on how low you were prepared to go. Not necessarily in price, but in design. Would he feel he was compromising his desire to be the best globemaker in the world if he agreed for them to be hinged at the equator and used as drinks cabinets? Would he be prepared to take advertising in the Pacific region? If he could get into first-class airline departure lounges, would he be happy layering his work with a web of route maps? (The answer to the last dilemma was: definitely yes, if only the right people at the airlines would return his calls.)

Earlier in the day he had finally commissioned the first *Churchill* ball at a cost of £2,800. He had considered aluminium, but when he learnt that it might degrade after a century he had plumped for the increased tolerance of fibreglass. It was being made by a man who did most of his moulding for Formula 1, while the base was being made by a man who worked for Aston Martin near Birmingham. Bellerby, who loves cars, had enjoyed these recent visits, but he was most excited about a visit that week to a company called Omnitrack. Omnitrack specialised in highly calibrated glorified castors, and provided the answer to one of his biggest dilemmas – how best to spin the *Churchill* without ruining the globe or getting a hernia.

He demonstrated how these castors would work by spinning a small globe he had made for the artist Yinka Shonibare. The globe was placed on three small triangular shaped struts, each with a small plastic ball at its centre. When Bellerby spun it with his fingers he did so with the sort of delight shown by a child snapping the string from a gyroscope. The globe made a gritty sound and it kept on spinning for much longer than expected. The phrase 'executive toy' sprung to mind, and Bellerby decided that small globes barely 20cm in diameter would be his next big enterprise.

His excitement soon manifested itself into tangible success. By June 2011, Bellerby had a waiting list of twenty-five for each of his globe sizes, the first *Churchill* had begun goring, and so he had moved into a new building, a warehouse in a nearby mews. The place was perhaps ten times as large as his old workshop, and was most recently a supply depot for hardware shops. It was leaky in the rain, but this was the price of success; within four years, Bellerby had gone a long way to realise his dream of building a bespoke globe factory, something that re-established a five-hundred-year tradition.

Pocket Map
Churchill's Map Room

Churchill's globe didn't win the Second World War – but his Map Room made sure he didn't lose it. The Map Room was at the heart of a fortified subterranean office complex near the back of Downing Street known officially as the Cabinet War Room and then the Churchill War Rooms, and if any single space could lay claim to being the leader's command post, this was it.

It was about as low-tech as you could get. Previously an Office of Works warren, where civil servants ordered administrative supplies, the bunker was converted for wartime use at great speed after the Munich crisis of September 1938. By the time it opened less than a year later it had proper bedrooms and bunks in the corridors, a BBC unit, a cabinet room and a map room, all beneath reinforced beams.

Between thirty and forty people were engaged here in the plotting of the war, with the Map Room issuing daily bulletins to Churchill, his heads of staff and the King. It had four things going for it: the brains of its staff, a bank of coloured phones known as 'the beauty chorus', a huge array of maps glued to the wall and laid in drawers, and, in compartmentalised trays, what may have been the most concentrated supply of coloured map pins in the world. These plotted every movement of every

British and Allied warship, merchant ship and convoy, and with their specific codes – Red for British, Brown for French, Yellow for Dutch, Yellow with a cross for Swiss, White for German – they made the wall into a literal game of *Risk*.

There were other symbols too: cardboard ships and dolphins, the latter pinned to the oceans when a gale was due. 'When a heavy attack developed I found nothing so heartrending as the constant reduction in the number of ships in a convoy,' remembered one Map Room officer. 'One had to take down the cardboard symbol from the chart, erase the scribbled total on it and substitute a lower figure, perhaps only to repeat the process within a short while.'

The maps were not confined to a single room; the main one in Churchill's bedroom concentrated on coastal defence, marked with felt-tip symbols demonstrating permanent and temporary look-outs and barricades, areas suitable for tanks and areas susceptible to sea swell. A large curtain covered the whole map from visitors, and when Churchill pulled it aside it resembled a window from which he must have once feared the sight of invasion.

Commander 'Tommy' Thompson, Churchill's personal assistant, reported that there wasn't one day in London when Churchill didn't spend time in the Map Room or its annexe, frequently calling in at 4 or 5 in the morning to receive information ahead of his generals. Churchill keenly demonstrated how little Englanders should pronounce the names of foreign locations on the map. When the Map Room chief, Commander Richard Pim, pronounced Walshavn as Valsharvern, Churchill was quick to correct him: 'Don't be so BBC,' he said. 'The place is WALLS-HAVEN.'

Pim was an experienced naval man who hardly left Churchill's side for the duration of the war, except in May 1940, when he took charge of several motor boats to bring back some 3,500 troops from Dunkirk. He had set up his first map tables in a library at the Admiralty before Churchill became Prime

Churchill and his map chief, Commander Richard Pim.

Minister, and afterwards established portable map rooms during Churchill's travels abroad. Apart from the intelligence displayed upon them, the maps themselves were unremarkable, many dating to the First World War. We are accustomed to hearing Churchill break news of the end of the war to us, but it was Pim who broke the news to him. In reply, the Prime Minister told him, 'For five years you've brought me bad news, sometimes worse than others. Now you've redeemed yourself.'

By the time Commander Pim set up his map room aboard ship en route to the Tehran Conference at the end of 1943 (a meeting between Churchill, Roosevelt and Stalin to discuss the opening of a second front in Europe), he estimated that Churchill had already covered 111,000 miles on his wartime travels. In four years he had spent 792 hours at sea and 339 in the air. When the war was over, the British geographer Frank A. de Vine Hunt constructed a unique map showing the journeys undertaken by Churchill between 1941 and 1945 – nineteen of them in all – and it is one of the most compelling and descriptive charts of the war. It works as a map, a story and a puzzle, the viewer being encouraged not only to follow the various numbered arrows detailing Churchill's journeys, but also to wonder why they were undertaken.

Today, the war rooms are open to visitors, and the Map Room is arguably the highlight. It retains a solemn air, as if the liberty of the west still hung in the balance, and is very much as Churchill and his commanders left it in August 1945. One man's rationed sugar cubes remain in their packet, boxes of map pins sit unopened in a cupboard, a 'Confidential' map of the Balkans from July 1944 is spread out on the table.

At the centre of one desk is an atlas that Churchill and his staff are likely to have consulted frequently in the final year, and it is noteworthy for providing an American view of the campaign. *Look at the World: The Fortune Atlas for World Strategy* was published in New York by Knopf in June 1944. The world it presented in its pages attempted, its editor declared, to show not just 'the strange places in which Americans are fighting and the distant islands and promontories that the trade routes pass' but specifically why the Americans are fighting in such strange places – an explanation, for instance, of troops in Greenland, Iceland and Alaska.

The maps were modelled on the 'complete azimuthal equidistant projection' centred on the Poles. The northern hemisphere thus featured North America near its centre, with Asia above it and Africa on its side on the upper right. The design of the double-page spreads highlighted the curvature of the earth, with its chief cartographer Richard Edes Harrison explaining in his introduction that the design reflects the new key instrument of the war, the aeroplane; it made perfect sense to construct a series of aerial war maps that amplified the complexities and vulnerability of troop movements on the ground. Harrison begged indulgence from readers who might find his unconventional projection disturbing, in the same way Mercator did some four hundred years before.

The commentary on each spread serves as a unique snapshot of the way the US viewed not only its role in the war, but its position on the globe. This was not an impartial atlas

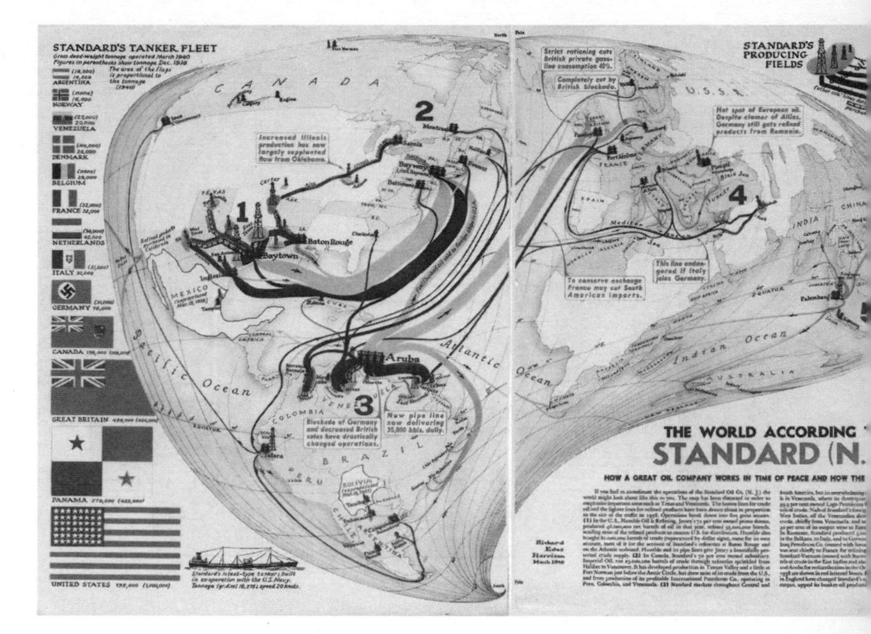

One of Harrison's wartime maps, showing the movements of tankers belonging to Standard Oil.

(if there ever was such a thing), but a judicious bit of geopolitical propaganda. 'The Mediterranean world, into which the Americans erupted with startling suddenness on November 7 1942, is the birthplace of Western civilisation,' the text proclaims at the foot of an arched map of Europe. The Alps 'may have figured largely in German ideas of a European fortress,' but an aeroplane 'makes light of Mediterranean mountains.'

Harrison hoped that his flat-map/globe combination gave the allies an advantage not available to its enemies. 'The Germans, in spite of excellence of execution, are notoriously traditional about maps,' he wrote. 'If they have "the geographical sense" neither their maps nor their strategy shows it'. No wonder Churchill found the atlas indispensable.

Chapter 19
The Biggest Map Dealer, the Biggest Map Thief

'Make me an offer!' W. Graham Arader III tells me after I had asked him about buying some of his most treasured possessions. 'Everything's for sale!'

This came as little surprise. W. Graham Arader III is the biggest map dealer in the world (the wealthiest, the most famous, the most combative and bombastic, the most feared, the most loathed) and I am in his bedroom, looking at the maps on his walls and getting the impression that in Arader III's world everything has really been for sale for ever, with the possible exception of his wife and seven children (one of whom is called W. Graham Arader IV). Every inch of Arader III's six-storey townhouse on the edge of Central Park is covered in maps – above his bed, above the fireplace, over and across his desk and over the doors. I think the walls are papered rather than painted, but it is not always possible to tell. The only places that haven't got maps on them are those covered with his other passion, rare natural history prints.

What sort of maps does he have? Every type of map! Or at least every type of map that has value, beauty and rarity, which means a concentration on nineteenth-century America and sixteenth-century Europe, all the great names. This is his Madison Avenue home and his showcase, but he also has four other galleries around the country. Between them they display a classic, framed history of cartography – Ortelius, Mercator, Blaeu, Visscher, Speed, Hondius, Ogilby, Cassini, John Senex, Carlton Osgood, Herman Moll and Lewis Evans – and the stories they portray serve as a neat summation of five centuries of trade and power. Here, etched large, is the Venetian silk route, the growth of the Dutch empire, the reigns of Suleiman the Magnificent and Philip II of Spain, the birth of America, the naval highpoint and subsequent decline of Great Britain.

Listening to Arader talk, you would soon believe that he has handled every map of significance in the history of the United States. 'Yes, I did own the map that Lewis and Clark used to plot their trip,' he says. 'I did own the Louisiana Purchase Proclamation signed by Jefferson and Madison ... Yeah, I'm number one. We've got a stock that is better than the next fifty dealers combined. It might be a billion dollars' worth of stuff, or five hundred million. It's really embarrassing that I have as much stuff as I do.'

It is mid-morning. Arader, who is a tall man with a considerable girth, has just been playing squash and is still wearing shorts. As he goes to shower I talk to Alex Kam, one of his strategic planners. He says that he is working hard on building up a new base of collectors, not least online, 'because if you only focus on the folks who have been spending a lot of money over the years, they get a bit older.' Kam finds that a lot of new US buyers are specifically interested in buying maps of their home states, 'those who want to look back at who they were.' One notable and obsessive exception was Steve Jobs, who used to pay Arader huge prices for botanical prints by the Belgian watercolourist Pierre-Joseph Redouté. 'He really loved Redouté roses,' Kam blogged after Jobs died in 2011, in the same month as one of these small nineteenth-century roses on vellum was on sale at Arader Galleries for $350,000. According to Kam, Jobs 'loved them so much he wanted us to gather up and buy every original Redouté rose in the known universe.'

When Arader returns he says he doesn't have much time, because he has a 'five million dollar sale' he needs to attend. He says he will be both buying and selling, but mostly selling. 'I have no choice really. The minute I lose sight of being a merchant it's over. I have overheads of $450,000 a month.'

The W in Arader's name stands for Walter, he tells me, which was also the name of his father and grandfather. His father was a navigator in the navy and studied maps professionally. When he became Secretary of Commerce of the Commonwealth of Pennsylvania in the 1960s, map collecting became a hobby, but he was interested, his son says, more in their beauty than in their purpose.

The young Arader began buying maps for him when he was eighteen and spent a year in England before starting at Yale. He bought from all the big names, including R.V. Tooley and Maggs Bros (one of the oldest London antiquarian dealers, achieving particular notoriety a century ago when they bought and displayed Napoleon Bonaparte's penis). Arader snapped up many pages from the big Dutch and Flemish atlases; in the 1970s these cost perhaps £80, compared with the £8,000 for a fine example today. At Yale he fell under the spell of Alexander Vietor, the head of the map department, and he began selling maps from his dorm room, predominantly, he says, to Jewish doctors at the Yale medical faculty. Then he started selling at antique fairs, where he was astonished to find himself the only map dealer present.

Arader discovered two things: that he had a talent for buying cheap and, after enough schmoozing and vamping with impressionable clients, selling dear; and that the map world was a sleepy place ripe for an alarm call. Thus he began

'Make me an offer!': W. Graham Arader III at home in his Atlas Room.

fulfilling what can only be called his eponymous calling: he became a raider. He bought what he claimed were 'criminally undervalued' maps and increased their price by ten.

It is unclear precisely how much the boom in the prices (one hesitates to say value) of rare maps in the 1970s and 1980s can be credited to Arader, and how much he was surfing a wave; the two probably fired each other. His tactics were to buy the very best and rarest, which meant few could compete with the quality of his stock (he has a habit of dismissing lesser items as 'common as dirt!'). He bought vast amounts, sometimes sweeping up entire auctions and estates, and he dreamed up a plan in which he and a few other leading dealers engineered a mass purchase of all the great maps in the world that weren't already in institutions.

This didn't happen in the coordinated way he planned, so he tried to do something similar on his own – and in so doing he became unpopular, very unpopular. 'Graham's pretty much at war with all of his colleagues,' another leading American dealer told me. 'He doesn't see much place in the world for anyone but himself, I guess. He adopts a very scorched-earth attitude towards everybody, and he's called me various names, including "a nest of vipers". Exactly how I personally manage to be a nest of vipers I'm not sure.'

It is easy to see how people might be offended by Arader's style. When I emailed him to request an interview he immediately put my letter on his blog. In his reply he said that he found almost all other map dealers 'dishonest, sacrilegious and wicked … I have been only able to find TWO honest map dealers other them myself in forty years.'

As one might expect from the map world's P.T. Barnum, Arader is not shy of publicity, and is a favourite of *Forbes* and *Fortune* magazine. He is a master of the soundbite: 'The Sun King Louis XIV – yes he built Versailles, but he was a jerk!' In 1987, a profile in the *New Yorker* characterised him as 'unscholarly yet imposingly knowledgeable,' which is spot-on. He tells me he reads a book a day. He picks one up from his coffee table, a book about Mount Desert Island off the coast of Maine (he's just bought the original manuscript map of it). He tells me he has more than 50,000 map reference books in his house, which he claims makes it the finest library of the history of cartography in the world, outgunning the British Library and Library of Congress (debatable, but he often makes the grand, sweeping statement that is difficult to check). 'When I buy something really important I will buy between a hundred and three hundred books on the subject, I will hire a professor from Columbia University to come and teach me.'

He looks around his bedroom. 'Very few people can tell you who the King of England was during the French and Indian wars, the American Revolution, the Treaty of Paris, the Louisiana Purchase and the War of 1812,' he says. 'That was George III. To have that as part of your core gives one a tremendous thrill for map collecting. You understand why Philip was able to subjugate Europe to the counter-reformation. The counter-revolution occurred because of his cash, his specie, this unbelievable bounty of gold from Mexico City. And these wars of religion, this festering fury, pretty much screwed up Europe for two hundred years ... This whole room is the story of what Philip's barrel of gold does.'

Arader parades his glossy knowledge frequently, but it only goes to reinforce what map lovers have always known: that to know about maps is to know about one's place in the world. 'I used to just see pretty maps, but now I see this whole cauldron of history,' he says. 'And I really know this history cold – I mean, you have to be a full professor somewhere to really keep toe-to-toe with me on this.'

Arader has a thing about professors, and, like so many successful and wealthy businessmen, he yearns to be respected not for his trading acumen but his scholarship. Accordingly, his ambitions are changing. 'Having all this is nice, but I hope to give it all away and die penniless,' he says. 'I'm sixty. By the time somebody's sixty, it's silly to be a person who focuses just on making money.' He shows me appreciative letters from the Dean of Northeastern University in Boston, thanking him for his generous donations of maps.

He then goes to his computer and searches the file titled 'Arader: My Dream'. 'Here it is. This is the course they're teaching – a different group of maps every day. My idea now is to take these maps and prints out of the museums and out of the libraries and out of the drawers, and use my influence and my wealth to teach the young.' He says he's giving them 'great big slugs of great material,' because he wants 'sixty bright-eyed kids every day to look at these maps and embrace history and geography and design and research.'

'Maps for education,' Alex Kam adds. 'That's where the sizzle is!'

The map world is not generally abrasive. I visited several map dealers and collectors, and found most of them civil, learned and passionate. But one thing they tended to have in common was a dislike of W. Graham Arader III.

Jonathan Potter, a mild-mannered London dealer who has known Arader for forty years, was among the most generous when he concluded, 'He's done a huge amount for the map business. He's brought a lot of collectors into the field. But I can't think of many people who haven't been bad-mouthed by him.'

Indeed. Most people with whom Arader was once friendly talk about him mainly in respect of their falling out. One such is William Reese, one of the world's leading experts on maps of the United States, whom we last met in connection with the Vinland map. Reese is a few years younger than Arader, and their paths crossed briefly at Yale (he was the man Arader called 'a nest of vipers'). When I visit him in his office in New Haven, Connecticut, in September 2011, he is cynical about Arader's new-found philanthropic urges. He tells me that when he had asked Arader to match a $100,000 donation he had given to Yale to help create a digital catalogue of its map collection, Arader had ridiculed his letter and posted it on his blog.

Reese is from Maryland, but a lot of his family grew up within walking distance of Yale. His professional interest in maps began as an adjunct to his interest in antiquarian books, but he says he has loved maps since he was young. 'I'm the sort of person who will take a coast-to-coast plane ride and follow our progress with a nationwide road map, as if we were doing it on the ground.'

His lucky break in the map world sounds like one of those mythical stories of finding a Rembrandt in the attic. In 1975 he was browsing through a big sale of a recently deceased book collector called Otto Fisher. Reese had originally gone to look at the picture frames, but then found something interesting in the rug sale. It was a map rolled up in brown butcher's paper. 'I felt very excited,' he remembers. 'I was pretty sure I knew what it was.' He had taken a course in Mesoamerican archaeology, and he thought that the map, which showed a Mexican valley from the sixteenth century, was on paper made from fig-bark. He paid $800 for it. When he got it home he narrowed it down to about 1540. He discovered that there was also another map on the other side. He decided to offer it to Yale, where he was in his second year studying American History. 'I showed it to them and we all sat there poker-faced. They asked, "Well what do you want for it?" It then cost $5,000 a year to go to Yale, so I said, "I'd like $15,000!" '

The college agreed. Reese instantly thought 'Damn!' It's now hanging in Yale's Beinecke Library, not far from the Gutenberg Bible. 'At the time it was probably worth about $25,000,' Reese thinks. 'Now it's probably worth two or three hundred thousand. But I don't care because it completely set the course of my life, and I walked out of there walking on air. I thought, "I can make a living out of this".'

I ask Reese how many serious (scholarly and moneyed) map collectors there are in the United States, and the answer surprised me: 'very few'. It was the 'You Are Here' factor again – most collectors collect only for a few years, predominantly interested in the area they were born or live in. Others only collect maps of the world. 'They start with the Mappa

Mundi as their first interest,' Reese says, 'and then go on from there. Most of the people who collect like that I would define as real amateurs, enthusiasts. Often their collecting is defined by their wall space – once they fill their walls, they're done.'

There are far fewer people who take it to the next level, where they store maps in drawers. Reese believes there are only 'a couple of dozen' very serious map collectors in the US, perhaps a couple of hundred worldwide, not including the institutions. There is one simple reason there aren't more: the scarcity of the really great items. 'With maps there's a risk that at the very, very top end there may soon cease to be a market,' Reese says. 'They've all been snapped up.'

But something strange happened to the supply of very fine maps in the first years of this century: items that were highly desirable but previously not around, began to appear. Bill Reese noticed it with rare American maps of the seventeenth century, particularly those drawn by the French explorer Samuel de Champlain, the first man to properly map the Great Lakes. Dealers suddenly had maps for sale that hadn't been seen for decades, and although they swiftly disappeared into private collections, they created a rare liquidity in the marketplace.

There was, of course, a reason for this: these maps had been stolen.

In September 2006 a fifty-year-old man called Edward Forbes Smiley III was sentenced to 42 months in jail and fined almost $2m after he admitted stealing 97 maps from Harvard, Yale, the British Library and other institutions. It was the biggest map theft anyone could remember, and it shocked libraries not only because of their losses (many of which they were unaware of until Smiley's arrest), but also because they had trusted him and enjoyed his company. Or at least they enjoyed

The unacceptable face of map-dealing: Edward Forbes Smiley III.

his connoisseurship: Smiley had been a dealer since the late-1970s and in the trade was considered 'one of us'.

But in fact he was really a common criminal. He used a blade from a craft knife to slice pages from books and atlases, which he would then disguise by trimming again. He favoured this method over another classic map thief's trick: after rolling a ball of cotton thread in your mouth, you unravel it against the bound edge of a map you wish to steal and close the volume; in a short while the enzymes in your saliva will thin the binding until it is weak enough to remove.

Bill Reese remembers dealing with Smiley in 1983 – the first and only time he trusted him. He sold him an American coastal atlas for $50,000, but his cheque bounced (he says he got the money eventually). Their paths crossed subsequently at auctions and map fairs, but their next proper encounter was in 2005, when Reese was brought in by Yale to assess the thefts from their collection. (Smiley had been caught at the university's Beinecke Library when a librarian spotted an X-acto blade on the floor; he was found with several rare maps in his briefcase and blazer pocket when he fled the building, including a map by Captain John Smith from 1631 – the first to mention New England.)

To enable Reese's audit, the library's map department was closed for an entire term. One of the problems he encountered was that the maps had not been fully catalogued electronically, something Forbes may have counted on when planning his targets. 'Smiley stole cards out of the card catalogue to cover his tracks,' Reese says, 'but it didn't work because they had a microfilm that had been made of the card catalogue in 1978, which he didn't know about.'

The Forbes Smiley case has had a modest impact on how leading research institutions protect their treasures. They tightened their security as best they could, and often added CCTV coverage. But many librarians felt uneasy at having to screen readers they had trusted for decades, and new security measures ran counter to the perspective of a library as a civil place designed for the free dissemination of knowledge. As one curator said at the time of the Smiley saga, 'we're in the business of being vulnerable.'

Small-scale map theft began at the same time as maps began; before they were valuable and decorative, they were just useful. But the significant heist stories in the last half of the twentieth century are compelling not only for their audacity and Cold War techniques, but also for the grand scale and serial nature of their crimes. Thieves kept on going back and back to the same rich well, as if a bank had left its vault open and the keys to the safe on the table.

In March 1963, for example, several colleges at Oxford and Cambridge found that for the previous ten months a man named Anthony John Scull had been razoring through their atlases with unfettered glee. He stole more than five hundred maps and prints, the majority from King's College, Cambridge. They included all the greats, at a time when all the greats were not under white-gloved supervision: Scull got away with Ptolemy, Mercator and Ortelius. In those days even rare maps didn't achieve great prices, but there is no doubt that dealers and even a couple of auction houses were grateful for Scull's supply. The total value of the stolen maps was estimated at £3,000 (perhaps £3m today) but the greater damage was to the broken atlases and books left behind.

In the United States, the most bizarre and cinematically heroic story may be the 'cassock crimes' of 1973, in which two American Benedictine priests stole atlases from leading college libraries and stored them in their monastery in Queens. Then there was the story of the Scandinavian thefts orchestrated by a British pair, Melvin Perry and Peter Bellwood, who thought – rightly as it turned out – that the best place to steal maps at the end of the 1990s would be the Royal Library in Copenhagen. Both Bellwood and Perry had previous convictions for thefts from the British Library and elsewhere when they started flying to Denmark. (Bellwood had a nice touch: on one visit he gained the librarians' trust by handing in a 500 Kroner note he had 'found on the floor'. He got away with Ortelius and Speed.)

Another famous case, again in the 1990s, was that of Gilbert Bland, who had stolen around two hundred and fifty maps before he was spotted by a reader ripping a page from an atlas at the Peabody Library at Johns Hopkins University in Baltimore. Bland sold his maps from a shop he ran with his wife in Florida, a stock he had purloined from nineteen libraries, from Delaware to British Columbia. His defence? 'I just wanted them.' Bland became the subject of an entertaining book, *The Island of Lost Maps*, by Miles Harvey, which also features a star turn from Graham Arader. Harvey called Bland the 'Al Capone of cartography, the greatest map thief in American history,' though that was before Forbes Smiley and his blade had appeared on the scene.

Where else would an eager person go for their cartographic education if not an increasingly fortified library? How about a series of tutorials run by W. Graham Arader III and his colleagues? Shortly before I visited him, Arader had begun placing advertisements for a summer course for high school and college students on which they could learn aspects of dealing. The course cost $1,200 a week and included what Arader had classified as the four key steps to making a sale: 'A: Introduction. B: The Artwork exists. C: The Artwork can be owned. D: The Sale of the Artwork.' Successful applicants would also receive lessons on maintaining relationships with clients and how to best use the Internet for trade. 'Lots of my clients pressure me into hiring their entitled children,' Arader told me. 'So now they'll have to pay for it. A kid comes in [to work with me], doesn't do anything the entire summer, gets about $5,000-worth of education – that's over!'

When I asked Arader for some of the key nuggets that he'd be offering his students, he directed me to his blog. Here he spoke of simple advice: compose handwritten thank-you letters; don't bombard your clients with emails but offer wise, measured and timely advice about items they might be particularly interested in.

There was also advice about spotting fraudulent maps that were masquerading with 'original colour'. In June 2011

Arader had taken a group of four interns (the last of the unpaying lot) up the road from his house to a viewing of an upcoming auction at Sotheby's. They settled on Lot 88, a classic seventeenth-century study of Africa with many maps by the Dutch geographer Olfert Dapper.

'It had the most amazing example of fake original colour imaginable,' Arader noted. 'We were all completely fooled – the greens oxidized through the paper almost perfectly. My first reaction was great excitement and lust for what looked like a magnificent book. But it was not.'

Arader's excitement soon turned to fury. He saw how cleverly the forger had tried to replicate the oxidisation of green that was usually the key indication of genuine contemporary colour. 'The only way to protect yourself is to look at atlases with original colour in the great libraries of the world,' he advised his interns, and vowed to expose the worst culprits. 'Hopefully one of these creeps will be lured into suing me so that we can use the courts to finally come to an answer. I am on the warpath! Be warned.'

Pocket Map
Women Can't Read Maps. Oh, Really?

In 1998, the Australian couple Barbara and Allan Pease self-published a funny and gentle book called *Why Men Don't Listen & Women Can't Read Maps: Beyond The Toilet Seat Being Up*. Within a year, the book had lost the bit about the toilet seat but had become a global hit (12 *million* copies), and not long after that it had become one of those books that people talked about at bus stops and at work. It was a war-of-the-sexes study a bit like John Gray's *Men Are From Mars, Women Are From Venus*, except this one took things further, veering into bonkers-land. It explained why men can't do more than one thing at a time, why women can't parallel park, and 'why men love erotic images and women aren't impressed.'

With reference to maps, their findings are unequivocal. 'Women don't have good spatial skills because they evolved chasing little else besides men,' they assert. 'Visit a multi-storey car park at any shopping mall and watch female shoppers wandering gloomily around trying to find their cars.' The Peases were really confirming a stereotype that had existed since Columbus laid out his sailing gear: men, nervous of asking a stranger the way, just get on better with directions that fold.

But is any of it true?

Several years before the Peases turned their success into a cottage industry, academics had begun publishing gender-related map studies of their own. Actually they'd been doing this for a century, but from the late-1970s they began appearing with an unusual urgency. In 1978 we had *Sex Differences In Spatial Abilities: Possible Environmental, Genetic and Neurological Factors* by J.L. Harris from the University of Kansas; in 1982 J. Maddux presented a paper at the Association of American Geographers entitled *Geography: The Science Most Affected by Existing Sex-Dimorphic Cognitive Abilities.*

Their approaches and findings varied, but yes, most psychological studies seemed to feel that when it came to such things as spatial skills, navigation and maps, men did seem to have the upper hand. It was thought likely that this would explain why the number of men doing PhDs in geography in the 1990s outnumbered women 4 to 1. It may also explain why, in 1973, the *Cartographic Journal* published a report by a man called Peter Stringer that stated he had recruited only women in his research into different background colours on maps because he 'expected that women would have greater difficulty than men' in reading them.

But what if there was a very simple explanation to all of this, beyond unfettered prejudice? What if men and women could each read maps perfectly well, but in different ways? What if the only reason women had trouble reading maps is because they were designed by men with men in mind? Could maps be designed differently to appeal to women's strengths?

In 1999, a project was carried out at University of California by academics in geography, psychology and anthropology. This involved an extensive review of the existing literature on spatial abilities and map-reading, well over a hundred papers by now, and also a new series of experiments involving 79 residents of Santa Barbara (43 women and 36 men, aged between 19 and 76).

The strongest conclusions in the literature – that men were better than women at judging the relative speeds of two moving images on a computer screen, and also at successfully judging the mental rotation of 2-D and 3-D figures – were thought to be not overly practical in real world situations. So the new experiments involved a bit of walking about and map sketching, as well as responding to verbal directions and learning to read fantasy maps. One of these showed an imaginary theme park named *Amusement Land*. It measured 8.5" x 11" and featured such landmarks as Python Pit, Purple Elephant Sculpture and Ice Cream Stand; having been given some moments to study it before it was taken away, participants then had to sketch the map for themselves with as many landmarks as possible. They were asked to perform a similar task with another fictional map entitled *Grand Forks, North Dakota*, which was actually a rotated map of the city of Santa Barbara. They were also led on a walking tour around a

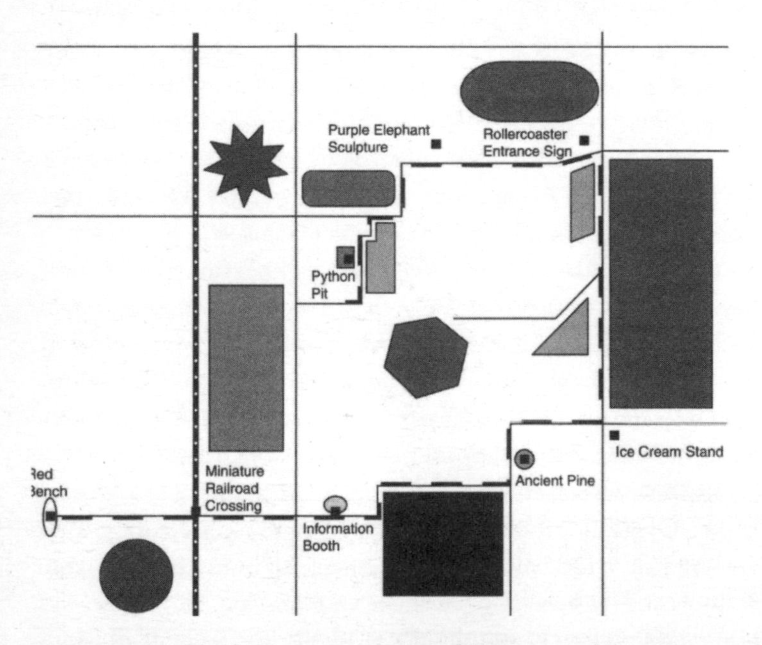

Amusement Land, where cartography, sex-differences and ice cream meet, at last.

part of the university campus, before being given a map of the area and asked to write in the route they had taken.

The authors concluded that although men were better at some tasks (estimating distances and defining traditional directional compass points), women were better at others (noting landmarks, some verbal description tests). When it came to map-use, both imaginary and real, women ruined the Peases' book title; they could read maps as well as men, only they read them slightly differently.

And there was growing evidence that they knew it. In 1977, the *Journal of Experimental Psychology* published an experiment that found that 20 of 28 male participants but only 8 of 17 females regarded themselves as having a 'good sense of direction'. But by 1999 this had shifted, at least in Santa Barbara, as residents of both sexes reported feelings of growing ability. Out of ten categories (including 'It's not important to me to know where I am', 'I don't confuse right and left much' and 'I am very good at giving directions') there was no statistically significant difference and high self-belief among both sexes. Men did appear to be more confident when in the category of 'I am very good at judging distances', but on the big issue of 'I am very good at reading maps' there was again no discernible difference.

What, then, is the perceived problem? The problem is, although women have no difficulties with navigation, the way they are told to navigate may be at fault. In December 1997, in an early British edition of *Condé Nast Traveller*, a writer called Timothy Nation wrote a brief essay wondering why, when we wander the streets of London it is far easier to get around by looking out for well-known landmarks than it is to rigidly follow a map or compass. This was because maps only follow the line of a street – they look down. But when we walk we tend to look up and around. The flat, two-dimensional, look-down approach is suited to cognitive strategies used by men, but it is one that generally puts women at a disadvantage.

Timothy Nation, whose real name was Malcolm Gladwell, the writer not yet known for his books *The Tipping Point* and *Blink*, then examined a famous experiment conducted at Columbia University in New York City in 1990 involving mazes and rats. This found that, when searching for food, male rats navigated differently to their female counterparts. When the geometry of their testing area was altered – dividers were introduced to create extra walls – the performance of male rats slowed down, while there was little effect on females. But the opposite was observed when the landmarks in the testing room – a table or a chair – were moved. Now the females became confused. This was the big news: males responded best to broad spatial cues (large areas and flat lines), whereas females relied on landmarks and fixtures.

Could this have been a freaky result? Possibly, but several other experiments in the last twenty years have produced similar findings. The most recent was in 2010, when the American Psychological Association reported an Anglo-Spanish experiment in which even more rats were placed in a triangular-shaped pool to find a hidden platform. Again a comparable result – female rats benefit from locational cues, whereas the males race by them.

Similar experiments have continued with men and women, again with comparable results. Few psychologists would now argue against these navigational differences. What is less certain is how these changes have come about. But we may well be back on the African plains with the hunter-gatherers. In this theory – and it's a plausible one – men and women's brains both developed through their navigational skills, but in different ways. Men sought the broad sweep of tracking and pursuit over large areas, while women tended to be down with the roots and berries, foraging skills aided by memory, and memory aided by landmarks.

The traditional map, however, a 2-D flat plane, is designed by hunters for hunters. Female gatherers don't get much of a look in. But with 3-D maps – either panoramic views with highlighted landmarks on paper, or digital renderings on screen – the road ahead becomes instantly more readable.

One further experiment, conducted in 1998 by social psychologist James Dabbs and colleagues at Georgia State University, found that the strategy differences of the sexes extend to verbal communication. Dabbs found that when men give directions they tend to use compass points such as north or south, whereas women focus on buildings and lists of other landmarks en route.

So perhaps Barbara and Allan Pease were right after all, or at least half-right. Men don't listen because they don't need to listen so much. And women can't read maps because they're the wrong sort of maps. What can possibly save this troublesome marriage? A plastic dashboard-mounted box perhaps?

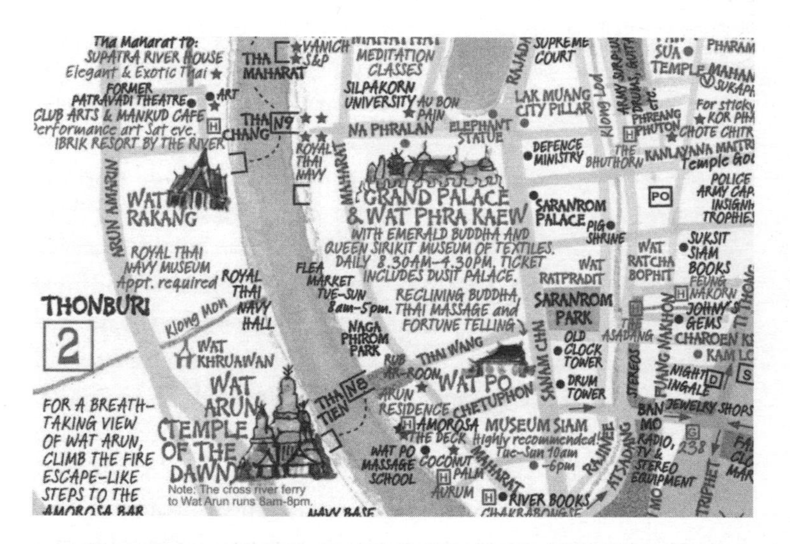

Maps for Women? Nancy Chandler, based in Bangkok, has been producing highly successful '3-D' maps of Thailand for the past two decades. They look crowded and a little chaotic but feature hand-drawn landmarks and useful text, as well as colour-coding for different attractions. And, yes, she notes that her maps are bought and used predominantly by women.

Chapter 20
Driving Into Lakes: How Sat Nav Put the World in a Box

It is the early 1980s, and you are travelling in an aeroplane at the beginning of your holiday. After take-off and drinks there is to be a movie. *Mr Bean* has not yet been made, and in-flight entertainment is still in its infancy, so the choice will be limited to one family film involving animals, shown on a row of small screens suspended from the ceiling. When the film is over there is another film, but it's one you've seen before: a stubby plane flying across the sky from your place of departure to your place of arrival. You glance at it half an hour later and it's still aloft. It's the most boring piece of information it is possible to watch.

The poet Simon Armitage has written of watching this film on a budget twelve-hour flight to Japan. The plane on the screen crawls over sea and land at a terrible pace, like a slow-motion shot on a putting green. Armitage watches with increasing desperation; after about three hours, a man behind him shouts 'GET IN THE HOLE!'

They're showing that movie again ...

The plane on the screen – not a film at all, of course, but a real-time miniaturisation of our point above the earth – is peculiarly compelling despite its deathly monotony. The screen also displays stats on ground speed and outside air temperature, but it is the plane on the map we can't stop watching, slightly jerky as it heads towards Stavanger and other places rarely in our thoughts, a map that serves to extend the tedium of flight rather than heightening its wonder. For many of us, travelling in the days before satellite navigation and phone apps, this was the first experience of watching a moving map.

The maps were first made available to passengers by an American company called Airshow Inc, but the fundamental technology that drives them – a system involving a calibration of gyroscope and accelerometer – originates from the aeroplane's cockpit. Known as inertial navigation, it is the map that has taken us on holiday for more than half a century.

Inertial navigation has its roots in military research, specifically in missile guidance. Before the Second World War, missiles used to be fired in much the same way as Horatio Nelson fired cannon balls – destructive velocity outgunning geographical felicity. But then the Germans and Americans devised a way of steering weapons by rudder and tailfin (the Germans used it on their V2s, a switch from blanket bombing to slightly more accurate targeting), and the technique advanced to such a degree that it became a way of describing things that were complicated: rocket science. But its basic science is something we can all understand. A fixed point is logged at take-off, and computers then maintain a route by adapting to airborne motion sensors. At its core, the system is not too far removed from the method of 'dead reckoning' employed by Columbus and Magellan to navigate the oceans.

That said, moving seat-back maps have come on a bit since the 1980s, and today offer 3-D graphics and instant satellite updates. The experience is now almost as good as looking out of the window. Inertial navigation is still used to guide our planes, although in the last decade it has been joined by another more accurate technology. It is the same piece of kit that has encouraged sea captains and car drivers to abandon their more traditional methods of navigation – lighthouses, paper maps, intuition – in favour of the indomitable, indisputable and surely failsafe method of advanced trilateration known as the Global Positioning System.

Over the years we have refined GPS to a point where almost anything going anywhere can be guided without a human being fouling things up. When the cruise ship *Costa Concordia* ran aground on a Tuscan shore in January 2012, killing thirty-two people, it wasn't GPS that was blamed; it was the captain for ignoring it. When the space shuttle *Discovery* took off for the International Space Station in 2006, GPS not only tracked its orbit, but also (for the first time) guided it home to Cape Canaveral. When people with a phone get lost anywhere in the world without a map, they are located and rescued thanks to GPS. And when you allow the 'enable location?' option on your phone app, it will be GPS that guides you to your nearest cashpoint.

What a wonder. What a potential disaster.

GPS is now such a significant part of our lives that the effects of failure would be catastrophic. Malfunction would be a blow not just to the digital cartographer and the iPhone user, it would be as if the world's entire harvest of electricity, oil and gas had run out at the same time. The loss of GPS would now affect all emergency services, all systems of traffic control including shipping and flight navigation, and all communications bar semaphore. It would affect the ability to keep accurate time and predict earthquakes. It would set the guidance and interception of ballistic missiles to haywire. What would begin with gridlock at road intersections would very rapidly turn the world dark, and then off. Everything would stop. We would be practically blind. We would not be able to stock our shops and feed ourselves. Only those who knew how to plough a field like they did in the middle ages would have a chance.

In the happy meantime, we have the pleasure of the TomTom Go Live 1005 World, a dashboard-mountable satellite navigation device that costs about £300 and contains detailed road maps of 66 countries. It is an astronomical, geographical and technological achievement of the greatest magnitude, and, given that it comes in a plastic and glass box with a five-inch HD touch-screen and live traffic reports, and, for an extra few pounds, the voice of Homer Simpson exclaiming 'You have reached your destination, and you can hold your head up high because you are a genius!', it is something that would have steered Ptolemy to the asylum.*

One buys a TomTom device (or a Garmin, Strabo, Mio, or GoogleMap-enabled phone) not for the maps but for the directions, which makes it a rather old-fashioned concept. Lost without a map in the pre-digital world, one would ask for turn-by-turn instructions from locals – left at the church and then left again at the town hall – and a sat nav performs precisely the same function on a grander scale. Sat nav is not attractive because it is a road map of course – Appleton, Michelin, the AA and Rand McNally could have sold you one of those a century ago. It is attractive because it is a language, a code, and for people who could never read a map before, a liberation.

Before we consider how it works, we should perhaps consider what happens when it doesn't. In 2010, a driver in Bavaria followed his sat nav when it told him to do a U-turn on a motorway, and crashed into a 1953 Rolls Royce Silver Dawn. The Rolls was was one of only 760 made, which may explain why its owner promptly had a heart attack. He recovered; his Rolls did not. A few months later there was the case of Robert Ziegler, a Swiss van driver who followed his sat nav up a narrow mountain goat track and, being unable to turn around or reverse, had to be rescued by helicopter. Then there was the sad tale of a mini-cab driver in Norfolk who followed his sat nav into a river. His boss gleefully told the newspapers 'He was in the car with his trousers rolled up. Fish were swimming around the headlights!'

There are so many of these examples that one has to wonder at which point drivers will stop entrusting their lives

* If you prefer, you can have Stephen Fry saying 'Now, when possible, could you turn around. Ideally, don't do it when impossible. Bless.' Or a character called All-The-Way Annie say, 'Take the ferry. Oooh, let's rock the boat, baby!' For a limited time you could also have bought a *Top Gear Edition* in which Jeremy Clarkson calls you a 'driverist' and observes, 'after 700 yards, assuming this car can make it that far, you have reached your destination, with the aid of 32 satellites and me. Well done!' but there were endorsement problems with BBC Worldwide and Clarkson is no longer available.

When sat nav fails, episode 97: Robert Ziegler's van, guided up a Swiss mountain goat track.

to the moving maps and start driving with their brains again. Is it when a journey that normally takes forty-five minutes through the back route starts taking an hour? Is it when a person who wants to go to New Haven, Connecticut, sets off in an undemanding way towards New Haven, Indiana? Or is it when we have grown tired of clogging up small villages with heavy goods vehicles and demand to be directed onto the path of an oncoming train?

Sat nav is military software for dummies. You turn it on, enter a postcode and it tells you, in certain words, precisely how to get there turn-by-turn. It takes both the fun and the anguish out of driving, and the challenge and rewards out of maps. The little box lets us forget just how big maps can (and, perhaps, ought to) be. Out on a hike, a simple GPS handheld has already begun to obliterate the pleasure and the exasperation of the massive windblown Ordnance Survey sheet, and the sat nav has done away with the large spiral-bound road atlas or A-Z. But at the same time, it can be harder for our brains to process the information. A large sheet map offers a perfect way to register where we have come from, where we are going, how we get there.

Satellite navigation as we understand it began in 1960 as an improved method of guiding intercontinental ballistic missiles. The project, officially called NAVSTAR, remained exclusively within the US Department of Defense* until the 1970s, when it was partially opened up to civilian use. Locational accuracy remained vague (to about 25 metres) until 2000, when Bill Clinton declared that since the Cold War was over it was high time for the last military restrictions to be lifted from GPS, and the modern era of electronic mapping began.

The way it worked was ingenious and insanely complicated, but at its core there was still a system of old-fashioned triangulation. A minimum of 18 orbiting satellites are required to cover the whole earth, although as many as 32 may be operational. The satellites beam their positions at precisely the same time via radio signals and electronic code, with each end user – be it rambler, pilot or driver – receiving the information via their (increasingly small) box of reckoning electronics. Several ground stations (in Hawaii, Ascension Island and elsewhere) receive the codes, and remotely monitor the satellites' health and accuracy.

* Like Antarctica, GPS is effectively run by the United States. But there are several other global navigation networks at various stages of execution, including systems in Russia (Glonass) and Europe (Galileo). The Chinese government, ever fearful of external reliance and influence, launched its own satellites to construct its own guidance systems, supplying regional coverage. This is now expanding into a network called Compass, extending beyond Asia to acknowledge a wider world.

The classic and simplest way of explaining the principles of GPS involves imagining yourself floating in a room at zero gravity. You measure your distance to the closest wall with a tape measure, then another wall the same way, and then your distance from the ground. You can calculate your position in relation to the fixed walls, floor and ceiling. The orbiting satellites and the GPS receiver are essentially a larger version of this, with radio signals taking the place of the tape measure.

The non-military concept of satellite navigation first became popular at the onset of the 1980s, but it was in ships not cars. The commercialisation of the US Navy's 'Transit' system, involving five orbiting satellites, was the talking point of the 1981 International Boat Show in London, when sets were on sale for under £2,000. Toyota announced its *Navicom* guidance gadget in 1983, although the system was initially based on inertial navigation (those gyroscopes and accelerometers) rather than GPS. Other Japanese firms introduced the new sat nav technology into cars at the beginning of the 1990s, with the early models running off CD-ROMS and usually a big box wired up in the boot.

But the watershed year was 2005, when a Dutch company that had sold 250,000 stand-alone devices the previous year, suddenly found that it could barely meet demand. It sold 1.7 million boxes in 2005, and within six years had sold more than 65 million. In-car digital mapping had finally arrived. People wished to be led, and they were willing to trust anonymous companies to take them to places they had previously managed to find by themselves. A similar transformation was taking place on mobile phones.

Not for the first time, the people at the forefront of a cartographical revolution were Dutch. For some reason that even the people involved can't explain, the Dutch have recently come to dominate the new digital mapping world in the same way as the Blaeu and Hondius dynasties had dominated the heavy paper one 350 years before.*

The heart of this new empire may be reached by turning right at the end of Damrak, following Prins Hendrikkade, passing the entrance of the waterfront tunnel, turning left into Kattenburgstraat, left again after the train tracks, following Piet Heinkade until the second set of traffic lights, turning right after Muziekgebouw aan het, turning right just before the viaduct, and then there you are at De Ruyterkade 154, the home of TomTom. Here you will find Harold Goddijn, CEO and one of the founders of the company in 1991.

Goddijn, early fifties, fastidious and manicured, used to work for the British handheld computer company Psion, where one of his responsibilities was developing software. He began with book-keeping and translating packages, utilities the busy executive bought on a floppy disc and would then sync from a computer to what was once called a personal digital assistant (PDA). The provision of mapping was a natural extension, but the problems were huge. Storage was expensive in the 1990s, and small devices couldn't provide the amount of processor power required to plan a random journey through a country. In the mid-1990s, prior to publicly accessible GPS, Goddijn bought early digital map files from a company called Automative Navigation Data (another Dutch company who had spent the previous few years digitising existing printed

* In the United States, the world of electronic navigable maps is dominated by Navteq, based in Chicago, the supplier of maps to *Garmin* and other brands, as well as *Bing Maps* and *MapQuest*. Navteq is now a subsidiary of the Finnish company Nokia, but from the early 1990s until 2007 it was owned by Philips, which is also, of course, Dutch. Etak, one of the car sat nav pioneers, and apparently named after the Polynesian method of navigating by celestial guidance, was also originally an American company, beginning in California in 1983. But in 2000 it was acquired by Tele Atlas, which is Dutch. And lastly, one of the key elements of sat nav, the ability to find the shortest route, relies on an algorithm developed in the 1950s by a man called Edsger W Dijkstra, who was, inevitably, Dutch.

'It's a very emotional topic': TomTom co-founder and CEO, Harold Goddijn, with a couple of his early products.

maps), and he supplied Psion users with a primitive method of plotting a manually scrollable route through the European road network.

These AND maps covered only about fifteen per cent of all roads, and the concept of real-time, turn-by-turn, spoken directions was still something for futuristic Hollywood movies, but Goddijn soon found that the market was eager for such things. A driver could plan a simple route between London and Paris from the written instructions on a PDA, and even though it would have been easier to do such a thing on an old-fashioned ring-bound road atlas, the idea of being on a portable machine seemed like the glittering future.

Which of course it was. Within a few years everything expanded – memory, processing power, route accuracy and GPS – and within a decade the first reliable stand-alone, voice-guided sat navs that knew where you were and how to take you somewhere else began appearing on the high street. 'We got emails coming out of our ears from people who couldn't stop telling us how liberated they were and how happy we made them,' Harold Goddijn claims. 'It's a very emotional topic. The feeling of getting lost in a car is very intense and the cause of misery. Everybody's been there. And the man and wife thing is ridiculed, but there is something in that. "Go left ... no ... I told you so." It's all real.'

Goddijn says he was astonished by the success of his product, and his supply had trouble keeping up. 'If you go to a bank and tell them, "It's now 2001, and my revenue is about seven million euros. But next year it's going to be forty million. And then 180 million the year after, and then 800 million and then 1.4 billion" – they just wouldn't believe you.'

But he is at a loss to explain precisely why the demand for these moving maps had suddenly begun to take hold in the world. It can't be reliability, because nothing was more reliable than a paper map. It can't be cost, because the old maps were far cheaper. It may be safety – the sat nav should be less distracting than paper maps, but it may be distracting in a different way, and the catalogue of mishaps with the devices increases each week. It may be the built-in speed cameras, which means sat nav motorway drivers in the UK now get a little warning beep if they deviate from the standard practice of driving at 77 mph in the middle lane on motorways. It may get you to a destination faster, but sat nav owners will tell you that most familiar routes they can drive faster without it. And surely it can't be the lure of Homer Simpson or Jeremy Clarkson.

So perhaps it is the fact that it limits the amount of route controversy in a car (in other words, arguments, usually involving a man and a woman). Or perhaps it is the fact that those who really believe they have absolutely no sense of direction and can't read maps can now find other things to worry about. Or perhaps it is just that we have lost the pleasure and challenge of traditional maps, and like to have someone do the map reading for us. In other words: laziness.

'It's true to a certain extent,' Goddijn concedes. 'There is a charm in reading maps, and there's more information on maps. There is a maritime equivalent: if you want to become a sailor you have to take a course in navigation and rightly so. You learn how to read a map and plot a course, how to do dead reckoning. A beautiful skill, but nobody uses it, because there's GPS and it works. So the skill becomes rusty, and people don't bother. And that's the same for car navigation. It's all very romantic to look at a map, but it's dangerous to use when driving. You always have to stop somewhere. I don't think it's realistic to assume that in ten or twenty years' time people will still use paper maps. Is that good or bad? I don't know. We're moving on.'

But there are already signs that we are not entirely satisfied with this new life. In 2008, the demand for sat navs began to slow down and even turn around, as the market edged towards saturation. TomTom faced increasing competition from free maps provided by Google, OpenStreetMap and others. To slow down the migration, TomTom devices have increasingly come laden with new innovations, some more gimmicky than others, including the ability to tweet your destination. Users may also benefit from what is called 'historical traffic' – anonymous data about routes and journey times gathered from drivers by central TomTom computers which is then used to improve future guidance.

In order to strengthen its hold on the market, TomTom bought its main supplier of digital maps Tele Atlas, another Dutch company, for almost US$3 billion. Tele Atlas still makes many of its maps in the old-fashioned way, by going out and recording what it sees, although now the maps are compiled not with quill and sextant but in a car battened down with roof cameras, lasers and 3-D imaging machines, all of which enable pinpoint locational accuracy but tend to miss landmarks set slightly away from the road that were significant in a pre-digital age, such as Stonehenge.

The resulting maps also have an effect on the way we learn to see things. When we're looking at maps on our dashboard or on phones as we walk, we tend not to look around or up so much. It is now entirely possible to travel many hundreds of miles – to the other end of a country, perhaps, or even a continent – without having the faintest clue about how we got there. A victory for sat nav, a loss for geography, history, navigation, maps, human communication and the sense of being connected to the world all around us.

Pocket Map
The Canals of Mars

Sir Patrick Moore's house is in Selsey, surrounded on three sides by sea. It is fifteen minutes' drive from Chichester in Sussex, and the cab driver says 'Oh yes –' when you climb in, '– it's the one with all the telescopes.'

Moore, the UK's best known and undeniably strangest astronomer, lives in a universe that revolves around his study. This houses his library of books on space and exploration, including about a hundred of his own – books on the mapping of the moon, books on Neptune and most other planets, novels he has written about many worlds other than his own. The room holds some small globes and telescopes, a vintage Woodstock typewriter and many medals, badges and other souvenirs from his travels. And then, occupying all other places, are his mementoes of *The Sky at Night*, the BBC programme he has hosted for more than 700 episodes, latterly with increasing help from other stargazers, including corkscrew-haired Queen guitarist Brian May.

Moore is aged eighty-eight when I call on him, the same number as there are constellations. 'I can no longer get into the garden to look at the sky,' he says mournfully, 'and I can no longer play the piano.' He is wearing a crimson toga. He suffers increasingly from a back injury (RAF, WW2), and his arthritis is painful. His hands and legs are swollen, the eye that usually takes a monocle is half-closed. He is, I fear, a dying planet, but from his padded swivel chair he is still a force around which things revolve: one day a month the *Sky at Night* team transform his office into something resembling a planetarium, bringing in additional globes, darkening the leaded windows and making Moore look like the squinty big-gussetted amateur we first encountered many moons ago when we chanced upon his eccentric enthusiasms on late-night television.

I had called on him to talk about the mapping of Mars, and within two minutes I was laden down with his own books on the subject, including *Mission To Mars, Peril on Mars, Can You Play Cricket on Mars?, The Domes of Mars,* and *The Voices of Mars.* Written over a fifty-year period, the books had one thing in common (apart from being about Mars): they almost all

Patrick Moore in his study at home, surrounded by celestial and terrestial globes – and more than a hundred of his own books.

contradicted each other. *Patrick Moore's Guide to Mars*, written in the late-1950s, was so different from *Patrick Moore on Mars*, written in the late-1990s, that it may as well have been about a parallel universe.

'Ah, the Red Planet!' Moore exclaims as I ask him about the changes he has seen in his lifetime. 'Before the Mariner spacecraft went there [in the 1960s and 70s] we thought we knew a lot, but we knew hardly anything. We had to change the maps and the names on the maps. I remember giving a talk at a university and almost everything that I told them turned out to be wrong!'

Moore was not alone. The entire atlas of the universe has changed fundamentally in the last century, as the power of telescopes and space rockets has taken us closer than ever before. But the cartographic history of Mars is a saga unlike any other, and not only because the planet is on average 140 million miles away, and more than seven months away by spacecraft (on average, the moon is only about 239,000 miles and four days away). This, and its relatively small size (about a third of the surface area of the earth), have made accurate observation difficult; when Galileo first looked through his telescope at the beginning of the seventeenth century he found Mars too unobservable to say anything interesting about. But it is not what astronomers couldn't see that makes Mars so intriguing, but what they could, or thought they could: canals, hundreds of them, perhaps thousands of them, and vegetation too, enough to feed a whole hungry nation of Martians. Life on Mars? That was a theory dreamt up not by science fiction writers or Hollywood; that was down to astronomical mappers.

In 1946, fresh from flying over Germany, Patrick Moore flew to Arizona. He was twenty-three and hooked, like many astronomers both before and since, on the story of the great observatory at Flagstaff, from where Pluto had first been seen sixteen years before. Although he didn't discover it himself (he was actually dead at the time), the PL in Pluto was named after Percival Lowell, the man who had set up the Flagstaff observatory and

Percival Lowell observing Mars from the observer's chair of the 24-inch refracting telescope in Flagstaff, Arizona.

installed its exciting 24-inch refractor telescope principally to study Mars. Flagstaff continues to be an important astronomical venue, but for a while it was regarded as a slightly nutty place, run by a man who, in Patrick Moore's words, 'claimed that Mars must undoubtedly be inhabited by beings capable of building a planet-wide irrigation system.'

Lowell wasn't a crackpot, he was a serious astronomer (he was a fellow of the American Academy of Arts and Sciences, and before that he was a career diplomat, representing US interests in Korea and Japan.) But in 1894 he became obsessed with a theory that went something like this: Mars was in trouble; it was running out of water; it was inhabited by intelligent beings; the reason we knew it was inhabited, and they were intelligent, is because they had constructed long straight canals to hold water and guide it down from its melting ice caps. Lowell began to publish these theories in 1895, and his maps appeared in national newspapers as the subject of serious debate. A Mars mania took hold. Science fiction had found its supposedly factual springboard, and the expansive imaginations of H.G. Wells, Ray Bradbury and others would find a fertile readership. Everyone, it seemed, wanted to believe in the prospect of life that the maps suggested; even the possibility of future colonisation.

Percival Lowell was the first astronomer to give the canals a proper living backstory, but he wasn't the first to see or map them. That honour goes to Giovanni Schiaparelli, the Italian who did more than anyone else to give every place on Mars a name. He too drew long linking straight lines on his maps of the planet, but he wasn't drawn on whether they were waterways or some other phenomenon; besides, he called them *canali*, which merely means a channel, and could easily have been a product of nature rather than a product of Martians with spades; some even believed that the 'canals' were nothing more than a telescopic reflection of veins in the viewer's bloodshot eye.

While the barely viable vision of life on Mars goes forth into the world – it lasted precisely seventy years, from Lowell's book *Mars* in 1895 to the day the Mariner 4 orbiter sent back its first photographs in 1965 and recorded a thin and very inhospitable atmosphere – let us consider how Mars was mapped before giant telescopes and space probes. It was predominantly a place of ghostly shapes and shadows, often obscured by dust and prone to seasonal transformations, distant and insignificant enough to absorb any delusions or fanciful nomenclature we may place upon its surface.

The celestial behaviour of Mars had been studied since before Ptolemy, while its orbital movements were carefully plotted by Copernicus and Tycho Brahe. But the first maps we know of were probably made by Francisco Fontana in Naples 1636,

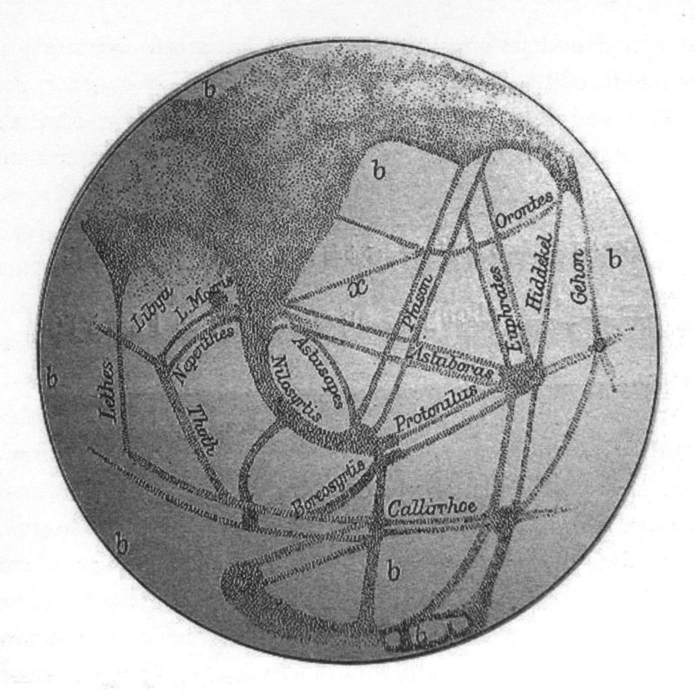

Giovanni Schiaparelli's map of Mars, made in Milan in 1877 – and introducing the planet's *canali*.

and they were fairly disastrous, consisting of little more than a shaded black dot in the middle of a sphere. Fontana called this shaded area a 'pill', but it turned out to be nothing but a common optical illusion. In 1659, however, there was genuine progress, as Christiaan Huygens, a Dutchman, drew a sketch of what we now recognise as Syrtis Major, a triangular shaped area roughly in the proportions of Africa. The polar caps of Mars were first detected by Giovanni Cassini (an Italian in France, the man who began the mapping dynasty responsible for the triangulation of France), and each decade brought more refined telescopes and more accurate drafting, until the early nineteenth-century German astronomers Wilhelm Beer and Johann von Mädler made the first attempt at a full map based on the Mercator projection, and set a prime meridian point at zero longitude in the middle of it.*

Beer and Mädler declined to name the key areas on their map, but others were less timid. A map by the British amateur Richard Proctor upheld the imperial tradition by naming the majority of what was commonly perceived as seas, islands and continents after prominent British astronomers, a system that held good until Schiaparelli constructed his own gridded map in 1877, attaching more than 300 names to the planet's surface, the majority of them inspired by earth's geography and classical myths so Proctor's Herschel II Strait became Sinus Sabaeus, and Burton Bay (named after Irish astronomer Charles Burton) became Mouth of the Indus Canal. 1877 was clearly a fine year for observation, with Mars close to both the earth and the sun; its two dwarf satellites, Phobos and Deimos, were first seen that year. Inevitably Schiaparelli's map was wide of the mark, and he had a particular problem with perspective; what we now know as volcanoes he called lakes. But it was based on scientific principles and the basic shape was roughly accurate. Intriguingly, its form looked not unlike the Victorian imagining of Eratosthenes map of the earth from 194 BC.

And then the canals came into view. In *The Worst Journey In the World*, Apsley Cherry-Garrard mentions that in 1893, just before the onset of golden Antarctic heroism, it was believed that 'we knew more about the planet Mars than about a large area of our own globe,' but this wasn't true; certainly the biggest uncertainty was still raging. Astronomers and some journalists travelled to Flagstaff with the hope of seeing what Schiaparelli and Lowell had seen, and some did indeed detect faint tracks and the hint of arid vegetation. But the most influential of them all, the Greek astronomer Eugenios Antoniadi, drew the most detailed pre-space age map of Mars in Paris in 1930 and concluded there was no sign of intelligent life. However, he did

* This was in the 1830s; the earth's prime meridian at the Greenwich Observatory was not internationally accepted for another half century.

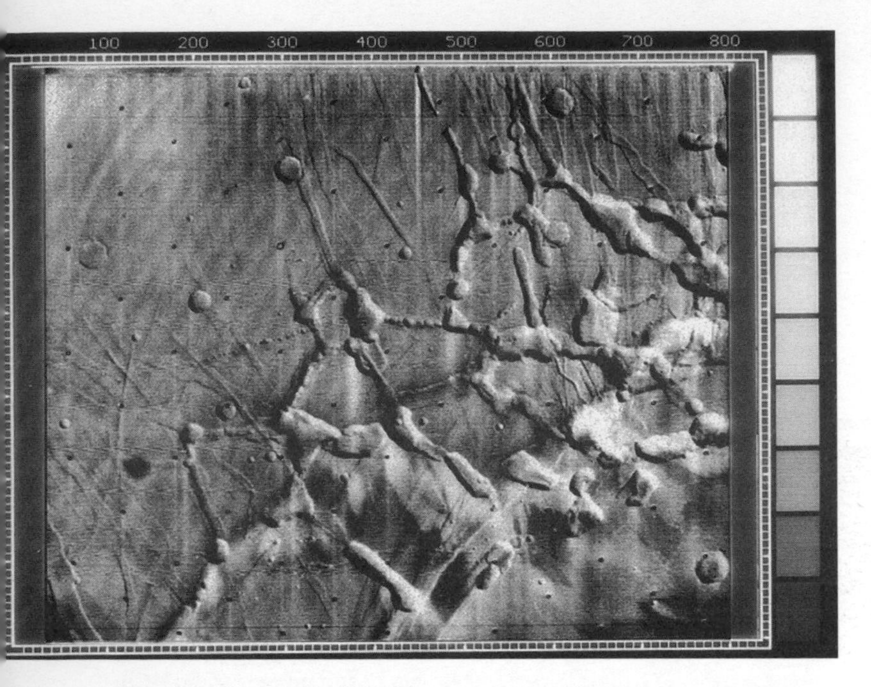

One of the first views of Mars from Mariner 9, showing grooves, craters and flat-topped mesas. The image is roughly 400 km across.

leave the door slightly ajar, stating that the canals 'have a basis of reality', for there were clear 'streaks' visible on repeated observations. And so the enticing prospect prevailed until 1965, when the Americans sent orbiting probes, and NASA began piecing together collaged maps based on grainy photographs that showed a barren rock-strewn landscape covered in a thin dust that invited no form of life and showed no form of canals.

NASA's first official *Atlas of Mars*, published in 1979, relied heavily on images from Mariner 9, the first spacecraft to completely orbit another planet in 1971–72, as well as the Viking craft that landed on the planet in the mid-70s. But it also depended on the airbrushed artistry of a cartographic team from the United States Geological Survey which had based itself at the Lowell Observatory at Flagstaff. No modern map, and certainly not those of the moon, relied so much on photo-mosaics and artistic interpretation for its 'true' accuracy, although with the mass of new images now available from the continual patrols of the Mars Exploration Rovers, the latest maps and Martian globes are stitched together by computer.*

'For some reason I regret that the canals are not there,' Patrick Moore told me. 'But that's science for you.' One can see his point. We are grateful for the digital accuracy and the new names, as well as the huge dark volcanoes that once appeared on early maps as dark seas. We have relearnt more about its atmosphere, its colour and its dust clouds in the past fifty years than anywhere else in the universe. And of course we should be happy to know this about a place we may one day visit despite the intolerable cold, a place that may indeed once have contained life and may yet conceal water below its surface. But the mapping of Mars had shown us the true romance of cartography, and perhaps only scientists have unreservedly welcomed its realities.

* In a cool nod to the sort of human exploration that had inspired many NASA staff at school, two of the recent Mars microprobes were named Amundsen and Scott. As for permanent names on the surface, these have been standardised: large craters are named after deceased scientists, while small craters take the name of small villages of the world.

Chapter 21
Pass Go and Proceed Direct to Skyrim

For the lucky few with time on their hands and a desire to escape their immediate surroundings, maps of Mars with deep canals can still be part of the daily routine. So may maps of the Moon, and maps of occupied France, and a map that recreates the experience of fighting in Iraq in 2003, and a map of a godforsaken urban landscape called Liberty City where you drive around doing terrible things to passers-by. For map fans interested in where the most intricate and beautiful maps have gone (now that museums and libraries have snapped up all the old ones and phone apps and live 3-D maps have done for the rest), this is where to start looking – in video games, the bold future of cartography.

How can this be? Aren't video games the object of scorn and derision, not least from fretful parents who fear their children are wasting the best part of their lives playing them? Aren't they addictive, mindless, repetitive and violent? All this may be true, although perhaps not quite as true as it was when video games took hold in the 1990s. For these days we may acknowledge other attributes, and, far from being a cultural nadir, you can make a decent case that video games are the most creative form of screen entertainment we have. Do they not stretch the young creative mind? In assigning a series of challenges, do they not demand new forms of exploration and problem solving, and a sense of achievement when levels are attained? Do they not also encourage perseverance and patience, and promote cooperation? And more to our point, for a young cartographer in the twenty-first century, is there a more demanding or defining industry within which to work?

Exhibit A: *Skyrim*. This is the fifth part of the *Elder Scrolls* series that began in 1994 and is the most popular digital role-playing game in history (ten million copies sold in the month of release in November 2011, with a sales revenue of $620m). It is an 'open-world' game that lets you either pursue a vast array of quests and skills, or just wander around without purpose, losing yourself in the lush dewy landscape of valleys and mountains, or in the ice of the tundras, encountering enveloping strangeness wherever you go. There is a story at the heart of it – your usual everyday battle against dragons and other foes in a dystopian Nordic-Medieval kingdom – but it is the geography that enthrals, a 3-D dreamworld both familiar and alien, a pixellated Mappa Mundi with a choice of pilgrimages and viewing angles. It is certainly not a place where one can function without an atlas.

Indeed, the game comes with a fold-out map, printed on textured faux parchment, but it's more of a mood-board than anything that will help you navigate beyond your first half-hour. Proper help comes in the form of a 660-page official game guide. This comes with 220-pages of maps, which gives you an idea of the complexity of the game, the hundreds of digital cartographers involved in its creation, and the endless

The 3-D dreamworld of Skyrim – a videogame whose appeal rests primarily on geography and maps.

days in which you may lose yourself in the Skyrim world (a world inevitably far more detailed than Cyrodiil, the country where *Elder Scrolls IV: Oblivion* took place, which was itself more detailed than Vvardenfell, the island at the centre of *Elder Scrolls III: Morrowind*).

Skyrim itself is a country in the continent of Tamriel on the planet Nirn (keep with this: after playing the game enough, it's Earth that becomes weird). Skyrim is divided into nine Holds, although their borders are inexact during gameplay. They have names such as Haafingar, The Reach and Eastmarch, and each Hold has its Primary Locations (large spaces requiring interior exploration, such as the vampire hideout Movarth's Lair), and Secondary Locations (seldom in need of further exploration, such as the Shrine of Zenithar in the Rift.) The locations are spotted with camps, mines, strongholds, dens, lairs and crypts, all with their own names and purpose – horse trading, food supplies, dangerous areas with enemies to be slain if one hopes to gain new skills and rejuvenate one's combative health.

To take one map alone: The Reach occupies the entire western edge of Skyrim, and judging by the ruins it was once a more populated and happier place, but something awful happened here. According to the guide, 'Karthwasten and Old Hroldan provide some degree of safety, and the Blades hideout known as Sky Haven Temple is another beacon of tranquillity surrounded by hard terrain and harder adversaries. Fort Sungard and Broken Tower Redoubt are both fortifications to explore, and two Orc Strongholds (Mor Khazgur and Dushnikh Yal) are also here for you to find. To the northeast is Hjaalmarch, but the majority of the Reach borders Whiterun.'

This is either your thing or it isn't, but the mapping of this imagination is original and impressive. As with a Blaeu atlas of old, the cartography of Skyrim should be credited to many hands, a team of perhaps thirty or forty employed at Bethesda Game Studios; the maps in the guide are credited to a firm called 99 Lives. If you were a mapper, why would you not want to meet this sort of challenge? And if you were a player, why would you not want to believe you were wandering alone in that free-form world with only a map and your wits to save you, an adventure in barely charted territory, one of the latest, greatest, and most underrated cartographical landmarks.

Until *Skyrim* came along the hottest map-within-a-game was *Grand Theft Auto 4* (*GTA4*). Its release in 2008 was an event anticipated by gamers in much the same way as a previous generation anticipated the new Beatles album, and the sales on the first day broke all records. Sales of the whole GTA series have surpassed 100 million copies, and its three main creators (one Scotsman and two Englishmen) have taken on the mantle of impossibly wealthy celebrities, a role they may have been aiming for when they named their company Rockstar Games not long after they opened for business in 1997.

It is clearly a thrilling game to play, the experience enhanced by the fact that everyone who isn't playing it is outraged by it (*Uniting Conservatives and Liberals in Hatred since 1988* as one of the GTA online trailers proudly has it). Certainly, the GTA series has generated a lot of bad press – real-life crimes were apparently inspired by it, there was a pornographic game hidden within its layers, questions were asked in Parliament – all of which boosted sales no end. It is indeed a violent adult game, but at its heart lies a simple pursuit of cops-and-robbers: you steal a car and outwit those chasing you. But as with *Skyrim*, the game is as much about navigation as it is about quests. As a player you are free to speed through the sequence of loosely disguised dystopian urban environments – London, New York, Miami, San Francisco and, coming soon with GTA 5, Los Angeles – as if you were truly at the wheel in a very vivid city.

Many of the cars in GTA 4 have satellite navigation installed, which one operates in much the same way as a real system – you put in the address and off you go. In the *Liberty City Guidebook* it states, 'GPS was invented because real men do not ask for directions. Now you can get automatically re-routed when you handbrake past that last turn at 150mph.' But you may also spend some time on foot and in the subway, which is where the provision of the large folding map comes in handy. This splits the city into five boroughs, each more unappealing than the last. The central one is Algonquin, a carbon copy of Manhattan with its Middle Park in place of Central Park and the Grand Eastern Terminal close to where Grand Central normally lives.

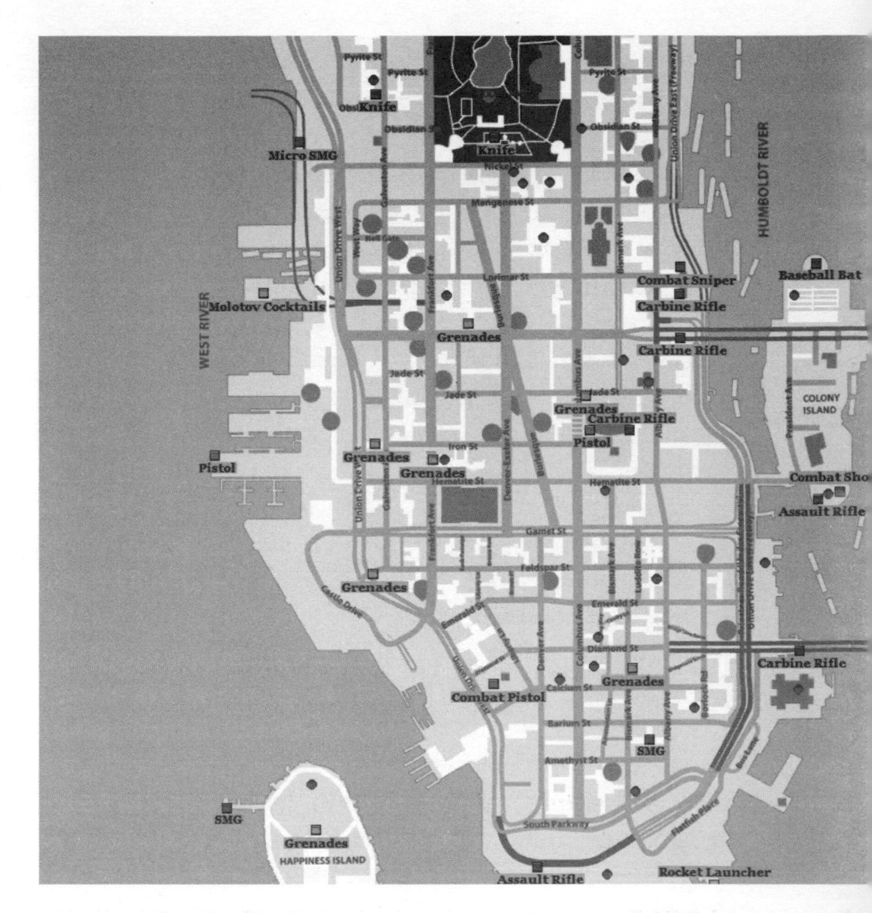

Liberty City can be a *very* bad place to make your way around: this is Algonquin, modelled on Manhattan.

The grid system remains, although the streets are all named after jewels and the numbered avenues have been replaced by Galveston, Frankfort, Bismarck and Albany. The designers know their mapping history: Columbus has been upgraded from a Circle to an Avenue of his own, an island-long stretch from Amethyst Street in the south village to Vespucci Circus in the heights. Of the other boroughs, you'd probably want to live in Broker, where the Brooklyn-style brownstones and leafy streets offer beaches and boardwalks and respite.

GTA navigation takes two forms – how to map your way through the various urban wastelands, and how to get around the architecture of the game itself. Both are handled by the controllers, which on the Sony PS3 means an entire knobbly dashboard with different buttons for accelerate, brake, steering, headlights, game radio station and mobile phone operation, and of course Fire Weapon. That's in the car. If you're on foot, there are buttons and sticks for walking, running, jumping, mounting a ladder and of course Fire Weapon. You begin by fumbling, but pretty soon you learn that a cautious mastery of maps is what you need to get you furthest. The canny player learns that the more you familiarise yourself with your environment, the more you benefit from it (the alley your pursuers don't know, the back-route that will shave eight seconds from your journey). It's a primitive skill, but is it taught as compellingly anywhere else?

Before computers, back in the analogue world, maps and games hit it off rather well too. The association stretches back at least to 1590, when the counties of England and Wales were displayed on a deck of cards (we can't be sure of the rules, but it may have been the very first example of Top Trumps). The top quarter of the card features the name of the country, the suit and the value; the middle section shows a map of that county, while the lower quarter shows the various properties, including length, breadth and distance from London. (Whether being nearer to London or further away was a game-boosting advantage is unclear.)

Another game, with the familiar 52 cards, was published in Paris in 1669 by Gilles de la Boissière. *Les tables geographiques reduites en un jeu de cartes* was truly international, featuring a small illustration of a country or state, including America,

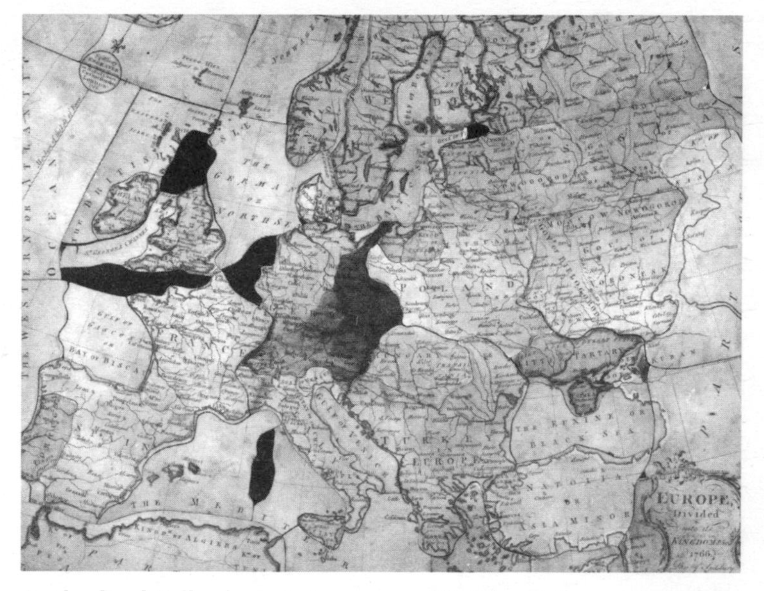

Scotland and Holland go missing on one of Spilsbury's jigsaws, from 1766.

Virginia, Florida, Mexico and Canada. A variation appeared a year later, now with each suit representing a continent: America is clubs, Asia is diamonds, Europe is hearts, and, in a category choice that would these days provoke wrath and hand-wringing, Africa was spades (this may have contributed to the racial slur 'as black as the ace of spades').

But the map would find no more natural home than two other types of pastime gaining favour in the middle of the eighteenth century: jigsaws and board games. The first ever jigsaw is believed to be an engraved map on wood made by the English cartographer John Spilsbury in the 1760s. The idea proved so popular, not least as a way of making school geography tolerable, that he printed and sawed his way not only through maps of the world, but also each of the four continents and the British Isles. A few years before, J. Jeffreys had produced a similarly learning-is-fun pastime with *A Journey Through Europe* or *The Play of Geography*, a map board over which you would advance with dice and rules.

We have continued to play derivations of this game for more than two hundred years, among them *Lincoln Highway* from 1926, in which players moved coloured pins coast to coast over a map of the US (the game was endorsed by the Automobile Club of America, the roads on the board apparently so accurate that they could have been used in a real journey), and *Hendrik van Loon's Wide World* from 1933, in which airplane and steamship pieces raced to complete remote voyages.

And then there was *La Conquête du Monde*, invented by a French film producer Albert Lamorisse in the mid-1950s, and renamed by a salesman at the US game maker Parker Brothers named Elwood Reeves. The word 'Conquest' was already on too many other games, so the salesman picked an initial from each of his grandchildren and named it – initially with an exclamation mark – *Risk!*

The rule book from an edition in the early 1960s has a simple claim – 'You are about to play the most unusual game that has appeared in many years' – and an equally simple purpose: 'The Object of the game is to occupy every territory on the board and in so doing eliminate all other players.' You had armies, you had dice, and gradually, if you had the time, it was hoped you would engulf the world. The game could take as long to set up as other games take to complete, while playing it could annex the kitchen table for both dinner and breakfast. The board was a large and colourful map of the world, although there was clearly something wrong with it. Six continents were each allocated a colour, and each contained several anomalous territories (Asia, for example, held Siberia, Yakutsk, Irkutsk, Afghanistan, China, Middle East and Kamchatka). 'The sizes and boundaries of the territories are not accurate,' the rule booklet explained. 'The territory marked Peru includes, in addition, the country of Bolivia ... It should be noted also that Greenland, Baffinland and a section of the Canadian mainland make up the territory marked Greenland.'

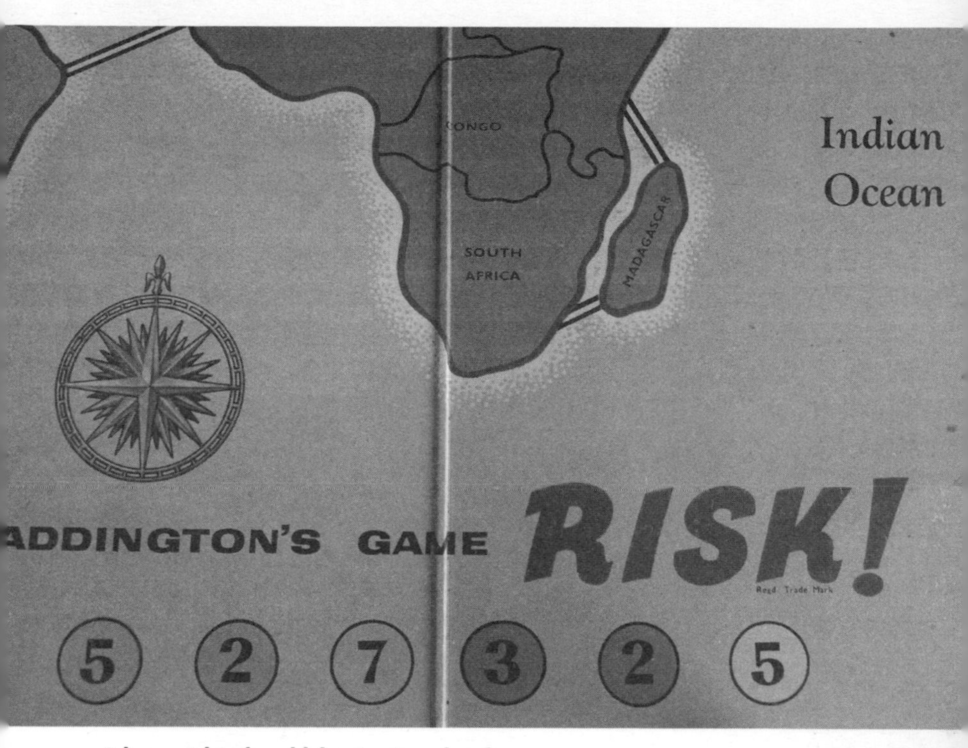

A long night of world domination ahead.

Risk! was a hit in many of the provinces it featured, although some countries got bored with its slow progression and speeded up play by amending the rules. In the UK the game was manufactured by Waddington's, the company that had enjoyed a special relationship with Parker Brothers since 1935, the year the British had licensed the word-play game *Lexicon* and the Americans had returned the favour by licensing something called *Monopoly*.

Designed by the Philadelphian Charles Brace Darrow, *Monopoly* began life in a modest white box with vaguely educational claims on empire building, but it wasn't long before it was bringing out the greedy worst in all of us. It also rapidly became the impressionable mind's proper introduction to urban geography. The game was a smash worldwide, the street names easily localised: out went St Charles Place, Boardwalk and other locations in or near Atlantic City, in came Mayfair, Rue de la Paix, Parco della Vittoria and Wenceslas Square. The board provided a shamelessly corrupt impression of the ease of both rent collection and travel, scrambling true and relative distances in rather the same way as the map of the London Underground. And as for Free Parking, it was clearly made in a simpler universe.

But the *Monopoly* board also has a secret map history, and it is one that may have transformed lives. In the late 1930s Waddington & Co was making more than playing cards and board games. It was also printing silk maps to fight the war. Air men and women would have them sewn into their jackets or concealed in the heels of their shoes as they flew over Europe: the maps wouldn't crease, spoil or betray themselves during searches, and they might help to get them home after a parachute mission or capture (the maps were based on world maps printed by Batholomew in Edinburgh and divided into country sheets as required).

The Americans were making similar maps, but it was only at Waddington that its maps and games divisions combined in this unique way. With the approval of the secret British 'escape and evasion' unit M19, silk maps were inserted between two pieces of the Monopoly board, the game pieces were adapted to include a compass, real money was shuffled into the playing notes, and the games were sent to European prisoner of war camps by dubious charity organisations such as the Licensed Victuallers Sports Association and the Prisoners' Leisure Hours Fund (the special access afforded the Red Cross was conditional on an agreement not to assist escape). Not all the games were so modified; the special ones were marked at strategic points on the board, and there must have been a great temptation to place a mark on the Get Out of Jail Free card.*

Inevitably, *Monopoly* is now available online, and you may play with people you will never meet from places you will never visit. But computer games with maps have been around since the end of 1961, when a group of young hackers at MIT were trying to find a way to show off the capabilities of a newly arrived machine, the PDP-1 from Digital Equipment Corporation. The word 'hacker' meant something else then, more like 'geek' today. The students were proud of their new machine's potential (and with a $120,000 price ticket this was just as well), but they were a little bored with its applications, including an early word processor. So they decided to build a game, something that's now regarded as the grandfather of all computer shoot'em-ups: *Spacewar!* (Precisely why so many

* This Monopoly story has assumed the tinge of myth and James Bond about it, but its ingenuity was certainly typical of MI9 and its American equivalent MIS-X. The key item in this mystery is a letter from an M19 agent, Captain Clayton Hutton, to Norman Watson, an executive at John Waddington, sent at the end of March 1941. It states: 'I shall be glad if you shall make me up games on the lines discussed today containing the maps as follows: One game must contain Norway, Sweden and Germany. One game must contain N France, Germany and frontiers. One game must contain Italy. I am also sending you a packet of small metal instruments. I should be glad if in each game you could manage to secrete one of these.'

There is no specific mention of Monopoly, but another letter refers to Free Parking (the space on the Monopoly board had been marked with a full stop to show that there was a map inside).

Monopoly had already annoyed the Germans before the war. Goebbels objected to the fact that the most expensive area on its Berlin board was Insel Schwanenwerder, where many Nazi leaders lived. Fearing that the Third Reich would be associated with capitalism and extravagance, the manufacturers Schmidt were advised to stop selling it; an allied bombing raid on the company subsequently destroyed any remaining copies. But the game is back in business in Germany, with Unification transforming the layout.

early names of games had exclamation marks after them is hard to say! Perhaps it has something to do with the esteem in which their creators held them.)

Spacewar! was a simple two-player idea featuring a couple of spaceships trying to blow each other up with missiles. A star added a gravitational pull that threatened to set the spaceships off course, but an early version had something missing – a realistic background that would provide a proper sense of dimension and velocity. So another program called *Expensive Planetarium* was crashed into the game, a map of the night sky above Massachusetts. The game was copied for other institutional owners of the PDP-1, and the addictive powers of computer gaming were experienced for the first time.

Thereafter almost all screen games required some sort of map for effective play – either a basic backdrop for 'shooters' such as *Space Invaders* and *Doom*, layouts for multi-level platform games such as *Super Mario* or *Prince of Persia*, a broad perimeter plan for simulation games such as *The Sims* or *Farmville*, or a cheat-sheet atlas to aid navigation through open-world challenges such as *Myst* or *Skyrim*. The maps sometimes come in the box with the game, but more often they *are* the game, and the cartographical interpretation of the landscape is the ultimate challenge. In this way, maps continue to tell a story in much the same way as they did with *Mappae Mundi*, and nowhere is this more true than in the mythical and magic world of *Dungeons & Dragons*.

D&D is a role-playing game involving Dungeon Masters, elves, wizards and the proud and relentless allegiance of those who play it, and no matter that the rest of the world regards them as one orc short of a moondance. Those whose imaginations were first sparked by the mental maps of *Lord of the Rings* will have little difficulty understanding the appeal or gameplay (gaining life experience, defeating opponents, mastering skills).

As with *Skyrim*, real-world abandonment is all, and unlike the tactical model-army war games that share the game's inspirational credit with fantasy fiction, the liberating pleasures of inhabiting an imaginary character may stretch well beyond the hours spent playing the game with friends (ultimately

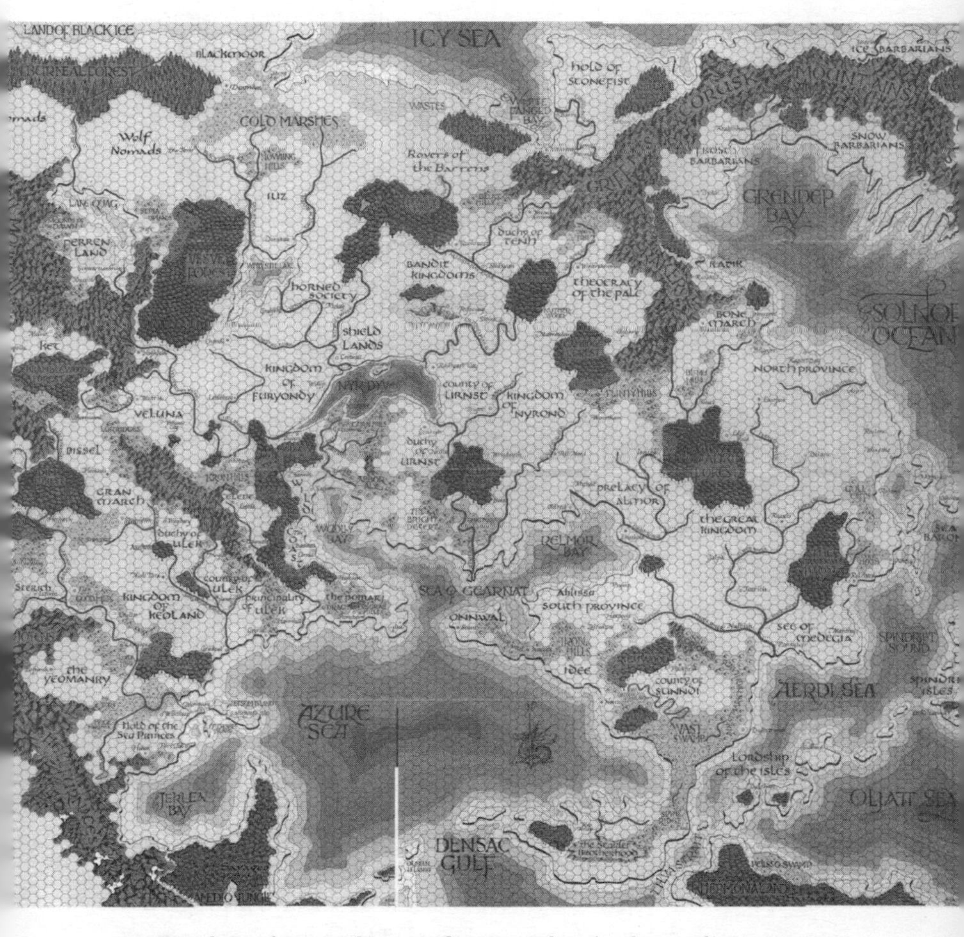

One Orc short of a moondance: welcome to Flanaess, home of mystery and the names of the creator's children.

perhaps into the online avatars of the once-huge *Second Life*). The basic (pre-computer) *D&D* game, created in 1974 by Gary Gygax and Dave Arneson, requires a map for successful gameplay, though it may be little more than an unlabelled grid over which polyhedral dice are rolled and representational game pieces moved.

In 1980, the American artist Darlene Pekul designed a 34-by-44-inch multi-sheet map of Flanaess, the easterly part of the Kingdom of Oerik, for the *D&D World of Greyhawk* campaign. Set on a one-centimetre hex grid, Pekul took Gary Gygax's original vision and created an entirely workable parallel universe. Regions such as Grand Duchy of Geoff contained the Valley of The Mage, and Oytwood and Hornwood forests, while the Kingdom of Keoland lies between the rivers Javan and Sheldomar, and has a distinctly aristocratic flavour with its many baronies, duchies and earldoms. The names on the map are rarely found within the *Times Atlas*, and are often anagrams or homophones of Gygax's children, friends or favourite things: Celene, Flen, Urnst, Linth, Nuthela. But the sea is still blue and the forests still green, and the margin of the map contains a key to the colour-coding and symbols that wouldn't look too out of place on an OS chart: a red dot for a castle, a red square for a walled town, three bars across a river denoting rapids.

The transition from dining room table to computers was a natural one, with programmers able to eliminate a lot of the game's more tiresome calculations and in so doing speed up the action. *Zork!* and *Alakabeth: World of Doom* were primitive 1970s combinations of role-playing games and treasure hunts, and they relied on punctuation marks and other text graphics rather than the 3-D images we are used to today. But the development was swift, and the black-and-white line drawings in such things as *Ultima* and *Wizardry* soon gave way to faster graphics and colour in *Tunnels of Doom*, the improvements aided by the code-writing possibilities of the first home computers, and the excited exchange of floppy discs in Ziploc bags at video and computer stores.

And so an entire generation of potential television viewers and scale model builders were lost to a more modern and exciting way to spend their time and money. And in this way, with no fanfare and very little resistance, a whole generation of parents were alienated, and (many years before mobile apps on phones) maps stealthily entered the lives of young people in an entirely new way. For what is *Skyrim* if not a huge, playable imaginary atlas? Would Ptolemy and Eratosthenes not have recognised it as a thing of wonder?

Chapter 22
Mapping the Brain

When Albert Einstein died in April 1955, a pathologist had his brain on a slab within a day. There was of course only one big question: would the brain of a genius look the same as the brain of a mortal? It turned out that some parts of his brain appeared narrower than the norm, while others were wider; some areas were almost non-existent, but these were compensated for with others that had clearly once pulsed with frantic activity. The findings caused quite a fuss at the time, because our understanding of the human brain was still in its infancy. We could master relativity and quantum theory without any firm understanding of how our brain managed it.

But this is gradually changing. Thanks to technology, brain mapping has entered an exciting phase, a phase where we can actually see things that twenty years ago were purely theoretical. Partly this is due to the work of Einstein himself. And one of the things we're beginning to grasp is how – and where – we are able to read a map.

It always amuses people to learn that Einstein couldn't drive; he probably had other things on his mind. But every time he took a cab – say from his office at Princeton to Newark Airport an hour away – there was one thing he could be relatively certain of: the person driving him had a brain bigger than he did. Or at least a certain part of it was bigger, the bit that successfully selected the quickest route, taking into account traffic conditions, the newest roadblock and the time of day. It was bigger because Einstein's cab drivers (okay, the better ones) had learnt a large map of the state of New Jersey, unknowingly broken it down into a system of molecules, cells and neurons, and reassembled them in just the right order to take their valuable cargo to his next assignment.

When Einstein came to London in the early 1930s to speak at the Royal Albert Hall it was the same: the cab driver would have the entire A-Z crammed up there. It was thought likely that this brain area in cab drivers would be larger than in those who, for example, were constantly getting lost on their way from their front door to the shops (apparently another Einstein occurrence). But it was only very recently that this theory was proven, in a scientifically elegant story that combines both everyday practical maps and the grander notion of the way we read and memorise them: the software and the hardware.

In 2000, a young woman called Eleanor Maguire and a group of colleagues at University College London published a paper in the *Proceedings of the National Academy of Sciences* that got its readers thinking about an obscure and vaguely mythical qualification called The Knowledge. London cabbies knew it only too well as the fiendishly frustrating series of routes or 'runs' they needed to learn before they could earn their licence. There used to be 400 runs to learn, and even though there are now only 320 (*Run 4: Pages Walk SW4 to St Martin's Theatre WC2*, perhaps, or *Run 65: St John's Wood Station NW8 to Brompton Oratory SW7*, each twisty enough to make you yearn for the Manhattan grid system) it takes an average of two to three years to learn them. Indeed, only about half of those who start on The Knowledge will stay the distance and get their badge (for as well as having to navigate some 25,000 streets, there are also about 20,000 'points of interest' to memorise.)

Maguire is a cognitive neuroscientist, and thus concerned with how learned behaviour affects the structure, function and passageways of the brain. But there was also a personal reason for her interest in cab drivers and mental maps. 'I am absolutely appalling at finding my way around,' she explained. 'I wondered, how are some people so good and I am so terrible? I still get lost in the Centre for Neuroimaging and I have been working here for fifteen years.'

Her breakthrough paper – *Navigation-Related Structural Change in the Hippocampi of Taxi Drivers* – produced headlines around the world for its key finding: that London cabbies who had the A-Z in their brains had a significantly larger right posterior hippocampus (the part responsible for spatial awareness and memory) than those who hadn't taken The Knowledge. This news was so handy, and perhaps surprising, for the Public Carriage Office (the body that licenced the black-cab drivers), that they started using it in their recruitment ads, the

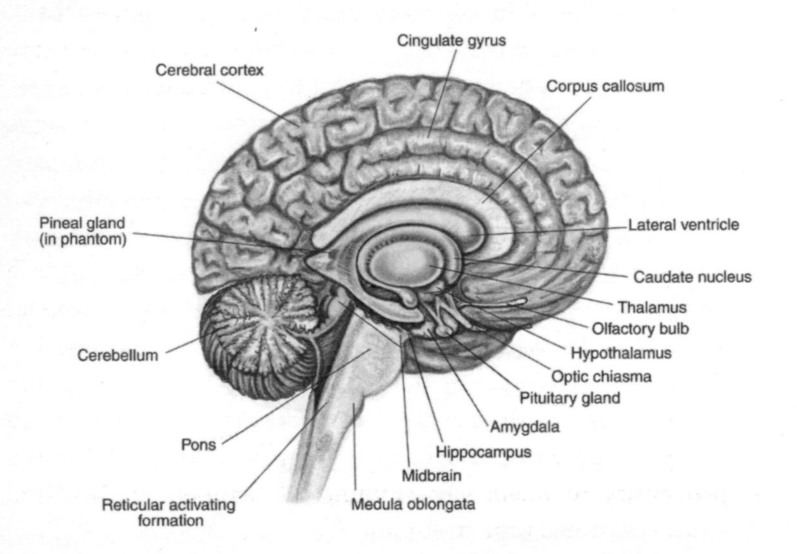

The brain mapped. If your right posterior hippocampus is a lot bigger than this, you're probably a London cab driver.

best boost since the cab driver Fred Housego won Mastermind in 1980. But the findings also provided hope to people unable to read a map or find their way around. Or, rather, to those who say they are unable to find their way around, like Eleanor Maguire. Her work suggested the opposite was true: spatial awareness and erudition is not an inherited trait, but a learnt one. Anyone with regular brain capacity and without brain disease can follow a compass, read a map, remember a route, and find their way back to their car. Learning a lot of maps showed that the brain was malleable plastic.

In 2001, a year after Maguire's research was published, a new study of two slides of Einstein's brain showed something equally intriguing. Einstein had significantly larger neurons on the left side of his hippocampus than on the right – that is, the opposite side to cab drivers. This suggested stronger nerve cell connections between the hippocampus and the neocortex, the part of the brain associated with analytical and innovative thinking, but no marked increase in cell growth on the part linked to the reinforcement of memory.

The methodology of the 'plastic brain' research did leave a few unanswered questions, however. Only sixteen cab drivers were used in the research – all male, right-handed, with a mean age of forty-four and a mean cab-driving duration of 14.3 years – and there was no way of being sure that they didn't become cab drivers because they already had a larger hippocampus before they began driving, and thus a propensity to retain vast amounts of mapping information and a vocational urge to exploit it.

And so, buoyed by the initial enthusiasm towards her work, Maguire and her colleagues at UCL designed further studies. In 2006, many of the doubts surrounding her initial survey were dispelled when she plotted the grey matter in the hippocampus of cab drivers against that of London bus drivers. Both had an aptitude for driving and stress, but the bus drivers were not required to memorise anything but relatively simple and repetitive routes. The bus drivers chosen as the control group had driven for the same number of years as the cab drivers. Again, only the cab drivers showed a significant enlargement in the right posterior hippocampus. The cab drivers also fared better than the bus drivers on memory tests on London landmarks (learned information), but less well on short-term recollection. This was reflected in the larger anterior hippocampus of bus drivers.

The implications of this work are great, and represent a potentially exciting advance in our understanding of spatial skills and memory. It opens doors to other areas, including the possibility of repairing memory loss caused by Alzheimer's, dementia and brain injury through accident. That is to say, the new mapping gives us structural knowledge of the brain and also functional knowledge, the possibility of clinical treatment. When it is complete, the mapping of a pulsing lump of protoplasm may hold the key to eradicating some of our most impenetrable diseases, and in so doing our greatest miseries.

There is a very deep history to this. We are cave men, and we have learnt to walk upright, and our brains have suddenly become very big. Somehow, in the last four million years, we have transformed ourselves from *Australopithecus* to *Homo habilis* to *Homo erectus* to ancient *Homo sapiens* to modern *Homo sapiens*, and towards the end of this run the size of our brains has swelled, probably more than any other creature, and the tasks we can perform as a result have increased greatly. We can, for instance, imagine other worlds beyond our own, and we can anticipate life before and after our own, and we can contemplate our role in the universe, and our own death. Not bad for something that weighs about three pounds, and (we think) unique among the animal world.

One of the other tasks we can perform is to speculate how this brain expansion came about. There are several theories, and prominent among them is the development of language. At some point we managed to make recognisable and repeatable sounds, and to assign these sounds a meaning and a vocabulary. Without knowing why, we developed grammar too. Even the most primitive form of communication would make the most basic tasks easier, and so our talent to make ourselves understood continued to expand (and obviously continues to do so). Our larynx would have had to expand in size and capability as well, and the power to accommodate these changes and possibilities would have caused the brain to expand, feeding on its own possibilities.

Another theory, popularised by the neurophysiologist William Calvin, wonders whether the growth spurt wasn't triggered by the physical, specifically the expansion of nervous tissue caused by our ability to throw and kill. The most successful hunter-gatherers were the ones who could lure their targets and dispose of them accurately and efficiently, and for this they needed a combination of strength, spatial awareness, cunning and timing. These are big things to lug around, and hence the need for more cranial computational space.

And there is a third theory, lucidly examined by Richard Dawkins in *Unweaving the Rainbow*, his celebration of the scientific imagination. Dawkins set out to find the *deus ex machina* that would have remodelled our brain capacity in the same way that the growth in personal computers coincided with the reduction in size and price of the transistor. Our brain capacity developed far more slowly than the capacity expansion of the computer, of course, but the metaphor is a hard one to resist:

Dawkins looks for a revolutionary event that was to the brain what the development of the mouse and Graphical User Interface was to the birth of the Apple Mac and Microsoft Windows. And he may have found one.

Back on the African plains with the hunter-gatherers, the skill of tracking is invaluable. The ability to read footprints, dung deposits and disturbed vegetation will lead to edible rewards, but this knowledge is insufficient in itself. The expert tracker needs expert spear chuckers, and an ability to communicate expert findings. If there was as yet no language, our tracker may mime his intentions to kill an antelope – a silent watch followed by a stealthy stalk and a sudden pounce – but miming precise location of the prey would be trickier. Dawkins suggests there was another way. 'He could point out objectives and planning manoeuvres on a map of the area.' A tracker would be 'fully accustomed to the idea of following a trail, and imagining it laid out on the ground as a life-size map and the temporal graph of the movements of an animal. What could be more natural than for the leader to seize a stick and draw in the dust a scale model of just such a temporal picture: a map of movement over a surface?'

This, of course, is also the beginning of cave paintings – humans and animals depicted in their daily round of survival, with representational figures standing for something else, and introducing the concept of scale and directional arrows and spatial difference.*

But as for the brain, we may have found the reason for expansion and sophistication. Richard Dawkins concludes with a question: 'Could it have been the drawing of maps that

* In 2009 archeologists from the University of Zaragoza revealed what they declared was the world's earliest map – a stone tablet found in a cave in Abauntz in northern Spain. Engravings on the stone, which dates from around 14,000 BC and measures around seven by five inches, seem to depict mountains, rivers and areas for foraging and hunting. 'We can say with certainty that it is a sketch, a map of the surrounding area,' said Pilar Utrilla, who led the research team. 'Whoever made it sought to capture in stone the flow of the watercourses, the mountains outside the cave and the animals found in the area.'

boosted our ancestors beyond the critical threshold which the other apes just failed to cross?'

In November 2010 Chris Clark, a colleague of Eleanor Maguire's at University College London, delivered a talk on brain mapping at the British Library, part of a series to accompany its Magnificent Maps exhibition.

Clark had trained as an accountant before making the considerable career switch to neuroscience and is now head of the Imaging and Biophysics Unit at UCL's Institute of Child Health, where he is concerned with a wide range of neurological diseases seen in children, including autism and cerebral palsy. The mapping of the brain – in particular the white matter consisting of connective tissue that links particular functions – may, he hopes, one day provide enough clues to explain the nature of why certain brain functions fail and allow us to understand how treatments might influence brain circuitry and ultimately restore function.

At the British Library, Clark began his presentation where modern scientific brain mapping started, with the Brodmann maps from 1909. Korbinian Brodmann was a German anatomist who, by examining stained sections of cortex under a microscope, managed to define 52 distinct regions according to their unique cellular make-up (cytoarchitectonics, as he called it). Area 4, for example, is the primary motor cortex, while Area 17 is the primary visual cortex. All of Brodmann's areas were numbered at the time but only some were named, and even fewer had a defined function. (The most notable and widely accepted was the language/speech centre named Broca's Area, the left frontal region named after a French anatomist who, during autopsy in the early 1860s, found lesions and other damage in speech-impaired patients; one of these patients was called Tan, the only word he could say.)

Brodmann's revolutionary demarcations had a popular natural precursor, albeit one rooted less in advanced neuroanatomy than in pseudoscience. Phrenology – in its baldest sense the study of regions on the surface of the skull as indicators of behaviour traits and personal qualities – had been all the rage in the alternative quarters of Victorian science, and the maps are as amusing to us today as they were once perceived to be revelatory. Once one accepted that all human thought and emotion was processed in some way within the brain (rather than the heart or perhaps an ethereal/religious channel), then it made sense to locate particular attributes and values to particular areas; this is what Brodmann was doing in a more sophisticated form. What made less sense was to believe that these attributes could be somehow measured, gauged and differentiated by bumps and lumps on the brain's bony casing – the equivalent of diagnosing a car engine by feeling the bonnet.*

That said, the Victorian maps popularised by leading proponents of phrenology, such as the German physiologist Franz Joseph Gall and the American Fowler brothers in New York, were complex, imaginative and crankily beguiling. The classic china bust of a skull now displayed sardonically in psychoanalysts' waiting rooms shows the simplest cranial elements: Domestic, Aspiring, Animal, Self-Perfecting, Moral, Reflectives and Perceptives. These resemble nothing less than countries on a world map (or perhaps areas in a Disney theme park), and are usually broken down into regions. So Perceptives contains Order, Individuality and the sinister Eventuality (which actually just means an ability to recall events),

* Such a primitive form of brain mapping would inevitably prove useful to those with a bent for racial purity and social cleansing; a sympathetic physician could suggest superiority with a bigger frontal or temporal lobe, the grim reality in the Third Reich and the earliest conflict between the Hutus and the Tutsis.

while the Self-Perfecting zone has Cautiousness, Self-Esteem and Firmness.

The leading American Phrenological proselytisers, Orson Squire Fowler and Lorenzo Niles Fowler, travelled the US, Britain and Ireland giving lectures and selling their *American Phrenological Journal* and books. These days they'd be driven out of town for their conjectures, but in 1876, when their *Illustrated New Self Instructor* reached its 11th edition, their readers clearly thought the Fowlers were onto something. The book may have been used to find your ideal husband or spot your local psychopath, and the task was made easier by more than a hundred engravings, showing various forms of disjunction. Like the Brodmann maps, each part of the head was assigned a numbered function. Unlike Brodmann, the Fowlers' diagrams had hair. Area number four for the Fowlers, about half way down the back of the head, represented Inhabitiveness,

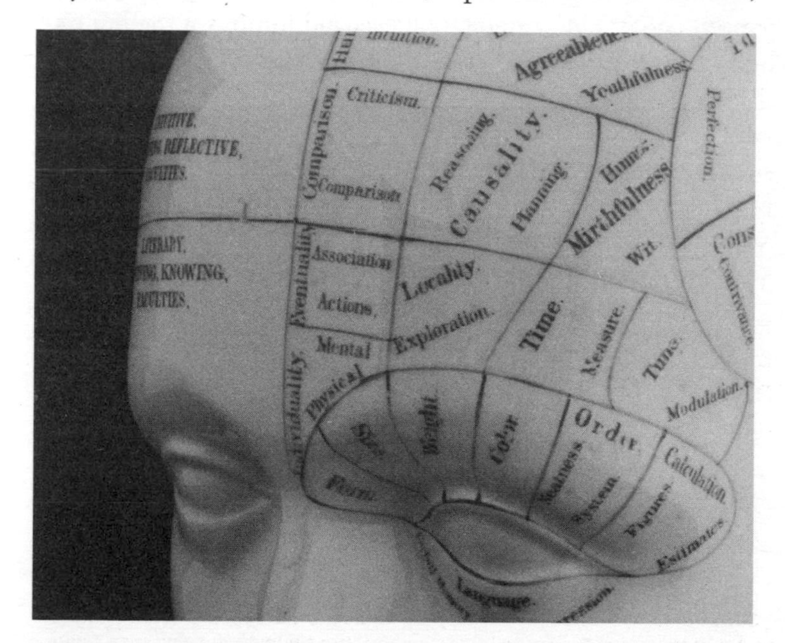

The Fowler brothers' phrenology bust with its map of the brain's perceived activity areas. Oddly, no space for sex or shopping.

the propensity to live near one's roots. A large bulge on the head indicated a large degree of domestic patriotism; but the lack of one indicated the life of a rambler. Similar markings were laid out for Amativeness at the very base of the skull (big sexual urges were revealed when this area was swollen, while a sunken area exposed modest frigidity).

Fortunately for medical science, it is Brodmann's work that has set the template for brain mapping for the last century, gradually refined into a neural jigsaw showing both brain function and connection (Broca's Area was assigned Areas 44 and 45). But it's only since the 1990s that we have been able to effectively map the function of these areas in a way that may prove clinically useful. The enabling technology – employed by Eleanor Maguire with her cab drivers and Chris Clark in his child health work – is that great forensic scanning tool Magnetic Resonance Imaging (MRI), and in particular the highly evolved specialisations known as Diffusion MRI and Functional MRI.

A few months after his talk I met Clark at his office in Bloomsbury. He shows me more slides – brain slices that display Ordnance Survey-style contours, images of thin long cylinders known as axons, fabulously coloured bundles of these axons known as tracts. And then there is a map of diffusion, an image that shows the passage of brain molecules moving through water in a random manner. And then an image of Einstein, who developed the coefficient establishing the 'time-dependent' process in which molecules move further and further away from their starting position as time increases.

Why is this significant? Because the movement of water through the underlying tissue structure – the slower the movement the darker the area on an image – suggests a concentration of structure that may be mapped over time. In the early 1990s diffusion MRI revolutionised the detection of brain damage only hours after a patient had suffered a stroke. Then tractography arrived to provide a method for mapping the connectivity of the brain, allowing the study of so-called disconnection syndromes such as Alzheimers disease and indeed normal aging.

The MRI scanner that lets us gather these images is an other-worldly-looking thing. But for clinical and experimental purposes it has one great advantage: unlike other forms of diagnostic imaging such as X-rays, it is thought to carry no risk of harm to the patient.

'You should have a look at the journal *Human Brain Mapping*,' Clark says. So I do. The March 2012 issue contains articles on changes in the hippocampus of bipolar patients who take (and do not take) lithium, and the results of localised brain activation in stroke patients after directed stimuli. These are not theoretical concerns, but may soon feed back into treatment programmes. Clark's work also already has practical applications. His unit is frequently called upon to provide diffusion images of a patient before surgery, particularly in cases of epilepsy where a patient isn't responding to drugs and requires the removal of a part of the temporal lobe. It's an effective operation but also a particularly delicate one, as the surgeon needs to avoid damaging the adjacent area known as Meyer's Loop for fear of causing a defect in the visual field. Tractography can play a crucial role in guiding the surgeon here, as it can with tumour removal; prior to the directional mapping of the neuro-pathways there was a much greater risk of cutting the motor cortex that attaches to the spinal cord. Clark says he does have 'a little bit of uncertainty' when patients submit themselves to the knife assisted by his imaging work. 'The culture in science is to constantly question what we do. Is this correct? What are the sources of error? Can we improve on what we are currently doing?'

Where are we heading with all this? Somewhere exciting. Advances in brain mapping have mirrored the development of

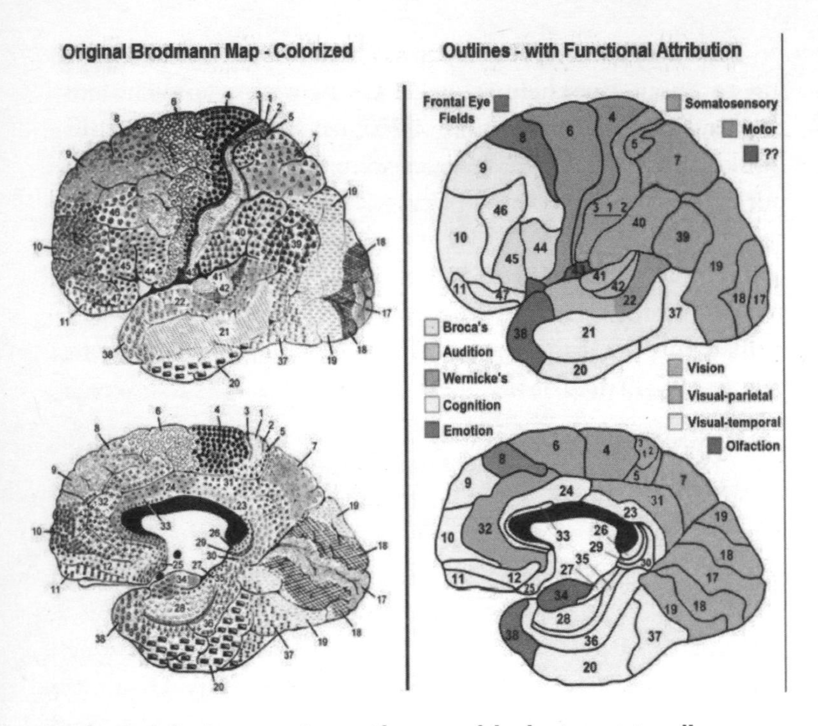

Brodmann's brain maps – in specific parts of the brain certain cell characteristics group together.

the human genome with something called the Human Connectome Project, a US-based enterprise that will eventually lead to an entire map of the brain's physical wiring. Unlike the Genome Project, which shows what makes us who we are, this neural ID will demonstrate how we process and store information, and why we behave the way we do. This 'anatomy of thought' involves capturing some 150 trillion neural connections, and to do this the neuroimaging division of Massachusetts General Hospital took possession of a new scanner at the end of 2011 with some excitement ('4 times the fieldcoils and 4 times as many water cooling layers!' as the machine it replaced). It was hoped that the brain mapping of 1,200 people would begin in the middle of 2012, about half of whom would be twins.

The study of connectivity is 'as hot as hot can get,' according to Susan Bookheimer, a UCLA neuropsychologist and the head of the Organisation of Human Brain Mapping. But Bookheimer and her colleagues were naturally reluctant to put a timeline on delivering practical applications from the completion of the Connectome. And of course even with completion there will be other, bigger, more contemplative questions to answer. The questions of consciousness and human purpose, for instance. The eternal question of how to produce a 3-D map of the planet.

Epilogue
The Instant, Always-On, Me-Mapping of Everywhere

The headquarters of Google Maps in Mountain View, California, contains the sort of diversions you'd expect in any standard office at the forefront of world domination – table football, table tennis, air hockey, a copious range of free quality snacks. The campus on which it stands, the Googleplex, also has a picnic area, a vegetable garden, cycle paths, massage rooms, a car wash, a dry cleaner, a crèche, a beach volleyball court, a dog parlour, a medical area with a dentist, a hairdresser and a carbon-free bus service to take you anywhere you'd like to go, personal security access permitting. There is visual humour too, such as the giant doughnut in an outdoor eating area, and the enormous red map pin outside the Google Maps building.

Inside the Google Maps building there are other jokes, including a green road sign that hangs above a cubicle. 'Welcome to Earth,' it reads in a Douglas Adams way. 'Mostly harmless!' On the reverse: 'Now Leaving Earth. Please check oxygen supply and radiation shielding before departing. Y'all Come Back Now!' And then there are the wooden signposts. There are two of them, each about five-feet tall, deliberately chipped and distressed to resemble something from a hundred years ago, something from a mid-west hiking trail perhaps, or a post you'd hitch your horse to. But they were made in the first years of the twenty-first century, the period in which Google Maps did away with the need for signposts altogether, and established itself as the newest revolution in cartography since – in fact, it is hard to think of a similar event since the Great Library of Alexandria opened for business around 330 BC.

The directions on the signpost point to the names of Googleplex meeting rooms, and read like the dead wood that they necessarily are: Eratosthenes, Marco Polo, Leif Ericsson, Sir Francis Drake, Ortelius, Vasco da Gama, Vespucci, Magellan, Livingstone, Stanley, Lewis and Clark, Shackleton, Amundsen and Buzz Aldrin. One day, perhaps, they will name a small plank on the signpost after Jens and Lars Rasmussen, the brothers who brought Google Maps to the world; or after Brian McClendon, the man chiefly responsible for inventing Google Earth.

I am standing with McClendon at Google's eight-screen map-wall, also known as the Liquid Galaxy. He swoops around the world he helped create with the same planet-on-a-string omnipotence that was once enjoyed by Mercator, or maybe God. 'This is really cool,' he says as he zooms out to the spinning blue-green earth, before zeroing in on a basketball court in Lawrence, Kansas, his home town.

When I visited in the spring of 2011 this was the big new thing – going inside buildings. It was still early days for internal mapping (McClendon said it would be a real palaver

The whole wide world at your control: the Liquid Galaxy map wall at the Googleplex.

getting permission to take photos inside private homes), but it signified the company's – and cartography's – intentions: to map every place on earth in more detail than anyone had ever managed before, and in more detail than most people had previously considered necessary. It brought to mind the absurd vision of Lewis Carroll in his last novel *Sylvie and Bruno Concluded*, where the ultimate map was on the scale of a mile to a mile, or Jorge Luis Borges' fantasy conceit *On Exactitude in Science*, published in 1946 but purportedly written in 1658, recalling a time when 'the Cartographers Guilds struck a Map of the Empire whose size was that of the Empire, and which coincided point for point with it'.

This time, of course, the entire map of the world, inside and out, wasn't being drawn to be big, but drawn to fit on a mobile phone. And it wasn't really being drawn, but photographed and computerised, rendered and pixelated from street level and satellite level. It was the product of applied science, and as such it was impersonal, unemotional, factual and more accurate and current than any map we've ever used.

And such a thing is useful: according to the research company comScore, about sixty million people per month used Google maps in the first quarter of 2012, and the company was by far the most dominant player, with 71 per cent of the market of online computer maps. On smartphones, Google had a 67 per cent share of the fifty million people who used mobile maps to get around. In the summer of 2012, Google estimated that about 75 per cent of the world's population had been covered with its high-resolution maps; about five billion people could say they were able to see their house. But there was still a long way to go.

Brian McClendon, a youthful-looking forty-eight in his Google uniform of polo shirt, jeans and trainers (he actually looked a bit like a young Bill Gates), was the first to admit that he was not a cartographer, despite the fact that he was in charge of Google Maps, Google Earth, Google Ocean, Google Sky, Google Moon and Google Mars. The clue was in his job title: Vice President of Engineering. His stated ambition was nothing less than a digital, instantly accessible live atlas of the entire world, something that would show not just the things that old-school atlases showed (major cities, geological facets, coastal contours, comparative data) but every house on every street and every car in every driveway. Then there would be the inside of buildings, enabling, say, a tour of the Louvre, and universal 3-D imaging with which to better judge distance and height, and then all that route-planning and live traffic information we've experienced with sat nav. Not forgetting all those apps that use Google to coordinate the other gimmicks on our computers and phones, such as photo location, the precise whereabouts of our friends, or (the ultimate dream of commerce) an app with the foresight to offer us a special deal just a few seconds before we walk past the shop where that special deal awaits. And that's just on the ground in our cities; the wilderness too will be fully mapped, the poles and the deserts, and names will be apportioned by Google where none existed before, just like cartography of old. And don't get McClendon started on the mapping of coral on the ocean floor, or the seamless recreation of craters of the Moon.

In such a way Google would not only show the entire world, but also have the power not to show it; it had the power to control information in ways that the most crazed eighteenth-century European despot could only dream about. This ambition, and this power, is less than a decade old, and is already very far removed from Google's original purpose when it began in 1998, which was to build a search engine which ranked web pages in order of popularity and usefulness (five years before there had been no need for such a company, for in 1993 there were only 100 websites in the world). But now 'search' wasn't the dominant thing any more; it was allying the results of search to maps.

In the spring of 2005, Sergey Brin, Google's co-founder, wrote a letter to his shareholders in which he made clear that the company was keen on new directions. Accordingly there were several new products that had been (or were just about to be) launched, including Gmail, Google Video and Google Scholar. There was also Google Maps, which would provide planning and driving directions from the web, and Google Earth, a downloadable program providing almost sixty million square miles of stitched satellite images. The first of these developments was nothing exceptional to those familiar with the services of AOL's MapQuest and MultiMap, although the increased speed at which the pre-rendered map tiles appeared on a screen and the integration of maps into Google search results was rather handy. But the introduction of Google Earth offered one of the Internet's 'hey wow' moments. No one, with the possible exception of Neil Armstrong and friends, had been able to see the earth in this way before, swooping and zooming from zero gravity, breathlessly zoning in on places we had visited on our holidays and places we would never visit if we lived for ever.

And where did people search first? The very same place they had looked for when they viewed the *Mappa Mundi* at the end of the thirteenth century, the place where they lived. 'Always,' Brian McClendon told me. 'And every new version, people go and say how does my town or my house look.' This is a part of human nature – the desire to know where we fit within the grander scheme of things. But it is also emblematic of the new form of cartography that Google and its digital counterparts represent: Me-Mapping, the placing of the user at the instant centre of everything.

Before it was bought by Google in 2004, the software that became Google Earth was previously known as Keyhole, which was co-founded by Brian McClendon. He says he knew he was onto something in the late-1990s, when he and colleagues at his previous company, Silicon Graphics, began combining images of the globe with images of the Matterhorn and nearby terrain, zooming in and out on a piece of hardware that cost $250,000. From the Matterhorn, Keyhole advanced to the Bay Area of San Francisco and zoomed into aerial images of a shopping centre. The commercial breakthrough occurred in 2003, when CNN began using the software in its coverage of the Iraq war. McClendon and his colleagues presented Keyhole to Google founders Sergey Brin and Larry Page in April 2004, and had an offer for the company within twenty-four hours.

Google's instincts were spot-on. There was so much public interest that when Google Earth launched its free service on 28 June 2005 the entire Google computer system almost melted (and new downloads were severely restricted in the first few days as new servers were deployed to meet demand). By the end of the year, Google Earth had become the key geographical tool on tens of millions of personal computers, and its early adopters would call their friends round so that they could feel slightly queasy together. An atlas had never been so much fun.

I asked McClendon what he thought the great explorers of the sixteenth century would have made of the world in zoomable form. 'Oh, they would have completely understood it.' In fact, they would have known slightly more about the world than early users of Google Maps did. In 2005, Google could only render maps of the United States and the UK, and there was no mainland Europe or Central or South America. The maps it did show were licensed from already well-established companies such as Tele Atlas and Navteq, and a few government agencies, but it had none of its own. The Google world only filled out slowly: in 2007 there was still no Pakistan or Argentina, and there were none of the places first touched by Amerigo Vespucci more than 500 years before. But by 2009 this had been rectified, and Google had captured almost the entire world, enhanced by the purchase of a vast cache of satellite imagery.

McClendon says there is an assumption that satellites have rendered human mapping obsolete, but they have strict limits: they can't track local details, they can't name things, they can't relate spatial awareness to real-world issues. Satellites may have seen the Antarctic, but they could never define its boundaries or obscurities. In the true wildernesses – the poles, the deserts, the jungles, even the unpopulated regions of fully developed countries – Google employees are increasingly setting off not in cars with cameras on the roof, but with cameras on backpacks and on the wings of planes, and they do well to arm themselves with the knowledge of border conflicts and the heated disputes over nomenclature.

As it becomes increasingly powerful, Google finds that it encounters obstacles it never anticipated, often geopolitical and social ones that seldom detained mappers with empire-building intent in centuries past. 'We have places that are named that are claimed to be owned by three different countries,'

McClendon says. 'And they have two or three names associated with them. We regularly get yelled at by countries. I didn't think we'd be that important. As it turns out we are more important to argue about than anything else. When the Nicaraguans invaded Costa Rica, they blamed Google Maps for doing it because our borders weren't right. They said that we just went to the land that Google had given us.'

On the day we met, McClendon said he had a new and seemingly less controversial passion: mapping the world's trees. It was a maniacal goal. By some estimates there are 400 billion trees in the world, and Google believes it has catalogued about one billion of them. 'So we have a long way to go - figuring out how to detect them, locate them, know their species.'

At the end of June 2012, McClendon gave a talk at the annual convention for Google developers and media in San Francisco. He began with a classic misconception: '*Hic sunt dracones*,' he said. 'This is what they wrote on old maps when they were drawing them and they didn't know where the borders were, to tell the people looking at the maps "don't go there, you might fall over the cliff". But our goal at Google has been to remove as many dragons from your maps as possible.'

I had been driven to Google by a colleague of McClendon's named Thor Mitchell. Mitchell began working for Google in 2006 after a long spell at Sun Microsystems, and he now managed a department called Google Maps API, which provided a set of tools to enable people outside the company to make software applications involving Google Maps. These could take the form of constructing location devices on your phone, or using maps on your website to show your restaurant or shoe store and boost commerce.

We had met at Where 2.0, a three-day conference in Santa Clara, near San Jose in California (certainly near enough to enable people at the conference to joke that they had known their way to San Jose). The conference attendees, who came from eighty countries, were all engaged in the business of maps and location, and the presentations they gave buzzed with phrases like 'proximity awareness', 'cross-platform realities' 'dataset layering' and 'rich context beyond the check-in'. There were contributions from many of the old big players, including Nokia, Facebook and IBM, and some of the relative newbies too, including Groupon and Foursquare (the concept of 'old' in the digital mapping world meant three years or more).

But the big news that week came not from the scheduled speakers, but from a surprise presentation from two attendees called Alasdair Allan and Pete Warden. Allan, a senior research fellow at the University of Exeter, had just finished some analytical work on the Fukushima nuclear disaster when he was 'looking for some cool stuff to do'. What he found, after some digging in the recesses of his MacBook Pro, was that every call he had made on his iPhone had been logged on his computer with coordinates for latitude and longitude. The information was not encrypted, and was available for everyone to see. He didn't suspect anything sinister on the part of Apple, but he was disturbed by the potential invasion of privacy. Phone carriers had by necessity logged customer calls to monitor usage and issue bills, but this was something else – the open tracking, for a period of almost a year, of an individual's whereabouts. Allan and Warden had no problem translating the recorded coordinates into maps, and one particularly striking screengrab from their presentation showed a train trip from Washington DC to New York City, with Allan's whereabouts being registered every few seconds. And of course Allan and Warden weren't alone: we were all being tracked, and all - potentially at least - being mapped.

The glowing promise of digital mapping has other downsides. Because the new digital cartography is really just an amalgamation of bits, atoms and algorithms, it should perhaps come as no surprise that all our WiFi and GPS devices send as well as receive. Some of this information we provide voluntarily when we enable the location option on our photo-sharing programs or apps, or when we let our sat nav feed back traffic information while on the road, but some of it is just sucked from us without our knowledge.

As we drove to the Googleplex, Thor Mitchell and I talked about the 3-D wonders of Google Street View, the hugely popular web application providing panoramic urban maps of the world. When it launched in 2007 it covered just five US cities, but by 2012 this had expanded to more than 3,000 cities in thirty-nine countries. Billions of drive-by photos had been stitched together to form a seamless cursor-led stroll or drive for its users, and Google cars had driven some five million unique miles to enhance the maps that it had licensed from other companies. But this too was now coming under scrutiny concerning privacy issues.

Between the beginning of 2008 and April 2010, the cars that had been gathering information for Google's maps had also been sweeping up personal information from the houses they passed. If you were on the Internet as one of Google's Subarus rolled by, Google logged the precise nature of your communications, be it emails, search activity or banking transactions. As well as taking photographs, the cars had been consciously equipped with a piece of code designed to reap information about local wireless services, purportedly to improve its local search provisions. But it went beyond this, as another program swept up what it called personal 'pay-load data' and led the Federal Communications Commission in the US and other bodies in Europe to investigate allegations of wiretapping. While there is no evidence that Google has made use of the personal information, a spokeswoman for the company did admit that 'it was a mistake for us to include code in our software that collected payload data'. At the beginning of its life, Google had one publicly stated mission: 'Don't be evil.'

But then there was another problem for Google Maps: something called Apple Maps. In June 2012, Apple announced that its forthcoming new mobile operating system would appear without Google Maps, which would be replaced with a service of its own. This would not actually consist predominantly of maps of its own making, for the company had already licensed Tele Atlas maps from TomTom. But its intent was clear: maps were the new battleground, and Apple no longer wanted to rely on or promote those of a rival.

But how would Apple's maps differ, and how would they hope to compete with such a giant in the field? Its big idea, it claimed, was try to bring new consumer joy to digital cartography, in the way it often did to other services. It promised greater ease of use, smooth integration with both its software and hardware, and an enhancement of such things as 3-D imaging, voice directions and live traffic. It would attempt to add real-time information allied to public transport, commercial buildings and entertainment venues, possibly enabling seat reservations and other purchases through iPhone credit.

Two things were happening at once here – integration and exclusivity. The technological capabilities of maps continued to astonish, and they were increasingly becoming what they had been in the age of the Spanish conquistadors – guarded, proprietary and inestimably valuable as routes to further riches. Google responded to Apple's withdrawal with a weary shrug that said 'good luck – it's a tough world out there'. It told a press conference that it invests hundreds of millions of dollars each year on its mapping services, and that in eight years it had built up an army of snowmobiles, boats and aeroplanes to achieve its aims. It promised a new feature called Tour Guide, enabling users to 'fly' over cities in 3-D. It also dramatically cut the prices of using its Google Maps tools on websites with heavy traffic, from $4 per 1,000 map loads to 50 cents per 1,000 map loads.

This wasn't the first time Google's mapping had faced serious competition. At the time of Apple's announcement, the online directory ProgrammableWeb counted 240 companies offering their own map platforms, more than double the amount in 2009. Some are bigger and more comprehensive than others. In 2009 Microsoft launched Bing Maps, an improvement on its earlier Virtual Earth with a refreshed 'bird's eye' view and an expanded global range (its base maps were supplied by Nokia's American subsidiary Navteq, which also supplied Yahoo Maps).

And a few days after Apple's announcement there was the prospect of another major player. Amazon's Kindle devices and an anticipated Amazon smartphone would both benefit from handheld mapping, and the leaked news in June 2012 that the company had recently purchased a 3-D mapping start-up suggested that the journey was well underway.

At Where 2.0, Blaise Aguera y Arcas, the principal architect of Microsoft's Bing Maps, promoted his product in a novel way. It was 'an information ecology', he claimed, which provided a 'spatial canvas ... a surface to which all sorts of different things can bind'.

In one sense this appeared to be a new artistic vision, but in another it was merely a new grand language to describe something that had been going on elsewhere for a while – the map mash-up. This had been happening in music, especially, the ability to take one bit of a song and crash it into another, an extreme form of sampling. The same was now happening with maps, and it was the hottest cartographical trend of the digital age.

Personalised, crowd-sourced additions may render a map subversive, satirical or simply newly useful. A list of the most popular mash-ups on ProgrammableWeb (in the middle of 2012 there were more than 6,700 to choose from) includes a map of where items on the BBC News are located in the world, the sites of the Top 50 medical schools in the US as tabled by US News (almost four-fifths east of Chicago) and many marine vessel and flight trackers (so you can point your phone at a boat or plane and find out what it is, where it's come from and where it is going).

And then there are others that are just timesucks: the vague location of the 'Top 99 Women' as voted by the drooling staff of *Askmen* Magazine (the red location markers, which are accompanied by photos and videos, are, predictably, mostly sited in California, but there are also top women in Germany, Brazil and the Czech Republic). Slightly more productive are several rock band road trips, on which you may follow the route of fans as they plot cross-country US drives listening to local artists en route (so pass your cursor over Baltimore, Maryland, and you'll hear Frank Zappa, Animal Collective, Misery Index and more). Most of these use Google Maps and Bing Maps as their base, and all would have been impossible even five years ago.

One of the most compelling is Twitter Trendsmap (trendsmap.com), a real-time projection of the world on which the most popular tweeting topics are overlayed in strips. Levels of activity vary according to what time of day one calls up the map, but one is usually guaranteed a lot of hashtags involving sport, political outrage and Bieber. To take a European morning in the summer of 2012, arsenal, vanpersie, wimbledon and shard were all trending in the UK, while Spain was covered with black tiles announcing *bankia*, higgs, *el-pais* and *particula*. India, meantime, was busy with secularism, olympics, bose and discovery, while a sleepy Brazil was covered with *casillas*, *buzinando*, *pacaembu* and *paulinos*.

The Twitter map is oddly reminiscent of a project from sixty years ago, when visitors to the Festival of Britain encountered a map called 'What Do They Talk About?', a regional survey of conversational habits in the British Isles. Elaborately designed by C. W. Bacon for the *Geographical Magazine* and Esso (lots of swirling banners of text and schoolbook illustrations, not that far from Matthew Paris in 1250), it stated that everyone talks about the weather, but if you travelled to Northern Ireland they also talked about No Surrender, while in Portsmouth it was Pompey's chances in the football league. If you went up the east coast of Scotland from Edinburgh to Aberdeen you would also be able to engage the locals with The New Pit, Golf, What the Bull Fetched and Philosophy, Divinity, Fish.

Such maps continue to thrive in the analogue world, where they are rightly categorised as art rather than engineering, and they have a rich history. We've already encountered some of the zoological classics (the eagles and octopuses, the London tube map's Great Bear) but there are examples in just about any field you can come up with. There are horticultural maps (Bohemia in the shape of a rose from 1677, by the Bavarian engraver Christoph Vetter, with Prague at the centre and Vienna at the root), and allegorical maps such as the *Paths of Life* (made by B. Johnson in Philadelphia in 1807, showing 'Humble District', 'Gaming Quicksands' and 'Poverty Maze'), and also amorous examples that became popular as Victorian postcards (one shows the course of the Truelove River, flowing through 'Fancy Free Plateau', 'Tenderness Crossing' and the 'Mountains of Melancholy' before settling at 'Altar Bay' and the 'Sea of Matrimony'). *

Perhaps the most celebrated of all is Saul Steinberg's Manhattanite's view of the world – a map that appeared on the cover of the *New Yorker* in 1976 and has been the subject of myriad variations on posters and postcards ever since. In some ways it was a precursor to digital 3-D and birds-eye maps, with the viewer flying over the bustle of 9th and 10th Avenues, crossing the Hudson River into Jersey, and then, with an absurdly telescoped perspective, leaping over Kansas City and Nebraska into the Pacific. A few vaguely significant locations dotted in cross-hatched wheat fields hove briefly into view – Las Vegas, Utah, Texas and Los Angeles to the west, Chicago to the east – and then far off in the distance, the small pink-tinged hallucinations of China, Japan and Russia. The message was simple: everything that happens, happens in self-obsessed New York. It was me-mapping before the iPhone made it *de rigueur*.

Steinberg's Manhattanite's view of the world.

The parody has been parodied many times, but the best modern parallel, and certainly the rudest, is to be found in the work of the much travelled Bulgarian graphic designer

* For a mouthwatering collection of similar work see *Strange Maps* by Frank Jacobs (Viking Studio, 2009), or his online blog: http://bigthink.com/blogs/strange-maps

Yanko Tsvetkov. Tsvetkov, who works under the name Alphadesigner, may well have constructed the most offensive and cynical atlas in the world, all of it stereotypical, some of it funny. His Mercator projection entitled *The World According to Americans* showed a Russia labelled simply 'Commies', and a Canada labelled 'Vegetarians'. He has also produced the *Ultimate Bigot's Supersize Calendar of the World*, which includes *Europe According to the Greeks*. In this one, the bulk of European citizens live in the 'Union of Stingy Workaholics', while the UK is categorised as 'George Michael'.

Despite the rigours of digital architecture we should be relieved that maps remain funny, inquisitive and poignant, and that it is often the quirky, inspired hand-drawn one-offs that reveal the greatest truths. A glorious map of the Glastonbury Festival created by *Word* magazine included locations labelled 'Man Selling Tequila Off a Blanket', 'Route of Aimless 4am Trudging (Contraflow)', Doorway to Narnia' and 'People Actually Having Sex'.

Or how about the *New Simplified Map of London* by a secretive (but one imagines local) hand going by the name of Nad, on a Flickr site devoted to 'Maps From Memory':

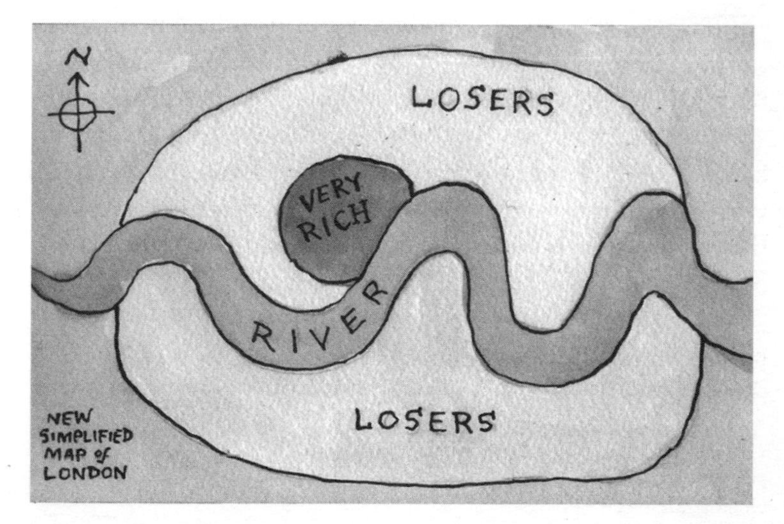

And perhaps there is another positive analogue reaction to the neatness and programmed ease of digital mapping. At the beginning of the twenty-first century, the modern art world has embraced cartography as never before, a trend heralded by Alighiero Boetti, Jenny Holzer, Jeremy Deller, Stanley Donwood and Paula Scher, and most passionately by the London-based artist and potter, Grayson Perry.

Grayson Perry's 2011 show at the British Museum, *The Tomb of the Unknown Craftsman*, contained pots, tapestries and drawings that suggested that we had entered a new golden age of hand-crafted wayfinding, albeit often of a mythical and highly autobiographical nature.

Perry had previously drawn a huge and complex modern-day *Mappa Mundi* called *Map of Nowhere*, complete with souls engaged on a Ruritanian pilgrimage to shrines labelled Microsoft and Starbucks, passing possible resting places marked 'binge drinking' and 'having-it-all'. Always religiously suspicious, Perry's map centres not on Jerusalem, but on 'Doubt'. His British Museum show went further, emphasising his love of maps that display the emotional and irrational, and maps as empirical objects displaying the commonplace.

The centerpiece was a tapestry more than 20ft long and 9ft in length entitled *Map of Truths and Beliefs*. At its heart was a depiction of the museum itself, with the main rooms each named for the afterlife (Heaven, Nirvana, Hell, Valhalla, Astral Plane, Avalon and, returning to cartography for the first time in 500 years, Paradise). The embroidery covered a collection of landmarks seldom seen together on other maps from any period: Nashville, Hiroshima, Monaco, Silicon Valley, Oxford, Angkor Wat and Wembley. There were figures from the artist's personal iconography, and symbols (walled cities,

Objects you could manipulate and upset: Grayson Perry in front of a section of his *Map of Truths and Beliefs*.

itinerant sailors, lonely citadels) that wouldn't have looked out of place on inked medieval calfskin, if only they weren't joined by helicopters, caravans and a nuclear power station. But they had as much mysterious right to be there as anything.

After an evening talk at the museum I asked Perry about his devotion to maps. He said that he thought all children shared this obsession before they lost their sense of wonder. 'I became aware of the possibility of maps as objects you could manipulate and upset,' he said, 'and the fact that they can tell personal stories rather than official ones.'

The gift shop offered a Grayson Perry map on a silk scarf, and it joined a growing list of map merchandise that has very little to do with getting around. A short walk from the British Museum will take you to Stanfords in Covent Garden, where the gift items suggest that cartography has reached unprecedented levels of hipness. Old-school paper maps appear as wrappers on 'It's A Small World' chocolates, on a global warming world mug (the coastlines disappear when you add hot liquid) and a giant eight-sheet pack of world wallpaper. The

World Map Shower Curtain proved so popular after appearances in *Friends* and *Sex and the City* that they brought out one showing the New York Subway map. And how to explain the distinct unusefulness of a map wrapped around a pencil, or the hip-flask with scraps of an old school atlas printed on it?

We should end not with the past but with the future, and this takes us back online. The biggest open-source mash-up of all is OpenStreetMap, which with its ambitions to cover the entire world with local contributors might consider itself more of a Mapipedia (though not to be confused with WikiMapia – another collaborative mapping project).

OSM began in 2004 as an alternative to Tele Atlas and Navteq, happy to be free of its regimented appearance and fees. It is truly a map of the people for the people, with volunteers tracking their area with a GPS device and a determination to plot not only the roads and landmarks that appear on other maps but also things the major players may not consider, or might regard as superfluous – a cluster of benches in a park, a newly opened store, a clever cycle route. It is often the most up-to-date map available, and increasingly benefits not just from personal on-the-ground additions but also from large datasets of aerial photography and officially licensed surveys. It's a goodwill map, and perhaps as close as we will get to a democratic map.

The same spirit, with greater urgency, invades the work of Ushahidi, a mapping platform that began as a method of monitoring violence in Kenya in 2008 and has since expanded to become the prime cartographic site for human rights work and emergency activism. Ushahidi's catchphrase is corny but true – Changing the World One Map at a Time – and as a mark of its influence the UN has employed the immediacy of

Ushahidi mapping in its crisis response to the killings in Syria and natural disasters in Japan and India.

Ushahidi's strength lies not only in its mapping tools, but in the ability of local people to employ them. Universal ease of use has been one of the great developments of digital cartography, and nowhere has this been more evident than in Africa, where inhabitants of the Kenyan slum region Kibera and villagers in the Congo rainforests have increased their global visibility, and with it their rights and heritage, with the aid of simple GPS and an indestructible platform with which they may place themselves on the map.

And so we are back where we started, the place where maps began to make us human. But Africa is no longer dark, the poles are no longer white, and we are fairly sure we live on a planet with more than three continents. More people use more maps than at any other time in human history, but we have not lost sight of their beauty, romance or inherent usefulness. And nor have we mislaid their stories.

There is, of course, still quite a lot to be said for getting lost. This is a harder task these days, but it's a downside we can tolerate. We may always turn off our phones, fairly safe in the knowledge that maps will still be there when we need them. We are searching souls, and the values we long ago entrusted in maps as guides and inspirations are still vibrant in the age of the Googleplex. For when we gaze at a map – any map, in any format, from any era – we still find nothing so much as history and ourselves.

Acknowledgements

I wish to express my great appreciation to everyone who has helped me with this book, and I am indebted in particular to those who gave up their time to be interviewed or answer what must have seemed very basic questions. Some appear in the text, but others do not, so thank you Bill Reese, the Very Reverend Peter Haynes, Dominic Harbour, Chris Clark, Graham Arader, Richard Green, Kate Berens, Peter Bellerby, Paul Lynam, Brian MacLendon, Thor Mitchell, Matt Galligan, Julia Grace, Norman Dennison, Tim Goodfellow, Harold Goddijn, Mark McConnell, Ian Griffin, Cressida Finch, Jonathan Potter, Alex Gross, Nicole Day, Francesca Thornberry, George Thierry Handja, Massimo De Martini.

I am also grateful to many people have provided additional advice and help: my agent Rosemary Scoular, Eleanor Farrell, Sara Wheeler, Bella Bathurst, Mark Ovenden, Max Roberts, Clare Morgan, Ralph and Patricia Kanter, Charlie Drew, Jack Drew, Tony Metzer, David Robson, Lucy Fleischmann, Suzanne Hodgart, Kristina Nilsson, Rosie Tickner, Deanna Yick, Nan Ross, Helen Francis, Carol Anderson, Diane Samuels and my children Ben and Jake Garfield.

A book such as this would be impossible without the patient attention and great knowledge of the librarians at The London Library, The British Library, the Royal Geographical

Society Foyle Reading Room and the New York Public Library. Their resources – literature as well as maps – lie at the core of this work.

These days, any student of cartography benefits inestimably from the wealth of material online, and two sites have been particularly useful to me in my research. The first, the David Rumsey Map Collection (www.davidrumsey.com), is a zoomable feast of more than 30,000 maps, and will provide many hours of delight. The second, Tony Campbell's www.maphistory.info, rightly bills itself as the gateway to cartography online, and you'll find not only fascinating stories but also links to other knowledgeable sites, societies, journals and conferences.

Several people read and commented on the draft copy of the book, and I have mentioned all but one above. But I would like to single out Andrew Bud, a great friend and faithful reader, for spotting some errors that would otherwise have caused sleepless nights, and for suggesting a couple of significant new directions.

I doubt whether any writer could wish for a more assured or painstaking editor than Mark Ellingham, and his work on the architecture of this book has been invaluable. The team at Profile have again been a pleasure to work with, and I would like to thank Andrew Franklin, Penny Daniel, Stephen Brough, Simon Shelmerdine, Peter Dyer, Niamh Murray, Claire Beaumont, Emily Orford, Anna-Marie Fitzgerald, Valentina Zanca, Ruth Killick and Rebecca Gray. Finally, I am hugely grateful for richly alluring jacket designed by Nathan Burton, and the endlessly imaginative design by James Alexander. The book would have been much the poorer without them.

Bibliography

Journals:
The Art Bulletin
Construction History Society Newsletter
The Cartographic Journal
The Geographical Journal (Royal Geographical Society)
Gesta
Imago Mundi
IMCoS Journal (International Map Collectors' Society)
The Map Collector
Transactions of the American Philosophical Society
The Wilson Quarterly

Alexander, Doris: *Creating Literature Out of Life*, Pennsylvania State University Press, Pennsylvania, 1996.

Auletta, Ken: *Googled: The End of the World as We Know It*, Penguin, New York, 2010

Baker, Daniel B (ed): *Explorers and Discoverers of the World*, Gale Research Inc, Detroit, 1993.

Barber, Peter and Harper, Tom: *Magnificent Maps: Power, Propaganda and Art*, British Library, 2010

Barber, Peter (ed): *The Map Book*, Weidenfeld & Nicolson, London, 2005

Barber, Peter and Board, Christopher: *Tales From the Map Room: Fact and Fiction About Maps and their Makers*, BBC Books, London 1993

Barrow, Ian J: *Making History, Drawing Territory: British Mapping in India c1756-1905*, OUP, 2003

Berthon, Simon and Robinson, Andrew: *The Shape of the World*, George Philip Ltd, London 1991

Binding, Paul: *Imagined Corners: Exploring the World's First Atlas*, Review, London, 2003

Booth, Charles: *Life and Labour of the People in London*, Macmillan, London, 1902

Booth, Charles: *The Streets of London: The Booth Notebooks*, Deptford Forum,London, 1997

Brotton, Jerry: *Trading Territories: Mapping the Early Modern World*, Reaktion Books, London, 1997

Carter, Rita: *Mapping the Mind*, Weidenfeld & Nicolson, London, 1998

Cherry-Garrard, Apsley: *The Worst Journey In The World, Antarctic 1910-1913*, Constable & Co Ltd, 1922

Christy, Miller: *The Silver Map of the World*, H Stevens, Son & Stiles, London 1900

Cosgrove, Denis (ed), *Mappings*, Reaktion Books, London, 1999

Crane, Nicholas: *Mercator: The Man Who Mapped The Planet*, Weidenfeld & Nicolson, London, 2002

Crossley, Robert: *Imagining Mars: A Literary History*, Wesleyan University Press, Connecticut, 2011

Dawkins, Richard: *Unweaving The Rainbow*, Allen Lane, London, 1998.

Dekker, Eli: *Globes From The Western World*, Zwemmer, London, 1993

Donovan, Tristan: Replay: *The History of Video Games*, Yellow Ant, East Sussex, 2010

Edson, Evelyn: *The World Map 1300-1492: The Persistence of Tradition and Transformation*, Johns Hopkins University Press, Baltimore, 2007

Fordham, Herbert George: *John Ogilby (1600-1676: His Britannia and the British Itineraries of the Eighteenth Century*, OUP, 1925
George, Wilma: *Animals and Maps*, Secker & Warburg, London, 1969
Goffart, Walter: *Historical Atlases*, University of Chicago Press, 2003
Goss, John: *The Mapmaker's Art: A History of Cartography*, Studio Editions, London, 1993
Harley, JB, Lewis GM & Woodward, David (eds): *The History of Cartography Vols 1-3*, University of Chicago Press, 1987-1998
Hartley, Sarah: *Mrs P's Journey*, Simon & Schuster, London, 2001
Harvey, Miles: *The Island of Lost Maps*, Random House, New York, 2000
Harvey, PDA: *Mappa Mundi: The Hereford World Map*, British Library, 2002
Hewitt, Rachel: *Map of a Nation: A Biography of the Ordnance Survey*, Granta, London, 2010
Jacobs, Frank: *Strange Maps*, Viking Studio, New York, 2009
Jennings, Ken: *Maphead: Charting the Wide, Weird World of Geography Wonks*, Scribner, New York, 2011
Keates, Jonathan, *The Portable Paradise*, Notting Hill Editions, London, 2011.
Knight, EF: *The Cruise of the Alerte* Longmans, Green and Co, London 1890.
Koch, Tom: *Disease Maps: Epidemics on the Ground*, University of Chicago Press, 2011.
Larner, John: *Marco Polo and the Discovery of the World*, Yale University Press, 1999
Lethem, Lawrence: *GPS Made Easy*, Cordee, Leicester, 1994.
Letley, Emma (ed): *Treasure Island*, OUP, 1998
Levy, Steven: *In The Plex: How Google Thinks, Works and Shapes Our Lives*, Simon & Schuster, New York, 2011
MacLeod, Roy (ed): *The Library of Alexandria: Centre of Learning in the Ancient World*, I.B. Tauris, London and New York, 2000
Markham, Clements: *Antarctic Obsession*, Erskine Press, Norfolk, 1986
McCorkle, Barbara B: *America Emergent* (catalogue), Yale University, 1985
Mollat du Jourdin, Michel and de la Ronciere, Monique et al: *Sea Charts of the Early Explorers*, Thames and Hudson, 1984
Monmonier, Mark: *Drawing The Line: Tales of Maps and Cartocontroversy*, Henry Holt, New York, 1995
Moore, Patrick: *On Mars*, Cassell, London, 1998
Morton, Oliver: *Mapping Mars: Science, Imagination and the Birth of a World*, Fourth Estate, London, 2002
National Maritime Museum: *Globes at Greenwich*, London 1999
Ogilby, John and Hyde, Ralph (introduction): *A-Z of Restoration London*, London Topographical Society, 1992
Ogilby, John: *Britannia, Volume The First*, A Duckham & Co, London, 1939
Ogilby, John: *London Survey'd*, London & Middlesex Archaelogical Society, 1895
Oliver, Richard: *Ordnance Survey Maps: A Concise Guide for Historians*, The Charles Close Society, London, 1994
Parker, Mike: *Map Addict: A Tale of Obsession, Fudge & The Ordnance Survey*, Collins, London, 2009
Parsons, Nicholas T: *Worth The Detour, A History of the Guidebook*, Sutton Publishing, Gloucestershire, 2007.
Pawle, Gerald: The War and Colonel Warden, Harrap & Co, London, 1963.
Pearsall, Phyllis: *From Bedsitter to Household Name*, Geographers' A-Z Map Company, Kent, 1990
Pease, Allan and Barbara: *Why Men Don't Listen and Women Can't Read Maps*, Pease Training International, Australia, 1998
Reeder, DA (Introduction): *Charles Booth's Descriptive Map of London Poverty*, London Topographical Society Reprint, 1984
Ross, James Clark: *A Voyage of Discovery and Research in the Southern and Antarctic Regions During the Years 1839-43*, John Murray, London, 1847
Ryan, Christoper and Jetha, Cacilda: *Sex At Dawn*, Harper Perennial, New York, 2010

Schwartz, Seymour: *The Mismapping of America*, University of Rochester Press, 2003
Seaver, Kirsten: *Map, Myths & Men: The Story of the Vinland Map*, Stanford University Press, 2004
Shephard, David: *John Snow*, Professional Press, Chapel Hill, North Carolina, 1995
Skelton, RA: *Explorers' Maps*, Routledge and Kegan Paul, London, 1958
Skelton, RA: *The Vinland Map and The Tartar Relation*, Yale University Press, 1995
Simkins, Peter: *Cabinet War Rooms*, Imperial War Museum, London, 1983
Stanley, HM: *The Exploration Diaries of HM Stanley*, W Kimber, London, 1961
Stevenson, RL: *Treasure Island*, Cassell, London, 1895
Stevenson, RL: *Essays in the Art of Writing*, Chatto & Windus, London, 1995
Tooley, RV: *Collectors' Guide to Maps of the African Continent and Southern Africa*, Carta, London, 1969
Tooley, RV: *Maps and Map-Makers*, Batsford, London, 1971
Tooley, RV: *The Mapping of America*, Holland Press, London, 1980
Tyacke, Sarah: *London Map-Sellers 1660-1720*, Map Collector Publications, Tring, 1978
Virga, Vincent: *Cartographia*, Little, Brown and Co. New York, 2007
Wallis, Helen M & Robinson, Arthur H (eds): *Cartographical Innovations: An International Handbook of Mapping Terms to 1900*, Map Collector Publications, 1987
Wheeler, Sara: *Terra Incognita: Travels in Antractica*, Jonathan Cape, London, 1996
Wheeler, Sara: *Cherry: A life of Apsley Cherry-Garrard*, Jonathan Cape, London, 2001
Whitfield, Peter: *The Image of the World*, The British Library, London. 1994
Whitfield, Peter: *The Mapmakers – A History of Stanfords*, Compendium, London, 2003
Wilford, John Noble: *The Mapmakers*, Junction Books, London, 1981
Williams, Kit: *Masquerade*, Jonathan Cape, London, 1979
Wilson, EA: *Diary of the Discovery Expedition*, Blandford Press, London, 1966.

Picture credits

p16 Courtesy of Facebook; **p28** Courtesy of the Trustees of the British Museum; **p44** © The Times, November 20th 1988; **p47** The Hereford Mappa Mundi reproduced by kind permission of the Dean and Chapter of Hereford and the Hereford Mappa Mundi Trust; **p48** The Folio Society digital facsimile of the Hereford Mappa Mundi, reproduced by kind permission of the Dean and Chapter of Hereford, the Hereford Mappa Mundi Trust and the Folio Society; **p53** The Folio Society digital facsimile of the Hereford Mappa Mundi, reproduced by kind permission of the Dean and Chapter of Hereford, the Hereford Mappa Mundi Trust and the Folio Society; **p59** Courtesy of the British Library; **p61** Courtesy of the British Library; **p66** Courtesy of the British Library; **p82** © Jim Siebold, Cartographic Images; **p84** © Jim Siebold, Cartographic Images; **p88** © Yale University Press; **p93** © Yale University Press; p111 Courtesy of Geography and Map Division, Library of Congress; **p136** Courtesy of Rare Book and Special Collections Division, Library of Congress; **p145** ©Leen Helmink antique maps, www.helmink.com; **p149** Courtesy of Arader Galleries; **p150** ©Leen Helmink antique maps, www.helmink.com; **p153** Courtesy of the David Rumsey Map Collection; **p155** Courtesy of the David Rumsey Map Collection; **p153** ©Janine Doyle; **p162** © Barry Lawrence Ruderman Antique Maps Inc; **p164** © Barry Lawrence Ruderman Antique Maps Inc; **p173** Courtesy of the Museum of London; **p206** Courtesy of the University of Texas Libraries, University of Texas at Austin; **p208** Courtesy of the David Rumsey Map Collection; **p212** Courtesy of the Lilly Library, Indiana University, Bloomington, Indiana; **p215** Courtesy of the Rare Books Division, Princeton University Library; **p281** ©Cassini Publishing Ltd; **p284** © Cassini Publishing Ltd; **p293** Courtesy of London Transport Museum; **p295** Beck's parody Courtesy of the London Transport Museum; **p295** the Daily Mail Moral Underground courtesy of The Poke; **p339** Courtesy of David Rumsey Map Collection; **p349** Courtesy of the Imperial War Museum, London; **p355** © Jessica Wager; **p371** Nancy Chandler's Map of Bangkok, reproduced courtesy of Nancy Chandler, www.nancychandler.net; **p373** Courtesy of Shutterstock; **p377** Courtesy of Europics; **p386** Courtesy of Radio Times, Immediate Media; **p388** © National Radio Astronomy Observatory; **p401** Courtesy of the British Library; **p426** ©Google Earth; **p438** ©Saul Steinberg; **p439** © Ellis Nadler; **p441** ©Olivia Harris/Reuters

While every effort has been made to contact copyright-holders of illustrations, the author and publishers would be grateful for information about any illustrations where they have been unable to trace them, and would be glad to make amendments in further editions.

Index

Figures in *italics* indicate captions to illustrations.

Vancouver
Calgary
Edmonton
Toronto
Ottawa
Montreal
Seattle
Portland
Sacramento
San Francisco
San José
Los Angeles
San Diego
Phoenix
Minneapolis
Chicago
Detroit
Cleveland
Buffalo
Boston
New York
Newark
Salt Lake City
Las Vegas
Denver
Houston
Monterrey
Guadalajara
Mexico City
Philadelphia
Pittsburgh
Baltimore
St. Louis
Washington
Atlanta
Charlotte
Dallas
Jacksonville
Miami
Santo Domingo
San Juan
Valencia
Caracas
Recife
Fortaleza
Salvador
Rio de Janeiro
Maracaibo
Medellin
Lima
Brasilia
Belo Horizonte
Valparaíso
Goiânia
Sao Paulo
Santiago
Porto Alegre
Montevideo
Buenos Aires
Edinburgh
Glasgow
Newcastle
Sheffield
Liverpool
Dublin
Nottingham
Manchester
Birmingham
London
Oslo
Helsinki
Gothenburg
Stockholm
Copenhagen
The Hague
Amsterdam
Rotterdam
Brussels
Antwerp
Bielefeld
Laon
Charleroi
Orleans
Paris
Lille
Rhine/Ruhr
Cologne
Bonn
Rouen
Strasbourg
Rennes
Ludwigshafen
Zurich
Nantes
Karlsruhe
Lausanne
Bordeaux
Stuttgart
Toulouse
Grenoble
Lyon
Milan
Porto
Bilbao
Montpellier
Turin
Genoa
Madrid
Marseille
Nice
Barcelona
Palma de Mallorca
Seville
Alicante
Palermo
Lisbon
Malaga
Valencia
Tunis
Algiers
Alexandria
Cairo
Under Construction
Lagos